Ergebnisse der Mathematik und ihrer Grenzgebiete

3. Folge · Band 19

A Series of Modern Surveys in Mathematics

Ivar Ekeland

Convexity Methods in Hamiltonian Mechanics

Springer-Verlag
Berlin Heidelberg New York
London Paris Tokyo Hong Kong

Ivar Ekeland
CEREMADE, Université Paris-Dauphine
F-75775 Paris Cedex 16, France

Mathematics Subject Classification (1980):
58Exx, 58Fxx, 70Hxx, 70Jxx, 70Kxx

ISBN 978-3-642-74333-7 ISBN 978-3-642-74331-3 (eBook)
DOI 10.1007/978-3-642-74331-3

Library of Congress Cataloging-in-Publication Data
Ekeland, I. (Ivar), 1944- Convexity methods in Hamiltonian mechanics / Ivar Ekeland, p. cm. - (Ergebnisse der Mathematik und ihrer Grenzgebiete; 3. Folge, Bd. 19)
Bibliography: p. Includes index.
ISBN-13: 978-3-642-74333-7
1. Hamiltonian systems. 2. Convex domains. I. Title II. Series.
QA614.83.E44 1990 514'.74-dc20 89-11405

Softcover reprint of the hardcover 1st edition 1990
2141/3140-543210 - Printed on acid-free paper

Table of Contents

Introduction

The study of periodic solutions of Hamiltonian systems has a long and prestigious history. It starts with the invention of calculus, when Isaac Newton wrote down the differential equations derived from the Hamiltonian $\frac{1}{2}p^2 - \frac{1}{\|q\|}$, and found their solution to be the Keplerian ellipses. The "Principia mathematica philosophiae naturalis", published in 1687, record this discovery and lay the foundations of western science. Two hundred years later appeared Henri Poincaré's famous treatise, the "Méthodes nouvelles de la mécanique céleste", which recognized for the first time the inherent complexity of nonintegrable Hamiltonian systems, and opened the way to the modern theory of dynamical systems.

In another book[1] I have tried to describe the intellectual posterities of Newton and of Poincaré. Their opposition appears already in their mathematics. Newton dealt with the Kepler problem, all the solutions of which are periodic (provided the energy level is negative). After him, attention focused on completely integrable systems, and on the regularities that could be found in the motion of mechanical systems. After two centuries of work by astronomers on the three-body problem, Poincaré proved that a complete description of the motion was impossible, and showed that most Hamiltonian systems would contain trajectories with wildly irregular behaviour. Attention then slowly started shifting to nonintegrable systems and to ergodic properties, culminating in today's interest in chaos.

Periodic solutions are relevant to both sides to the story. On the one hand, they are paradigms of regularity, the simplest solutions one can find apart from equilibria. On the other hand, a tubular neighbourhood of a periodic solution may contain examples of all kinds of trajectories, as illustrated by today's computer pictures, which show elliptic islands surrounded by a chaotic sea. The interest in periodic solutions to Hamiltonian systems has been unabated for three centuries now. In an oft-quoted sentence[2], Poincaré considers them to be "the only opening through which we can force our way into an otherwise impenetrable citadel".

---

1 "Mathematics and the unexpected", Chicago University Press, 1988.
2 "Méthodes nouvelles", t. 1, p. 82.

In the case of completely integrable systems, periodic solutions are found by inspection. For nonintegrable systems, such as the three-body problem in celestial mechanics, they are found by perturbation theory: there is a small parameter ϵ in the problem, the mass of the perturbing body for instance, and for $\epsilon = 0$ the system becomes completely integrable. One then tries to show that its periodic solutions will subsist for $\epsilon \neq 0$ small enough. Poincaré also introduced global methods, relying on the topological properties of the flow, and the fact that it preserves the 2-form $\sum_{i=1}^{n} dp_i \wedge dq_i$. The most celebrated result he obtained in this direction is his last geometric theorem, which states that an area-preserving map of the annulus which rotates the inner circle and the outer circle in opposite directions must have two fixed points.

And now another ancient theme appear: the least action principle. It states that the periodic solutions of a Hamiltonian system are extremals of a suitable integral over closed curves. In other words, the problem is variational. This fact was known to Fermat, and Maupertuis put it in the Hamiltonian formalism.

In spite of its great aesthetic appeal, the least action principle has had little impact in Hamiltonian mechanics. There is, of course, one exception, Emmy Noether's theorem, which relates integrals of the motion to symmetries of the equations. But until recently, no periodic solution had ever been found by variational methods.

This was because the least action principle is in fact a misnomer. Periodic solutions do not minimize the action; they are just extremals, i.e. curves for which the first-order variations vanish. So the classical methods of the calculus of variations, which are used for instance to find minimizing geodesics, do not apply to this case, and progress on this question had to wait till suitable mathematical tools had been developped.

It is one of the achievements of modern nonlinear analysis that it is now possible to find periodic solutions of Hamiltonian systems by variational methods. Since the pioneering paper of Paul Rabinowitz in 1978, this problem has served as a trying ground for the various methods of critical point theory in function spaces, Morse and Liusternik-Schnirelman theory, linking and saddle-point theorems. It is also the hope of all of us who have been working in this field that the variational techniques which have been honed on finite-dimensional Hamiltonian systems will prove instrumental in solving even more difficult problems of mathematical physics, such as the study of nonlinear waves and fields.

This book is decidedly mathematical in character. Its aim is to give the state of the art concerning nonlinear systems with convex Hamiltonian, that is, when $H(t, p, q)$ is convex in the (p, q) variables.

Physically, such systems behave like nonlinear springs, and one will therefore expect the existence of nonlinear modes of vibrations – that is, periodic solutions of the corresponding equations... Mathematically, a device originating with Frank Clarke dramatically improves the situation and makes the problem amenable to by now classical techniques of minimization, mountain-pass theorem, and Liusternik-Schnirelman theory. Many, if not most, important Hamiltonians are non-convex, for instance Newton's, $H(p,q) = \frac{1}{2}p^2 - \frac{1}{\|q\|}$. Such problems with singular problems can also be treated by variational methods, and are the subject of much active research nowadays. I feel, however, that the unity and simplicity the book gains by focusing on convex Hamiltonians more than offsets the loss in scope. In addition, some of the most interesting results in the convex theory, for instance the multiplicity results in Chapter V, have no counterpart in the non-convex case.

This book is written with two purposes in mind:

(a) to serve as an introduction to the subject of convex Hamiltonian system, and more generally to nonlinear analysis and critical point theory.

(b) to serve as a reference for the properties of the index of periodic solutions to convex Hamiltonian systems.

The first task has been greatly facilited by the recent appearance of the excellent book by Jean Mawhin and Michel Willem[3]. By giving a concise and complete account of critical point theory, they have enabled me to concentrate on the various aspects of index theory.

The index, as we define it in this book, is an integer associated with a linear system on a time interval, provided the Hamiltonian is positive definite. Other definitions exist, valid for any quadratic Hamiltonian, but here again the convexity assumption enables us to give a purely analytic description, without having to go into the geometry of the symplectic group.

The advantage of convexity are not limited to ease of exposition. In the final analysis, it is because the index is a non-decreasing function of the interval in the positive definite case (in other words, all conjugate points contribute positively to the index), that we are able to prove the minimal period result of Chapter IV and the multiplicity theorems of Chapter V. No such results are known in non-convex situations.

I have intended Chapter I to be a complete exposition of index theory. In so doing, I have encountered the need for a reliable exposition of Krein theory, adapted to my particular needs. This beautiful theory stems from the study of stability of linear Hamiltonian systems. It is, of course, the creation of Mark G. Krein, but went unnoticed in the West, so that Jürgen Moser independently rediscovered it a few years later. So the first three sections of Chapter I contain an account of Krein theory, which will be used in later sections to prove iteration formulas for the index.

Chapter II is a self-contained account of convex analysis, and gives the abstract duality principle which is the setting for the remainder of the book.

The last Chapters, III, IV and V, contain our results on periodic solutions of nonlinear Hamiltonian systems. They can be read separately, provided the reader is prepared to refer back to Chapters I and II occasionally. The need for the index theory of Chapter I does not occur before Section 3 of Chapter IV.

---

[3] "Critical point theory and Hamiltonian systems", Springer-Verlag 1989.

Difficulty increases from Chapter III (fairly elementary; requires no critical point theory, only a good grasp of the functional analysis in Chapter II) through Chapter IV (fairly technical; relies on the mountain-pass theorem of Ambrosetti and Rabinowitz; full proofs are given) to Chapter V (relies heavily on Liusternik-Schnirelman theory in the context of group actions; technicalities are refered to the original papers). Chapters III and IV are mostly concerned with nonautonomous problems: the fundamental period T is prescribed. We show the existence of solutions with the given period T (fundamental mode), and also of solutions with higher period kT (subharmonics). Chapter V deals with autonomous problems: the energy of the system is prescribed, and we seek periodic solutions (of any period) lying on that energy level. The culminating point is theorem V.3.15, which shows that there are at least two such solutions on each energy level, and infinitely many in general (except in the trivial case where there is just one degree of freedom, $n = 1$).

In each section, propositions are numbered sequentially, together with definitions, lemmas, corollaries and theorems. Within the same section, they are refered to by this number. If they are in another section, or another chapter, the number of the section, or the chapter, is recalled also. For instance, Lemma 1 means Lemma 1 in the present section, while Lemma 1.1 means Lemma 1 in Section 1 of the present chapter, and Lemma I. 1.1 means Lemma 1 of Section 1 in Chapter I. Formulas are numbered according to the same principles.

The results in Chapters III, IV and V have been obtained in the ten years of intense activity which followed Rabinowitz's landmark result and Clarke's duality principle. These years have been made extremely gratifying by the openness and good fellowship which prevail among those who have concerned themselves with this theory. I wish to thank particularly my friends and collaborators, Antonio Ambrosetti, Nassif Ghoussoub, Helmut Hofer, Jean-Michel Lasry and Pierre-Louis Lions, who have taught me so much.

Paris, March 1989 I. Ekeland

Chapter I. Linear Hamiltonian Systems

1. Floquet Theory and Stability

Consider a system of m linear equations with continuous T-periodic coefficients:

$$\dot{x} = M(t)x \tag{1}$$

where $M(t)$ is a real $m \times m$ matrix, depending continuously on $t \in \mathbb{R}$ such that:

$$M(t+T) = M(t) . \tag{2}$$

The solutions to the initial value problem:

$$\dot{x} = M(t)x , \quad x(0) = \xi \in \mathbb{R}^m \tag{3}$$

are given by:

$$x(t) = R(t)\xi , \tag{4}$$

where the matrix $R(t)$ solves the initial-value problem:

$$\frac{d}{dt}R(t) = M(t)R(t) , \quad R(0) = I . \tag{5}$$

We shall refer to $R(t)$ as the *matrizant* of system (1). By the general theory of linear systems, the matrizant is invertible for every t, with $R(t)^{-1} = R(-t)$. In the case of a system with periodic coefficients, such as this one, Floquet theory gives us some more information.

Indeed, note that if $R(t)$ solves problem (5), then $R_T(t) := R(t+T)$ solves problem:

$$\frac{d}{dt}R_T(t) = M(t)R_T(t) , \quad R_T(0) = R(T) , \tag{6}$$

so that $R_T(t) = R(t+T) = R(t)R(T)$.

Since $R(T)$ is invertible, the set of eigenvalues $\operatorname{Spec} R(T)$ does not contain 0. Choose a simply connected domain Ω, and a determination of the logarithm, $\log : \Omega \to \mathbb{C}$, such that $\operatorname{Spec} R(T)$ is contained in Ω. By standard results on Banach algebras, $\log A$ is well-defined for all matrices A whose spectrum is contained in Ω, and it is a holomorphic function of A. In fact, we have the formula:

$$f(A) = (2i\pi)^{-1} \int (zI - A)^{-1} \log z \, dz \tag{7}$$

where the integral is taken over any closed curve in Ω winding once around $\operatorname{Spec} A$.

Define a matrix C with complex coefficients by $C := T^{-1} \log R(T)$. We have

$$R(T) = \exp CT \tag{8}$$

$$\operatorname{Spec} C = T^{-1} \log\{\operatorname{Spec} R(T)\} \tag{9}$$

(10) C and $R(T)$ have the same invariant subspaces.

We now understand system (1) as a differential equation in $\mathbb{C}^m$. We have two linear systems to compare:

$$\dot{x} = M(t)x \, , \quad \text{with matrizant } R(t) \tag{11}$$

$$\dot{y} = Cy \, , \quad \text{with matrizant } \exp(Ct) \, . \tag{12}$$

Clearly, the time-dependent change of variables $x = P(t)y$, with $P(t) = R(t) \exp(-Ct)$, brings the first one into the second. It is remarkable that the matrix $P(t)$ is T-periodic:

$$\begin{aligned} P(t+T) &= R(t+T) \exp(-CT) \exp(-Ct) \\ &= R(t)R(T) \exp(-CT) \exp(-Ct) = R(t) \exp(-Ct) = P(t) \, . \end{aligned} \tag{13}$$

This is the content of Floquet theory: there is a T-periodic linear change of variables in $\mathbb{C}^m$ which changes the original equation into an equation with constant coefficients.

On may ask whether there is a *real* change of variables which does the same job. By formula (7), the answer will be yes if we can take for log the standard determination of the logarithm, that is, if $R(T)$ has no real negative eigenvalue. Even if $R(T)$ does have negative eigenvalues, note that $R(2T) = R(T)^2$ never does, so that one will always be able to find a real $2T$-periodic change of variables which turns system (11) into a system with constant coefficients, namely $\dot{y} = \log R(2T)y$.

The eigenvalues of $R(T)$ are called the *Floquet multipliers*: they are uniquely defined. The eigenvalues of C are called the *Floquet exponents*: they depend on the particular choice of C, which is related to $R(T)$ by the equation

$R(T) = \exp(CT)$. If we denote by λ_i the Floquet multipliers, and by ω_i the Floquet exponents, properly ordered, we have the obvious relation:

$$\lambda_i = \exp(\omega_i) \qquad \text{for } 1 \leq i \leq m$$

so that the Floquet multipliers give the Floquet exponents modulo $2\pi i T^{-1}$. Note for instance that Eq. (1) has a periodic solution if and only if 1 is a Floquet multiplier. More generally, the equation $\dot{x} = M(t)x$ has a kT-periodic solution if and only if one of the Floquet multipliers is a k-th root of unity.

We now turn to question of stability. The system $\dot{x} = M(t)x$ is called *positively* (resp. *negatively*) *stable* if all its real solutions remain bounded for all $t > 0$ (resp. $t < 0$). It is called *stable* if it is both positively and negatively stable, that is, its real solutions are bounded for all times $t \in R$.

The problem then is to find a criterion for stability which requires knowledge of the solution on a finite interval only – one period for instance. Such a criterion is provided by Floquet theory. Denote by $\mathcal{D}$ the unit disk in $\mathbb{C}$, and by $\mathcal{U}$ the unit circle:

(14) $$\mathcal{D} := \{y \,||z| \leq 1\}$$

(15) $$\mathcal{U} : \{y \,||z| = 1\} \ .$$

Proposition 1. *The system $\dot{x} = M(t)x$ is positively stable if and only if $R(T)$ is diagonalizable and its spectrum lies entirely in $\mathcal{D}$. It is stable if and only if it is diagonalizable and its spectrum lies entirely on $\mathcal{U}$.*

Proof. Solutions x of Eq. (8) and solutions y of Eq. (9) are related by the T-periodic change of variables $x(t) = P(t)y(t)$. Clearly, $x(t)$ is bounded if and only if $y(t)$ is bounded. In addition, x is a solution of (8) in $\mathbb{C}^m$ if and only if its real and imaginary parts are solutions of (1) in $\mathbb{R}^m$. Finally, the system $\dot{x} = M(t)x$ is positively stable if and only if all solutions of $\dot{y} = Cy$ remain bounded in $\mathbb{C}^m$ when $t \to +\infty$.

The matrices C and $R(T)$ commute, so they have the same invariant subspaces, and $R(T)$ is diagonalizable if and only if C is diagonalizable.

If C is diagonalizable, let ξ_i, $1 \leq i \leq m$, be a basis of eigenvectors, corresponding to the eigenvalues ω_i. Then the y_i defined by $y_i(t) = \exp(\omega_i t)\xi_i$ span the spaces of solutions, and the remain bounded if and only if all the Floquet exponents ω_i have non-positive real part, that is, all the Floquet multipliers $\lambda_i = \exp(\omega_i)$ lie in the unit disk.

If C is not diagonalizable, it is known that there is an eigenvector ξ, associated with an eigenvalue ω, and a solution y which can be written $y(t) = p(t)\exp(\omega t)\xi$, with $p(t)$ a non-constant polynominal. Such a solution is unbounded when $t \to +\infty$.

Stability is characterized in the same way; this time the Floquet exponents ω_i must be zero $\left(\text{modulo } 2\pi i T^{-1}\right)$. □

The question of stability becomes more delicate if additional restrictions are put on the system – namely, that it must be Hamiltonian.

Definition 2. The linear system (1) is called *Hamiltonian* if its dimension is even, $m = 2n$, and we have $M(t) = JA(t)$, where $A(t)$ is a symmetric matrix and the matrix J is given by:

$$\begin{pmatrix} 0 & I_n \\ -I_n & 0 \end{pmatrix} .$$

The matrix J has the obvious properties:

$$J^* = -J = J^{-1} . \tag{16}$$

It follows for instance that $(J\xi, \xi) = 0$ for all ξ in $\mathbb{R}^{2n}$.

From now on, we shall consider Hamiltonian systems only. We are given a matrix $A(t)$, symmetric, T-periodic, depending continuously on t:

$$A(t) = A^*(t) \tag{17}$$

$$A(t+T) = A(t) \tag{18}$$

and we are interested in the linear differential system:

$$\dot{x} = JA(t)x . \tag{19}$$

The fundamental property of such a system is that its matrizant is symplectic. Recall that a matrix M is called *symplectic* if it preserves the 2-form associated with J:

$$(J\xi, \zeta) = (JM\xi, M\zeta) \qquad \text{for all } (\xi, \zeta) . \tag{20}$$

In other words, M is symplectic if $M^*JM = J$. Symplectic matrices in $\mathbb{R}^{2n}$ form a group, which we denote by $\operatorname{Sp}(\mathbb{R}^{2n})$.

Proposition 3. *The matrizant $R(t)$ of a linear Hamiltonian system is symplectic for all t.*

Proof. We have to check the relation:

$$R(t)^*JR(t) = J \quad \forall t . \tag{21}$$

This is obviously true for $t = 0$, since $R(0) = I$. We differentiate the left-hand side with respect to time. Using formula (5), with $M = JA$ we get:

$$\frac{d}{dt}R^*JR = (JAR)^*JR + R^*J(JAR) . \tag{22}$$

Using the properties of J, and the fact that A is symmetric, this becomes:

$$\frac{d}{dt}R^*JR = R^*A(-JJ)R + R^*(JJ)AR = R^*AR - R^*AR = 0 . \tag{23}$$

So $R^*(t)JR(t)$ is a constant matrix, and it must be equal to its value for $t = 0$. □

Note that there is a converse:

Proposition 4. *Let $R(t)$ be a symplectic matrix, a C^1 function of t in $[0,T]$ such that $R(0) = I$. Then there is a symmetric matrix $A(t)$, depending continuously on t, such that $R(t)$ is the matrizant associated with the linear Hamiltonian system $\dot{x} = JA(t)x$, namely*

$$A(t) := -JR'(t)R(t)^{-1} . \tag{24}$$

Proof. Defining $A(t)$ in this way, we have immediately the equation $R'(t) = JA(t)R(t)$, which, together with $R(0) = I$, defines the matrizant. All we have to do is to show that $A(t)$ is symmetric.

We have:

$$A^*(t) = (R(t)^*)^{-1} R'(t)^* J . \tag{25}$$

Since $R(t)$ is symplectic, we have $R(t)^*JR(t) = J$. This yields

$$(R(t)^*)^{-1} = -JR(t)J \tag{26}$$

and also, by differentiating:

$$R'(t)^*JR(t) + R(t)^*JR'(t) = 0 . \tag{27}$$

Replacing in formula (25) both $R(t)^*$ and $R'(t)^*J$ by their values taken from the preceding equalities, we get

$$A^*(t) = JR(t)JR(t)^*JR'(t)R(t)^{-1} \tag{28}$$

and the right-hand side is equal to $A(t)$ since $R(t)JR^*(t) = J$. □

This result will be useful later on. Meanwhile, let us go back to Proposition 3, and investigate its consequences.

Corollary 5. *$R(t)$ preserves volume and orientation:*

$$\operatorname{Det} R(t) = 1 . \tag{29}$$

Proof. Relation (21) gives $[\operatorname{Det} R(t)]^2 = 1$, so that $\operatorname{Det} R(t) = \pm 1$. For $t = 0$, we get $\operatorname{Det} I = 1$; since $\operatorname{Det} R(t)$ is obviously a continuous function of t, the result follows. □

Corollary 6. *If λ is a Floquet multiplier, so are its inverse λ^{-1}, its complex conjugate $\overline{\lambda}$, and $\overline{\lambda}^{-1}$. They all have the same multiplicity. If 1 or -1 is a Floquet multiplier, it must have even multiplicity.*

Proof. Since $R(T)$ is real, if λ is an eigenvalue, so is $\overline{\lambda}$, and the multiplicity is the same. All we have to worry about is λ^{-1}.

By relation (19), we have

$$-JR(T)^*J = R(T)^{-1} \tag{30}$$

and hence:

$$R(T)^{-1} - \lambda I = J^{-1}\left(R(T)^* - \lambda I\right) J \, . \tag{31}$$

This shows that if λ is an eigenvalue of $R(T)^*$, it is an eigenvalue of $R(T)^{-1}$ with the same multiplicity. But the eigenvalues of $R(T)^{-1}$ are the inverses of the eigenvalues of $R(T)$, while the eigenvalues of $R(T)^*$ are the eigenvalues of $R(T)$, with the same multiplicities. The result follows for $\lambda \neq \pm 1$.

If -1 is an eigenvalue of $R(T)$, it must have even multiplicity, otherwise $\operatorname{Det} R(T)$, which is the product of the eigenvalues, would be negative, contradicting Corollary 5. If 1 is an eigenvalue, all other eigenvalues having even multiplicity, and the dimension of the space being even, 1 itself must have even multiplicity. □

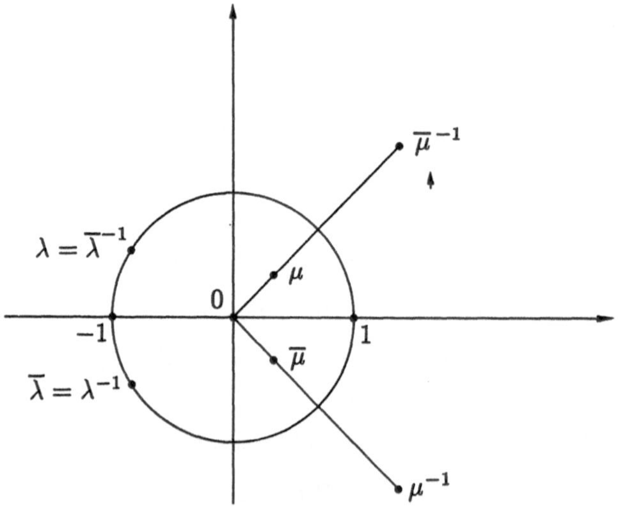

Fig. 1. Non real eigenvalues on the unit circle come in pairs $(\lambda, \overline{\lambda})$; off the unit circle, they come in quadruplets $(\mu, \overline{\mu}, \mu^{-1}, \overline{\mu}^{-1})$.

There are four different types of Floquet multipliers λ :

$\lambda = +1$ or $\lambda = -1$. Then λ has even multiplicity.

$\lambda \in \mathbb{R}$ and $\lambda \neq \pm 1$. In other words, λ is a real Floquet multiplier with $|\lambda| \neq 1$. Then $\lambda = \overline{\lambda}$, and λ again belongs to a pair of Floquet multipliers, namely $\{\lambda, \lambda^{-1}\}$.

$|\lambda| = 1$ and $\lambda \neq \pm 1$. In other words, λ is a non-real Floquet multiplier of modulus one. Then $\lambda^{-1} = \overline{\lambda}$ and λ belongs to a pair of distinct Floquet multipliers, namely $\{\lambda, \overline{\lambda}\}$.

$|\lambda| \neq 1$ and $\lambda \notin \mathbb{R}$. In other words, λ is non-real Floquet multiplier away from the unit circle. Then λ belongs to a family of four distinct Floquet multipliers, namely $\left\{\lambda, \overline{\lambda}, \lambda^{-1}, \overline{\lambda}^{-1}\right\}$.

This fact has remarkable consequences. First, we see that the Floquet multipliers cannot all lie in the interior of the unit disk: if $|\lambda| < 1$ then $|\lambda^{-1}| > 1$. This means that, for linear Hamiltonian systems, positive stability implies negative stability, and hence stability.

Secondly, suppose all the Floquet multipliers are simple and lie on $\mathcal{U}$. Then they are of type III, and they can be numbered $\left(\lambda_1, \overline{\lambda}_1\right), \ldots, \left(\lambda_n, \overline{\lambda}_n\right)$. Then all symplectic matrices M which are close enough to $R(T)$ will also have simple eigenvalues $(\mu_1, \overline{\mu}_1), \ldots, (\mu_n, \overline{\mu}_n)$ close to the preceding ones. As a consequence, none of these can stray away from the unit circle: if for instance we had $|\mu_1| \neq 1$, there would be not one but two eigenvalues for M close to λ_1, namely μ_1 and $\overline{\mu}_1^{-1}$.

This leads us to the notion of strong stability, which is unique to Hamiltonian systems, and which we shall investigate in the following section.

Notes and Comments. The material in this section is classical, and goes back to the work of Poincaré [Poi2] and Liapounov [Lia]. It will be found in many books, such as the ones by Hale [Hal], Rouché and Mahwin [RouM], or Roseau [Ro2].

2. Krein Theory and Strong Stability

Consider the Hamiltonian system in $\mathbb{R}^{2n}$

$$\dot{x} = JA(t)x \tag{1}$$

with $A(t) = A(t)^*$, and $A(t+T) = A(t)$ for all t.

Definition 1. The system (1) is called *stable* if all its solutions remain bounded when $t \in \mathbb{R}$. It is *strongly stable* if there exists some $\epsilon > 0$ such that, whenever $B(t)$ is a symmetric T-periodic matrix, depending continuously on t, with:

$$\|B(t) - A(t)\| \leq \epsilon \quad \forall t \tag{2}$$

the system $\dot{x} = JB(t)x$ is stable.

This definition ties in immediately with the following one:

Definition 2. A symplectic matrix M is called *stable* if all its iterates M^k remain bounded when $k \in \mathbb{Z}$. It is called *strongly stable* if there is some $\epsilon > 0$ such that all symplectic matrices N such that $\|M - N\| \leq \epsilon$ are stable.

Floquet theory tells us that all complex solutions of system (1) can be written as $x(t) = P(t)R(T)^{t/T}\xi$, for some $\xi \epsilon \mathbb{C}^{2n}$, where $R(t)$ is the matrizant and $P(t)$ is T-periodic, so that system (1) is stable if and only if the matrix $R(T)$ is stable.

Proposition 3. *System* (1) *is strongly stable if and only if the matrix* $R(T)$ *is strongly stable.*

Proof. Any Hamiltonian system which is close to system (1) must have a matrizant $S(t)$ which is uniformly close to $R(t)$. If $R(T)$ is strongly stable, $S(T)$ will also be stable, and so will the corresponding Hamiltonian system.

Conversely, assume $R(T)$ is not strongly stable. Then there is a sequence M_n of unstable symplectic matrices converging to $R(T)$. Then $R(T)^{-1}M_n$ is a symplectic matrix, close to the identity. Set

$$C_n := \log\left[R(T)^{-1}M_n\right] , \tag{3}$$

and define $R_n(t)$ by

$$R_n(t) := R(t)\exp\left[C_n t/T\right] . \tag{4}$$

By the following lemma, $R_n(t)$ is a symplectic matrix, and $R_n \to R$ in the C^1 topology when $n \to \infty$. By Proposition 1.3, R_n is the matrizant of the Hamiltonian system:

$$\dot{x} = JA_n(t)x \tag{5}$$

with $A_n(t) = -JR_n'(t)R_n(t)$, so that $A_n \to A$ uniformly when $n \to \infty$. But $R_n(T) = M_n$, so that system (5) is unstable. □

Lemma. *Let* C *and* M *be two matrices such that* $M = \exp C$ *with* $\|C\|$ *small. Then* M *is symplectic if and only if* JC *is symmetric.*

Proof. If M is symplectic, we have:

$$\exp C = J^{-1}\left(\exp C^*\right)^{-1} = \exp\left\{-J^{-1}C^*J\right\} . \tag{6}$$

It is known that the standard determination of the logarithm near the identity is the inverse of the exponential map near 0. From the preceding equality we deduce $C = -J^{-1}C^*J$, and hence $JC + C^*J = 0$, which is the desired result.

The converse is clear by exponentiating. □

Questions about the strong stability of Hamiltonian systems are therefore reduced to questions about the strong stability of symplectic matrices. By the arguments developed at the end of the preceding section, any symplectic matrix whose eigenvalues are simple and lie on the unit circle is strongly stable. What happens in the case of multiple eigenvalues is the subject of Krein theory, into which we now proceed.

Note that, since J is antisymmetric, $-iJ$ is Hermitian. Denote it by G and endow $\mathbb{C}^{2n}$ with the nondegenerate Hermitian structure associated with G:

$$(Gx, y) := (-iJx, y) = i \left\{ \sum_{k=1}^{n} \left(x_k \overline{y}_{n+k} - x_{n+k} \overline{y}_k \right) \right\} . \tag{7}$$

Note that all real vectors are G-isotropic:

$$(Gx, x) = 0 \quad \forall x \in \mathbb{R}^{2n} . \tag{8}$$

Let M be a real symplectic matrix. By definition, we have $(Gx, y) = (GMx, My)$, that is, M is G-unitary.

Proposition 4. *Let λ and μ be eigenvalues of M. If $\lambda\overline{\mu} \neq 1$, the two corresponding eigenspaces are G-orthogonal.*

Proof. If $Mx = \lambda x$ and $My = \mu y$, we get:

$$(Gx, y) = (GMx, My) = (\lambda x, \mu y) = \lambda\overline{\mu}\,(x, y) \tag{9}$$

which leads to $\lambda\overline{\mu} = 1$ or $(Gx, y) = 0$. □

A similar result holds for the invariant subspaces. Recall that E is an *invariant subspace* for M if $ME \subset E$. With each eigenvalue λ of M is associated an invariant subspace E_λ, defined by:

$$E_\lambda = \bigcup_{m \geq 1} \text{Ker } (M - \lambda I)^m . \tag{10}$$

In fact, $E_\lambda = \text{Ker } (M - \lambda I)^m$ for some $m \leq m_\lambda$, where m_λ is the multiplicity of λ. That is, λ is a root of order m_λ of the characteristic polynomial Det $(M - zI)$. We have the splitting:

$$\mathbb{C}^{2n} = \bigoplus_{\lambda \in \Lambda} E_\lambda \tag{11}$$

where Λ is the set of eigenvalues.

An invariant subspace E is *irreducible* if it cannot be split into two non-trivial invariant subspaces. Any irreducible invariant subspace is contained in one of the E_λ, and each E_λ splits (non-uniquely) into a direct sum of irreducible invariant subspaces.

We shall say that an eigenvalue λ is *semi-simple* if each irreducible invariant subspace of E_λ is one-dimensional, that is, the invariant subspace is just the eigenspace:

$$E_\lambda = \text{Ker } (M - \lambda I) . \tag{12}$$

M is diagonalizable if and only if all its eigenvalues are semi-simple.

We now extend Proposition 2:

Proposition 5. *Let λ and μ be eigenvalues of M. If $\lambda\overline{\mu} \neq 1$, the associated invariant subspaces E_λ and E_μ are G-orthogonal.*

Proof. Pick two vectors $x \in E_\lambda$ and $y \in E_\mu$. We have $(M - \lambda I)^p x = 0$ and $(M - \mu I)^q y = 0$ for some integers $1 \le p \le m_\lambda$ and $1 \le q \le m_\mu$. Set $p + q = m$ and argue by induction on m.

For $m = 2$, we have $p = q = 1$, so that x and y are eigenvectors, and the result follows from Proposition 4.

Assume the result holds for $m \le \overline{m}$ and set $m = \overline{m} + 1$. Define two new vectors $x' := (M - \lambda I)\, x$ and $y' := (M - \mu I)\, y$. We have $(M - \lambda I)^{p-1} x' = 0$ and $(M - \mu I)^{q-1} y' = 0$. It follows from the assumption that:

$$(Gx', y) = (Gx, y') = (Gx', y') = 0 \; . \tag{13}$$

Replacing x' and y' by their values, we get three equations:

$$(GMx, y) = \lambda\,(Gx, y) \tag{14}$$

$$(Gx, My) = \overline{\mu}\,(Gx, y) \tag{15}$$

$$(GMx, My) - \lambda\,(Gx, My) - \overline{\mu}\,(GMx, y) + \lambda\overline{\mu}\,(Gx, y) = 0 \; . \tag{16}$$

Writing the two first equations into the third, and using the fact that M is G-unitary, we get:

$$(1 - \lambda\overline{\mu})\,(Gx, y) = 0 \; . \tag{17}$$

So $\lambda\overline{\mu} = 1$ or $(Gx, y) = 0$. □

Corollary 6. *If λ is an eigenvalue of M away from the unit circle, $|\lambda| \ne 1$, then the invariant subspace E_λ is G-isotropic:*

$$(Gx, y) = 0 \qquad \text{for all } \; x \text{ and } y \text{ in } E_\lambda \; . \tag{18}$$

Proof. Take $\lambda = \mu$ in Proposition 5. □

Note that, if $\lambda = \pm 1$, there are real eigenvectors, and they must be G-isotropic. We also have the following:

Proposition 7. *If λ is an eigenvalue of M on the unit circle, and λ is not semi-simple, then there must be a G-isotropic eigenvector.*

Proof. By the theory of the Jordan form, we can find an eigenvector x and another vector $y \in E_\lambda$ such that:

$$My = \lambda y + x \; .$$

Hence:

$$(GMx, My) = (\lambda Gx, \lambda y + x) = |\lambda|^2\,(Gx, y) + \lambda\,(Gx, x) \; .$$

But the left-hand side is just (Gx, y) since M is G-unitary. Since $|\lambda|^2 = 1$, we are left with $(Gx, x) = 0$, as desired. □

We are now in a position to state the central definitions of Krein theory. Recall that a nondegenerate Hermitian form G can always be written as a sum of squares by a linear change of variables; the number p of positive squares and the number q of negative squares are invariant, that is, they do not depend on the particular basis in which G has been diagonalized. The pair (p, q) is called the *signature* of G.

Definition 8. Let λ be an eigenvalue of M on the unit circle. The restriction of G to the invariant subspace E_λ must be nondegenerate. The signature (p, q) of G on E_λ is called the *Krein type* of λ. If $q = 0$, that is, G is positive definite on E_λ, we say that λ is *Krein-positive.* If $q = 0$ or $p = 0$, we say that λ is *Krein-definite*; otherwise, we say that λ is *Krein-indefinite.*

Denote by G_λ the restriction of G to E_λ. The fact that G_λ is nondegenerate follows readily from the orthogonal decomposition

$$\mathbb{C}^{2n} = E_\lambda \bigoplus F_\lambda \qquad \text{with } F_\lambda = \bigoplus_{\mu \neq \lambda} E_\mu \tag{19}$$

and the two subspaces E_λ and F_λ are G-orthogonal by Proposition 5.

Note a simple lemma:

Lemma 9. *If λ has Krein type (p, q), then $\overline{\lambda}$ has Krein type (q, p). As a particular case, if $\lambda = \pm 1$ is an eigenvalue, we must have $p = q$.*

Proof. Since the real part of (Jx, x) is zero, we have $(J\overline{x}, \overline{x}) = -(Jx, x)$. Hence:

$$(G\overline{x}, \overline{x}) = -i\,(J\overline{x}, \overline{x}) = -(Gx, x)\ . \tag{20}$$

If $[\xi_1, \dots \xi_m]$ is a G-orthogonal basis for E_λ, then $[\overline{\xi}_1, \dots \overline{\xi}_m]$ is a G-orthogonal basis for $E_{\overline{\lambda}}$, with $(G\xi_k, \xi_k) = -\left(G\overline{\xi}_k, \overline{\xi}_k\right)$. The result follows for the case when λ is not real.

If $\lambda = \pm 1$, the invariant subspace E_λ is even-dimensional and real. Starting from a G-orthogonal basis $[\xi_1, \dots \xi_{2m}]$, which we normalize by $(G\xi_k, \xi_k) = \pm 1$, we get a new G-orthogonal basis $[\overline{\xi}_1, \dots \overline{\xi}_{2m}]$ where the signs of all the squares have been changed. Since the number of positive and negative squares is an invariant, there must be equally many of them. □

If λ is Krein-definite, there can be no G-isotropic vector in E_λ. It then follows from Proposition 7 and the preceding remark that if 1 or -1 are eigenvalues, they must be Krein-indefinite, and so must be all the eigenvalues which are not semi-simple.

These simple remarks lead us to a characterization of strong stability:

Theorem 10. *M is strongly stable if and only if it is stable and all its eigenvalues are Krein-definite.*

Proof. Assume M is stable and all its eigenvalues are Krein-definite. I claim M is strongly stable. Otherwise, there would exist a sequence M_n of unstable symplectic matrices converging to M. Either M_n has an eigenvalue outside the unit circle, or M_n has an eigenvalue on the unit circle which is not semi-simple. In either case, there is a G-isotropic eigenvector x_n:

$$M_n x_n = \lambda_n x_n \quad \text{and} \quad \|x_n\| = 1 \tag{21}$$

$$(Gx_n, x_n) = 0 \; . \tag{22}$$

Since λ_n is a root of Det $(M_n - zI)$, we can extract from the sequence λ_n a subsequence converging to a root of Det $(M - zI)$, that is, an eigenvalue of M. By compactness of the unit ball, the x_n also have a convergent subsequence; let x be its limit. Taking limits in the preceding equalities, we see that x is an isotropic eigenvector of M. This is impossible since all eigenvalues of M are Krein-definite. So M must be strongly stable.

Conversely, assume M is strongly stable. Then M is stable, so that all its eigenvalues lie on the circle and are semi-simple.

For every eigenvalue λ with positive imaginary part, we choose in the eigenspace Ker $(M - \lambda I)$ a G-orthogonal basis, say $[\xi_1, \dots, \xi_m]$, which we normalize by the condition $(G\xi_k, \xi_k) = \pm 1$. This is possible since G is non-degenerate. We then take $\left[\overline{\xi}_1, \dots, \overline{\xi}_m\right]$ as a basis for the conjugate eigenspace Ker $\left(M - \overline{\lambda} I\right)$. If ± 1 is an eigenvalue, the corresponding eigenspace is real and even-dimensional; we can choose its G-orthogonal basis to be $\left[\xi_1, \dots, \xi_m, \overline{\xi}_1, \dots, \overline{\xi}_m\right]$, with $(G\xi_k, \xi_k) = 1$ and $\left(G\overline{\xi}_k, \overline{\xi}_k\right) = -1$. Putting everything together, we get a G-orthogonal basis of eigenvectors $[\xi_1, \dots, \xi_n, \overline{\xi}_1, \dots, \overline{\xi}_n]$ for $\mathbb{C}^{2n}$. We have $(G\xi_k, \xi_k) = -\left(G\overline{\xi}_k, \xi_k\right)$, and by rearranging the basis, we can always assume that $(G\xi_k, \xi_k) = 1$ for $1 \leq k \leq n$.

Assume there is an eigenvalue λ which is not definite. It must have two eigenvectors with opposite G-norms, say ξ_1 and $\overline{\xi}_1$ if $\lambda = \pm 1$, and ξ_1 and ξ_2 if $\lambda \neq \pm 1$. Define a linear transformation M_ϵ by setting:

$$M_\epsilon \xi_1 = \lambda \left(\xi_1 \cosh \epsilon + \overline{\xi}_1 \sinh \epsilon\right) \tag{23}$$

$$M_\epsilon \overline{\xi}_1 = \lambda \left(\xi_1 \sinh \epsilon + \overline{\xi}_1 \cosh \epsilon\right) \tag{24}$$

if $\lambda = \pm 1$, and

$$M_\epsilon \xi_1 = \lambda \left(\xi_1 \cosh \epsilon + \xi_2 \sinh \epsilon\right) \tag{25}$$

$$M_\epsilon \xi_2 = \lambda \left(\xi_1 \sinh \epsilon + \xi_2 \cosh \epsilon\right) \tag{26}$$

$$M_\epsilon \overline{\xi}_1 = \overline{\lambda} \left(\overline{\xi}_1 \cosh \epsilon + \overline{\xi}_2 \sinh \epsilon\right) \tag{27}$$

$$M_\epsilon \overline{\xi}_2 = \overline{\lambda} \left(\overline{\xi}_1 \sinh \epsilon + \overline{\xi}_2 \cosh \epsilon\right) \tag{28}$$

if $\lambda \neq \pm 1$, and $M_\epsilon = M$ on the invariant subspace generated by the other ξ_k.

By construction, M_ϵ is real $\left(\text{that is, } M_\epsilon \mathbb{R}^{2n} \subset \mathbb{R}^{2n}\right)$, and symplectic, that is, $(G\xi, \zeta) = (GM_\epsilon \xi, M_\epsilon \zeta)$, as we readily check. On the other hand, ξ_1 (if $\lambda = \pm 1$) or $\xi_1 + \xi_2$ (if $\lambda \neq \pm 1$) is an eigenvector of M_ϵ, the corresponding eigenvalue being λe^ϵ, which is outside the unit circle if $\epsilon > 0$. So M_ϵ is not stable, and $M_\epsilon \to M$ when $\epsilon \to 0$. This contradicts the fact that M is strongly stable. □

We draw an interesting cosequence: a "normal form" for strongly stable linear Hamiltonian systems.

Proposition 11. *The linear Hamiltonian system* (1) *is strongly stable if and only if there is a real T-periodic symplectic change of coordinates $x = P(t)z$ which puts in the form $\dot{z} = JH'(z)$, where $H(z)$ is the quadratic form given by:*

$$H(z) = \frac{1}{2}\sum_{i=1}^{n} \alpha_i z_i^2 \ , \tag{29}$$

and where $\alpha_i + \alpha_j \neq 2k\pi/T$ for $1 \leq i,j \leq n$ and all $k \in \mathbb{Z}$. Then $e^{i\alpha_1 T}, \dots, e^{i\alpha_n T}$, are the Floquet multipliers of positive type, repeated according to multiplicity.

Proof. Assume $R(T)$ is strongly stable, and write its eigenvalues as $e^{i\alpha_1 T}, \dots, e^{i\alpha_n T}, e^{-i\alpha_1 T}, \dots, e^{-i\alpha_n T}$, the ones of positive type first. Since all the eigenvalues must be Krein-definite, $\alpha_i + \alpha_j \neq 2k\pi T$ for all i, j and k; in particular, $\alpha_i \neq k\pi$, so ± 1 cannot be an eigenvalue. Choose as before a G-orthogonal, basis $[\xi_1, \dots, \xi_n, \bar{\xi}_1, \dots, \bar{\xi}_n]$ of eigenvectors, the first ones being of positive type, $(G\xi_k, \xi_k) = 1$. Define a new basis $[\zeta_1, \dots, \zeta_{2n}]$ of $\mathbb{C}^{2n}$, consisting of real vectors, by:

$$\zeta_k = \frac{\xi_k - \bar{\xi}_k}{i\sqrt{2}} \tag{30}$$

$$\zeta_{k+n} = \frac{\xi_k + \bar{\xi}_k}{\sqrt{2}} \ . \tag{31}$$

We check that $(J\zeta_k, \zeta_j) = 0$ unless $|k - j| = n$, and $(J\zeta_k, \zeta_{k+n}) = -1$. Denote by Z the transition matrix from the canonical basis of $\mathbb{R}^{2n}$ to the ζ-basis, that is, the real matrix whose columns are $\zeta_1, \dots, \zeta_{2n}$. We have $Z^* JZ = J$, so that Z is symplectic.

We obtain from $R(T)\xi_k = e^{i\theta_k}\xi_k$ (setting $\theta_k := \alpha_k T$)

$$R(T)\zeta_k = \cos\theta_k \zeta_k + \sin\theta_k \zeta_{n+k} \tag{32}$$

$$R(T)\zeta_{n+k} = -\sin\theta_k \zeta_k + \cos\theta_k \zeta_{n+k} \ . \tag{33}$$

This means that the restriction of $R(T)$ to the invariant subspace generated by ζ_k and ζ_{n+k} is a rotation of angle θ_k: it can be written as $e^{JT\alpha_k}$. So $R(T)$ can be written as e^{JCT}, where C is a real diagonal matrix in the ζ-basis, with eigenvalues α_k, $1 \leq k \leq n$, as it was defined in formula (29).

The symplectic change of variables $x = P(t)z$, with $P(t) = R(t)Ze^{-JCt}$ then brings the original system into the form $\dot{z} = JCz$.

Conversely, assume that the system can be brought into the above form by a T-periodic symplectic change of variables. Then it is obviously stable, and the Floquet multipliers are the $e^{i\alpha_k T}$. All we have to show is that the eigenvectors are Krein-definite. Denote by e_k, $1 \leq k \leq n$, the standard basis of $\mathbb{R}^{2n}$, set $\zeta_k := P(t)e_k$, so that $R(T)$ is written as e^{JAT} in the ζ-basis. Note that the vectors:

$$\xi_k = \frac{\zeta_{n+k} + i\zeta_k}{\sqrt{2}} \ , \quad \xi_{n+k} = \overline{\xi}_k \ , \tag{34}$$

for $1 \leq k \leq n$, are G-orthogonal and satisfy:

$$(G\xi_k, \xi_k) = 1 \ , \quad (G\overline{\xi}_k, \overline{\xi}_k) = -1 \tag{35}$$

$$R(T)\xi_k = e^{i\alpha_k T}\xi_k \ , \quad R(T)\overline{\xi}_k = e^{-i\alpha_k T}\overline{\xi}_k \ . \tag{36}$$

Now let E be an eigenspace of $R(T)$, associated with an eigenvalue $e^{i\alpha T}$. Either $\alpha = \alpha_k \pmod{2\pi/T}$ for some k, and E is spanned by the corresponding ξ_k, or $\alpha = -\alpha_k \pmod{2\pi/T}$ for some k, and E is spanned by the corresponding $\overline{\xi}_k$. Since $\alpha_i + \alpha_j$ is never a multiple of $2\pi/T$ the two situations are mutually exclusive. Relations (35) then show that the restriction of G to E is positive definite in the first situation, and negative definite in the second. □

Let us warn the reader that it would be a mistake to believe that if a T-periodic Hamiltonian system is stable (that is, $R(T)$ is stable), then $R(t)$ will be stable for all t. In fact, the eigenvalues of $R(t)$ will usually have plenty of opportunities to leave the unit circle $\mathcal{U}$ as t increases from 0, but all the spot checks conducted at the times $t = kT$, $k \in \mathbb{N}$, will find them safely back on $\mathcal{U}$. More about this in the next section.

Notes and Comments. The characterization of strong stability (Theorem 10) was formulated by Krein in 1950 [Kre1] to [Kre4], and rediscovered independently by Moser in 1958 [Mo1]. It covers a number of well-known situations, where a small perturbation applied to a stable linear system will cause instability; this phenomenon is usually referred to a *parametric resonance* (see [Arn2] § 25). The most famous example is the swing, that is, a vertical pendulum the length of which varies periodically. This is modelled by Mathieu's equation:

$$\ddot{x} + (1 + \epsilon \cos \omega t)x = 0 \ ,$$

and the system is destablized for $\omega = 2/k$, $0 \neq k \in \mathbb{Z}$, with the strongest effect taking place at $\omega = \pm 2$, that is, when the destabilizing period is one-half the natural period. This is easy to understand via the analysis we conducted in this chapter. The unperturbed system is a linear, positive definite, Hamiltonian system, and the eigenvalues of $R(t)$ stay on the unit circle. They go through $(-1)^k$ at the time $t_k = k\pi$, and so they are strongly stable for all $t \neq k\pi$. The perturbed system is $2\pi/\omega$-periodic, and its matrizant $R_\epsilon(t)$ will be close to $R(t)$. So $R_\epsilon(2\pi/\omega)$ will be close to $R(2\pi/\omega)$, and hence stable unless $\omega = 2/k$. There is an abundant literature on Mathieu's equation and its generalization, Hill's equation: see [MW] and the references therein. The salient facts will be found in the textbooks we already quoted, [Arn2], [Hal], [JorS], [RouM], [Rou2], and a thorough discussion in the second volume of the beautiful treatise by Yakoubovitz and Starzhinskii [YakS]. The first volume contains an extensive account of Krein theory, its application to a variety of stability problems, and a complete bibliography.

Moser's approach is different, the emphasis being on normal forms, of which Proposition 11 is but a very simple example. This enables him to extend the scope of strong stability to the nonlinear case, by proving the following result.

Suppose a (nonlinear) Hamiltonian $H(x,t)$ is given as a formal power series:

$$H(x,t) = H_2(x,t) + H_3(x,t) + \dots$$

where $H_k(x,t)$ is a homogeneous polynomial in x of degree k, with T-periodic coefficients. Clearly, $x(t) = 0$ for all t is a solution of the associated Hamiltonian system $\dot{x} = JH'(t,x)$. Suppose the linearized system $\dot{y} = JH_2''(t,x(t))\,y$, which has T-periodic coefficients, is strongly stable. Then there is a formal power series with T-periodic coefficients

$$G(x,t) = G_2(x,t) + G_4(x,t) + \dots$$

which is positive definite (there are even terms only, and $G_{2k}(x,t) > 0$ for $x \neq 0$) and such that all coefficients in the formal power series

$$\frac{d}{dt}G(t,x(t)) = (G'(x), JH'(x)) + \frac{\partial}{\partial t}G(t,x(t))$$

vanish. If the series for G were convergent, this would imply that the nonlinear system is stable in the sense of Liapounov. In general this series will diverge, but Moser's result at least shows that perturbation theory will be unable to account for any instability, and leads us to believe that such a system will be considered stable for most practical purposes.

3. Time-Dependence of the Eigenvalues of R(t)

In the particular case of a *positive definite* system, we can describe in more detail the behaviour of the eigenvalues of $R(t)$ as t increases from 0.

Consider the Hamiltonian system in $\mathbb{R}^{2n}$

$$\dot{x} = JA(t)x \tag{1}$$

with $A(t) = A(t)^*$, and $A(t+T) = A(t)$ for all t, and assume that $A(t)$ is positive definite for all t:

$$(A(t)\xi, \xi) > 0\,. \tag{2}$$

We first investigate the behaviour of the eigenvalues of $R(t)$ for t close to 0, that is, $R(t)$ close to the identity:

Proposition 1. *Consider the system* (1), (2). *Then there is an* $\epsilon > 0$ *such that, whenever* $|t| \leq \epsilon$, $R(t)$ *has all its eigenvalues on* $\mathcal{U}$ *and* $\neq \pm 1$. *For* $0 < t < \epsilon$, *the* n *eigenvalues in the upper half-circle are all Krein-positive, and for* $\epsilon < t < 0$ *they are all Krein-negative.*

Proof. We have $R(0) = I$, so that all the eigenvalues start at 1 when $t = 0$.

Let us first show that the eigenvalues stay on U and are different from 1 when t is small. By Corollary 2.6, and the following note, it is enough to show that $R(t)$ has no G-isotropic eigenvector for $|t| \neq 0$ small. Suppose otherwise; then there is a sequence $t_n \to 0$, with $t_n \neq 0$ for all n, and sequences $\lambda_n \in \mathbb{C}$ and $x_n \in \mathbb{C}^{2n}$ with:

$$(3) \qquad R(t_n)\, x_n = \lambda_n x_n\ , \quad (Gx_n, x_n) = 0\ , \quad (x_n, x_n) = 1\ .$$

Hence:

$$(4) \qquad (G(R(t_n) - I)\, x_n, x_n) = (\lambda_n - 1)\,(Gx_n, x_n) = 0\ .$$

Divide by t_n and let $n \to \infty$. By compactness, we can assume that the sequence x_n converges to some unit vector x. Remember also that $\dot{R} = JA(t)R$. We get:

$$(5) \qquad (GJA(0)x, x) = 0\ .$$

Remember now that $G = -iJ$ and that $A(0)$ is positive definite. The left-hand side of the above equality is nonzero, pure imaginary in fact, and we have a contradiction. The first part of the proposition is proved. We have even more: all eigenvalues must be Krein-definite.

So the eigenvalues are either Krein-positive or Krein-negative. Let us see what happens to the Krein-positive ones for small positive t. Take sequences $t_n \to 0$ in $\mathbb{R}^+$, $\lambda_n \in \mathbb{C}$ and $x_n \in \mathbb{C}^{2n}$ such that:

$$(6) \qquad R(t_n)\, x_n = \lambda_n x_n\ , \quad (Gx_n, x_n) > 0\ , \quad (x_n, x_n) = 1\ .$$

Then:

$$(7) \qquad (G(R(t_n) - I)\, x_n, x_n) = (\lambda_n - 1)\,(Gx_n, x_n)\ .$$

When $n \to \infty$, we can assume that x_n converges to some unit vector x by compaactness. We get:

$$(8) \qquad (G(R(t_n) - I)\, x_n, x_n)\,/t_n \longrightarrow (GJA(0)x, x) = i\,(A(0)x, x)\ ,$$

and hence:

$$(9) \qquad \liminf -i\,(\lambda_n - 1)\,/t_n > 0\ .$$

This implies that the imaginary part of λ_n must eventually be positive, and concludes the proof. □

We now investigate the behaviour or Krein-definite eigenvalues:

Proposition 2. *Consider again the system (1), (2). Assume $R(t_0)$ has a Krein-positive eigenvalue $e^{i\theta_0}$ on $\mathcal{U}$, with eigenspace E_0. Then there is an $\epsilon > 0$ and an $\eta > 0$ such that, for $|t-t_0| < \epsilon$, $R(t)$ has at least one eigenvalue of the form $e^{i\theta}$, with $|\theta-\theta_0| < \eta$; they are all Krein-positive, and the sum of their multiplicities is equal to $\dim E_0$. If $t \to t_0$ and if $e^{i\theta}$ is an eigenvalue of $R(t)$ with $|\theta-\theta_0| < \eta$, then $\theta \to \theta_0$ and*

(10) $\liminf (\theta-\theta_0)(t-t_0)^{-1} \geq \mathrm{Min}\,\{(A(t_0)x,x)/(Gx,x)\,|x \in E_0\} > 0$

(11) $\limsup (\theta-\theta_0)(t-t_0)^{-1} \leq \mathrm{Max}\,\{(A(t_0)x,x)/(Gx,x)\,|x \in E_0\} < \infty\,.$

Proof. Denote by $E_0 \subset \mathbb{C}^{2n}$ the eigenspace associated with the eigenvalue $e^{i\theta_0}$. Since the eigenvalue $e^{i\theta_0}$ is semi-simple, E_0 is also the invariant subspace associated with $e^{i\theta_0}$, that is:

$$E_0 = \mathrm{Ker}\,\left(R(t_0) - e^{i\theta_0} I\right)^{2n}\,. \tag{12}$$

By standard compactness arguments, when $t \to t_0$, the eigenvalues of $R(t)$ converge to the eigenvalues of $R(t_0)$, and if λ is an eigenvalue of $R(t)$ converging to an eigenvalue λ_0 of $R(t_0)$, the corresponding invariant subspace E_λ of $R(t)$ converges to a limit which is contained in the invariant subspace of $R(t_0)$ associated with the eigenvalue λ_0. Denote by $\Lambda(t)$ the spectrum of $R(t)$, and consider the G-orthogonal decomposition given by Proposition 2.5:

$$\mathbb{C}^{2n} = \bigoplus_{\lambda \in \Lambda(t)} E_\lambda\,. \tag{13}$$

Counting the dimensions, we see that E_0 must be spanned by the limits of the E_λ, for $t \to t_0$, $\lambda \in \Lambda(t)$ and $\lambda \to e^{i\theta_0}$. Since the restriction of G to E_0 is positive definite, so is the restriction of G to the E_λ. The result follows, except for formulas (10) and (11) which we now prove.

Take sequences $t_n \to t_0$, $\theta_n \to \theta_0$ and $x_n \in \mathbb{C}^{2n}$ such that:

$$R(t_n)\,x_n = e^{i\theta_n} x_n\,, \quad (Gx_n, x_n) = 1\,. \tag{14}$$

By continuity, G is uniformly positive definite on the E_λ since $E_\lambda \to E_0$. So the sequence x_n is bounded, and we may assume that it converges to some unit vector x_0 with $R(t_0)\,x_0 = e^{i\theta_0} x_0$ and $(Gx_0, x_0) = 1$. Thus:

$$\left(R(t_n) - e^{i\theta_n}\right)(x_n - x_0) = \left(R(t_0) - R(t_n) - e^{i\theta_0} + e^{i\theta_n}\right) x_0\,. \tag{15}$$

Take the inner product with Jx_n on both sides. The left-hand side vanishes since:

$$\begin{aligned}(R(t_n)(x_n-x_0), Jx_n) &= \left(x_n - x_0, R(t_n)^* Jx_n\right)\\ &= \left(x_n - x_0, JR(t_n)^{-1} x_n\right)\\ &= \left(x_n - x_0, Je^{-i\theta_n} x_n\right) = e^{i\theta_n}(x_n - x_0, Jx_n)\,.\end{aligned} \tag{16}$$

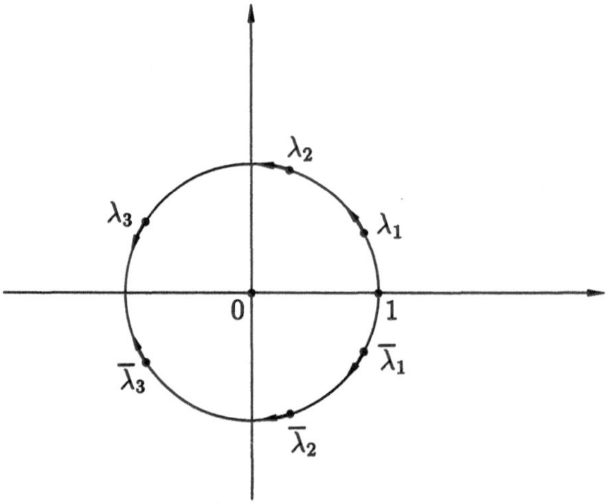

Fig. 2. ($n = 3$) The eigenvalues of $R(t)$ emanate from 1 for $t = 0$ and move on the unit circle, the positive ones on the upper half-circle and counterclockwise, the negative ones on the lower half-circle and clockwise. When $\lambda_3(t)$ and $\overline{\lambda}_3(t)$ meet at -1, they may branch off on the real line, and $R(t)$ will then become unstable.

Equation (15) then yields:

$$([R(t_0) - R(t_n)]\, x_0, Jx_n) = \left(e^{i\theta_0} - e^{i\theta_n}\right)(x_0, Jx_n) \ . \tag{17}$$

Divide by $t_n - t_0$ and let $n \to \infty$. Then $i\,(x_0, Jx_n) = (Gx_0, x_n)$ converges to $(Gx_0, x_0) = 1$. Hence:

$$\begin{aligned} \lim\,(\theta_n - \theta_0)\,(t_n - t_0)^{-1} &= (JA\,(t_0)\,R\,(t_0)\,x_0, Jx_0)\,e^{-i\theta_0} \\ &= (A\,(t_0)\,x_0, x_0) \ . \end{aligned} \tag{18}$$

Formulas (10) and (11) follow immediately. □

As an immediate consequence, we get:

Corollary 3. *If $e^{i\theta_0}$ is a simple eigenvalue of $R\,(t_0)$, there is an $\epsilon > 0$, a neighbourhood $\mathcal{V}$ of $e^{i\theta_0}$ and a C^1 function $t \to \theta(t)$ defined on the interval $(\theta_0 - \epsilon, \theta_0 + \epsilon)$ such that $e^{i\theta(t)}$ is the only eigenvalue of $R(t)$ in $\mathcal{V}$. We have:*

$$\frac{d\theta}{dt}(t_0) = \frac{(A\,(t_0)\,x, x)}{(Gx, x)} \tag{19}$$

where $x \in \mathbb{C}^{2n}$ is an eigenvector of $R\,(t_0)$. □

In other words, Krein-negative eigenvalues move clockwise on the unit circle, and Krein-positive eigenvalues move counterclockwise, as long as they don't collide with eigenvalues of opposite sign.

The behaviour of the eigenvalues as t increases from 0 is now clear, always provided the system is positive definite. At $t = 0$ all the eigenvalues of $R(t)$ are bunched together at $\lambda = 1$. As t increases, they separate, the Krein-positive ones moving counterclockwise on the upper half-circle, and the Krein-negative ones clockwise on the lower half-circle. Instability can occur only when two eigenvalues of different Krein-sign meet, and this happens for the first time at $\lambda = -1$.

Say we have for $t < t_1$ a simple Krein-positive eigenvalue $\lambda_1(t)$, which meets its mate $\lambda_{n+1}(t) = \overline{\lambda}_1(t)$ at $\lambda_1(t_1) = \overline{\lambda}_1(t_1) = -1$. The eigenvalues can cross each other and continue their motion on the unit circle. In most cases, however, the eigenvalues will leave the unit circle and start moving on the negative real half-line, where they are related by $\lambda_{n+1}(t) = \lambda_1(t)^{-1}$. Eventually, they will both come back to -1 at some later time t_2, where they will resume their motion on the unit circle, with the Krein-negative one now moving clockwise on the upper half-circle. It will then interfere with the remaining Krein-positive eigenvalues, and perhaps leave the unit circle again. All this can happen for a stable T-periodic system: the eigenvalues must only be back on the unit circle at the times $t = kT$, $k \in \mathbb{Z}$.

To have a more complete picture, we now turn to Krein-indefinite eigenvalues. As the following results show, they immediately split up into Krein-definite eigenvalues, and eigenvalues which leave the unit circle.

Proposition 4. *Consider again the system* (1), (2). *The set of points t such that $R(t)$ has a Krein-indefinite eigenvalue on the unit circle is discrete; that is, there are only finitely many such points in any bounded interval.*

Proof. Suppose $R(t)$ has a Krein-indefinite eigenvalue $\lambda \in \mathcal{U}$. If λ is not semi-simple, apply Proposition 2.7. If λ is semi-simple, the eigenspace $\operatorname{Ker}(M - \lambda I)$ coincides with the invariant subspace $\operatorname{Ker}(M - \lambda I)^m$, on which the Hermitian form G is indefinite. In either case, there is a G-isotropic eigenvector.

Now argue by contradiction. Say there are infinitely many such points t_k in a certain interval. Then there are sequences $t_k \to t$, with $t_k \neq t$, and $\xi_k \in \mathbb{C}^{2n}$, with $\|\xi_k\| = 1$, such that:

$$R(t_k)\,\xi_k = \lambda_k \xi_k \tag{20}$$

$$(G\xi_k, \xi_k) = 0\ . \tag{21}$$

Arguing by compactness, we may assume that $\xi_k \to \xi$ and $\lambda_k \to \lambda$ with:

$$R(t)\xi = \lambda\xi \tag{22}$$

$$\|\xi\| = 1\ . \tag{23}$$

Start from the relation:

$$(R(t) - \lambda)(\xi_k - \xi) = (R(t) - R(t_k) - \lambda + \lambda_k)\,\xi_k\ . \tag{24}$$

Take the inner product of both sides with $J\xi_k$. The left-hand side vanishes:

$$(25)\qquad \begin{aligned}(R(t)(\xi_k-\xi), J\xi_k) &= (\xi_k-\xi, R(t)^* J\xi_k) = \left(\xi_k-\xi, JR(t)^{-1}\xi_k\right)\\ &= \left(\xi_k-\xi, J\bar{\lambda}\xi_k\right) = \lambda(\xi_k-\xi, J\xi_k)\end{aligned}$$

and we are left with:

$$(26)\qquad ((R(t)-R(t_k))\,\xi_k, J\xi_k) = (\lambda-\lambda_k)(\xi_k, J\xi_k)\ .$$

The right-hand side vanishes by relation (21) since $G = iJ$. Divide by $(t-t_k)\to 0$, and take limits. We get:

$$(27)\qquad (A(t_0)\,\xi,\xi) = 0$$

which contradicts the fact that $A(t_0)$ is positive definite. □

Corollary 5. *Let $\lambda\in\mathcal{U}$ be an eigenvalue of $R(t_0)$ with Krein type (p_0, q_o). Then there is some open interval $\mathcal{N}$ around t_0 and some neighbourhood $\mathcal{V}$ of λ in $\mathbb{C}^{2n}$ such that, if $t_0\neq t\in\mathcal{N}$, $R(t)$ has only Krein-definite eigenvalues in $\mathcal{V}\cap\mathcal{U}$. If $t<T$, the Krein-positive eigenvalues are on the negative side of λ_0 in $\mathcal{V}\cap\mathcal{U}$, while the Krein-negative ones are on the positive side; their positions are interchanged when $t>T$.*

Denote by p_t (resp. q_t) the sum of the multiplicities of the Krein-positive (resp. Krein-negative) eigenvalues contained in $\mathcal{V}\cap\mathcal{U}$. Then p_t and q_t are constant on each side of t_0; we set

$$(28)\qquad p_0^- := p_t\ ,\quad q_0^- := q_t \qquad \text{for } t<T$$

$$(29)\qquad q_0^+ := q_t\ ,\quad p_0^+ := p_t \qquad \text{for } t>T\ .$$

We have:

$$(30)\qquad p_0^+ - q_0^+ = p_0 - q_0 = p_0^- - q_0^-\ .$$

Proof. By Proposition 4, there is some open interval $\mathcal{N}$ around t_0 such that, for $t_0\neq t\in\mathcal{N}$, $R(t)$ has only Krein-definite eigenvalues on $\mathcal{U}$. Choose a neighbourhood $\mathcal{V}$ of λ in $\mathbb{C}^{2n}$ such that λ is the only eigenvalue of $R(t_0)$ in $\mathcal{V}$. Remember now that the eigenvalues are the roots of the algebraic equation Det $(R(t)-\omega I)=0$; taking a smaller $\mathcal{N}$ if need be, we may assume that the sum of the multiplicities of the eigenvalues of $R(t)$ contained in $\mathcal{V}$ is p_0+q_0, and that they all converge to λ when $t\to t_0$.

By Proposition 2, we know that Krein-positive eigenvalues move positively as t increases: since they converge to λ, they must be on the negative side of λ for $t<T$ and on the positive side for $t>T$. The Krein-negative eigenvalues lie on the other side of λ. So, for $t\neq T$, Krein-positive and Krein-negative eigenvalues cannot interfere. They have to stay on the unit circle, and their number stays constant as long as $t\neq T$.

Everything is now proved except formula (30). Denote by $E(t_0)$ the invariant subspace of $R(t_0)$ associated with the eigenvalue λ, and by $E(t)$, for $t\in\mathcal{N}$, the direct sum of the three following subspaces associated with $R(t)$:

$E_+(t)$ is the direct sum of the invariant subspaces associated with the Krein-positive eigenvalues in $\mathcal{V} \cap \mathcal{U}$.

$E_-(t)$ is the direct sum of the invariant subspaces associated with the Krein-negative eigenvalues in $\mathcal{V} \cap \mathcal{U}$.

$E_0(t)$ is the direct sum of the invariant subspaces associated with the eigenvalues in $\mathcal{V}$ which lie away from the unit circle.

When $t \to t_0$, the subspace $E(t)$ converges to the subspace $E(t_0)$. The Hermitian form G is nondegenerate on $E(t_0)$, with p_0 positive squares and q_0 negative ones. If t is close enough to t_0, that is, if s has been chosen small enough, the restriction of G to $E(t)$ will be nondegenerate with signature (p_0, q_0).

The subspaces $E_+(t)$, $E_-(t)$, $E_0(t)$ are G-orthogonal. We have $p_t = \dim E_+(t)$ and $q_t = \dim E_-(t)$. I claim that the restriction of G to $E_0(t)$ has signature (r_t, r_t) where $2r_t = \dim E_0(t)$. It then follows that:

$$p_0 = p_t + r_t \quad \text{and} \quad q_0 = q_t + r_t \tag{31}$$

and we get formula (30) by substraction.

Note first that, by Proposition 2.5, the restriction of G to $E_0(t)$ is nondegenerate. Now approximate $R(t)$ by a sequence $M_k \to R(t)$ of diagonalizable symplectic matrices. Then $E_0(t)$ is the limit of invariant subspaces E_k associated with M_k. Each E_k splits into one-dimensional eigenspaces corresponding to eigenvalues λ of M_k away from the unit circle. Group these eigenspaces in pairs $F_\lambda := E_\lambda \oplus E_{1/\bar{\lambda}}$; the F_λ are pairwise orthogonal by Proposition 2.5, and they contain isotropic vectors by Corollary 2.6. The restriction of G to F_λ must therefore have signature (1,1). So the restriction of G to E_k must have signature (r_t, r_t), where r_t is the number of such pairs F_λ. Since $E_k \to E_0(t)$, we have $2r_t = \dim E_0(t)$, and since the restriction of G to $E_0(t)$ is nondegenerate, it must also have signature (r_t, r_t). □

In other words, Krein-positive and Krein-negative eigenvalues leave the unit circle in pairs, while the remaining ones continue their motion on $\mathcal{U}$ in the direction prescribed by their Krein-sign. A Krein-indefinite eigenvalue is the place where such a collision occurs. We formalize this idea by a definition:

Definition 6. The number r_t is constant on each side of t_0:

$$r_0^- := r_t \quad \text{for } t < T \tag{32}$$

$$r_0^+ := r_t \quad \text{for } t > T \tag{33}$$

We refer to $2r_0^-$ as *the number of eigenvalues which arrive on the unit circle at* λ, and to $2r_0^+$ as *the number of eigenvalues which leave.* □

Denote by m the multiplicity of the eigenvalue λ. It follows from formulas (30) and (31) that:

$$2r_0^- + p_0^- + q_0^- = m \quad \text{and} \quad 2r_0^+ + p_0^+ + q_0^+ = m \tag{34}$$

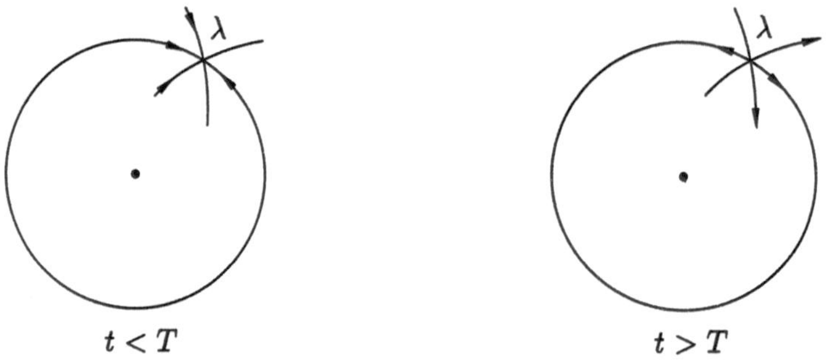

Fig. 3. At time $t = T$, there is a collision between four simple eigenvalues of $R(t)$ indicated by arrows resulting in an eigenvalue λ on the unit circle with multiplicity 4 and Krein type (2,2). In the notations of Corollary 5, we have $p_0^+ = q_0^+ = \overline{p}_0^- = \overline{q}_0^- = 1$, and $r_0^+ = p_0 - p_0^+ = 1 = p_0^- - p_0^- = r_0^-$. By Definition 6, this means that 2 eigenvalues arrive on the unit circle at λ, and 2 eigenvalues leave.

(35) $$p_0 = p_0^- + r_0^- \quad \text{and} \quad q_0 = q_0^- + r_0^-$$

(36) $$p_0 = p_0^+ + r_0^+ \quad \text{and} \quad q_0 = q_0^+ + r_0^+ .$$

Later on (Proposition 5.13), we shall prove that

(37) $$r_0^+ + r_0^- = m - \dim \operatorname{Ker} (R(T) - \lambda I) .$$

We conclude this section by a general result, which ensures that eigenvalues of $R(t)$ have to move with t. It is a consequence of Propositions 4 and 2, but we prefer to state it independently.

Proposition 7. *Consider again system* (1), (2). *Let* $\lambda \in \mathcal{U}$ *be an eigenvalue of* $R(t_0)$. *There is some open interval* $\mathcal{N}$ *around* t_0 *such that, if* $t_0 \neq t \in \mathcal{N}$, *then* λ *is not an eigenvalue of* $R(t)$.

Proof. Argue by contradiction. Say there are sequences $h_k \to 0$, with $h_k \neq 0$, and $\xi_k \in \mathbb{C}^{2n}$, with $\|\xi_k\| = 1$, such that:

(38) $$R(t_0 + h_k)\,\xi_k = \lambda_k \xi_k .$$

Arguing by compactness, we may assume that $\xi_k \to \xi$, where:

(39) $$R(t_0)\,\xi = \lambda\xi .$$

Start from the relation:

(40) $$(R(t_0 + h_k) - \lambda)(\xi_k - \xi) = (R(t_0) - R(t_0 + h_k))\,\xi .$$

Take the inner product of both sides with $J\xi_k$. The left-hand side vanishes – see the calculations (16) – and we are left with:

(41) $$((R(t_0) - R(t_0 + h_k))\,\xi, J\xi_k) = 0 .$$

Divide by $h_k \to 0$, and take limits. We get:

$$(A(t_0)\xi, \xi) = 0 \tag{42}$$

which contradicts the fact that $A(t_0)$ is positive definite. □

Notes and Comments. It is a standard feature of Krein theory to investigate the behaviour of the Floquet multipliers under perturbations.

Consider the Hamiltonian system:

$$\dot{x} = J(A_0(t) + \mu A_1(t))\, x$$

where A_0 and A_1 are given and μ is a parameter. It is assumed that A_1 is positive definite. Krein proved that, for real μ, the Krein-positive eigenvalues move positively on $\mathcal{U}$, while Gelfand and Lidskii investigated complex μ and defined Krein-positive eigenvalues to be those which move into the unit disk as μ moves into the upper half-plane [GeLi]. For the same system, the Krein-Liubarskii theorem ([YakS], ch. 3 § 4) gives a detailed analysis of Krein-indefinite eigenvalues. It states that if $e^{i\theta_0}$ is an eigenvalue of $R(\mu_0, T)$ with an irreducible invariant subspace of dimension $d > 1$, then there is an $\epsilon > 0$ and a holomorphic map φ from a neighbourhood of 0 in $\mathbb{C}$ to a neighbourhood of $e^{i\theta_0}$, with $\varphi'(0) \neq 0$, such that Det $(R(\mu, T) - \varphi(z)I) = 0$ whenever $|\mu - \mu_0| < \epsilon$ and $z^d = (\mu - \mu_0)$. It follows that the set of $\omega \in \mathbb{C}$ such that Det $(R(\mu, T) - \omega I) = 0$ for some $\mu \in \mathbb{R}$ contains d real analytic curves $\omega_1(s), \dots \omega_d(s)$, defined for $|s|$ small, with $\omega_j(0) = e^{i\theta_0}$ for $1 \leq j \leq d$; one of them has to lie on the unit circle, say $\omega_1(s) \in \mathcal{U}$ for all s, and the angle between the tangents to $\omega_j(s)$ and $\omega_{j+1}(s)$ at 0 is $2\pi/d$.

Unfortunately, all this analysis has been carried out under the assumption that everything depends analytically on the parameter λ, which rules out the possibility of studying time-dependence by this method (that is, taking T as parameter). This is why the results in this section are not covered directly by those we just quoted. However, the spirit is the same. For every d-dimensional irreducible invariant subspace of $R(t)$, associated with the eigenvalue $\lambda \in \mathcal{U}$, we will get $(d-1)$ curves bifurcating from the unit circle at λ, each of which carries a pair of eigenvalues of $R(t)$. The total number of such curves is $r_0^+ + r_0^-$; the r_0^+ counts the "outgoing" curves (those which carry a pair of eigenvalues of $R(t)$ for $t \geq T$; they start at λ for $t = T$ and separate for $t > T$, one inside the unit disk and one outside), and r_0^- the "ingoing" ones.

4. Index Theory for Positive Definite Systems

We continue our investigation of T-periodic linear Hamiltonian systems in $\mathbb{R}^{2n}$:

$$\dot{x} = JA(t)x\ , \tag{1}$$

where $A^*(t) = A(t) = A(t+T)$, under the added assumption that the matrix $A(t)$ is positive definite:

$$(A(t)x, x) > 0 \quad \forall x \neq 0 \, . \tag{2}$$

Since A depends continuously on t, we can find positive constants a and b such that:

$$a \|x\|^2 \geq (A(t)x, x) \geq b^{-1} \|x\|^2 \quad \forall (t, x) \tag{3}$$

and $A(t)$ has an inverse $B(t) := A(t)^{-1}$, which is also continuous, T-periodic and positive definite. By the preceding inequalities, we have:

$$b \|x\|^2 \geq (B(t)x, x) \geq a^{-1} \|x\|^2 \quad \forall (t, x) \, . \tag{4}$$

With every $s > 0$ we associate a real Hilbert space E_s and a quadratic form Q_s on E_s defined by:

$$E_s := \left\{ x \in W^{1,2} \left(0, s; \mathbb{R}^{2n}\right) \middle| x(0) = x(s) \right\} \tag{5}$$

$$Q_s(x, x) := \frac{1}{2} \int_o^s [(Jx, \dot{x}) + (B(t)J\dot{x}, J\dot{x})] \, dt \, . \tag{6}$$

Note that, because of the periodicity condition on x:

$$Q_s(x + \xi, x + \xi) = Q_s(x, x) \quad \forall \xi \in \mathbb{R}^{2n} \, , \tag{7}$$

the true variable in formula (6) is $\dot{x}$, and we can choose for x any primitive we like. The only prerequisite on $\dot{x}$ is that:

$$\dot{x} \in L^2 \left(0, s; \mathbb{R}^{2n}\right) \quad \text{and} \quad \int_o^s \dot{x} \, dt = 0 \, . \tag{8}$$

In other words, we have

$$Q_s(x, x) = q_s(\dot{x}, \dot{x}) \, , \tag{9}$$

where q_s is a quadratic form on the Hilbert space $L^2_o(0, s)$ defined by:

$$L^2_o(0, s) = \left\{ u \in L^2 \left(0, s; \mathbb{R}^{2n}\right) \middle| \int_o^s u \, dt = 0 \right\} \tag{10}$$

$$q_s(u, u) = \frac{1}{2} \int_o^s [(Ju, \Pi_s u) + (B(t)Ju, Ju)] \, dt \, . \tag{11}$$

Here Π_s denotes the primitive of u with mean value zero:

$$\frac{d}{dt} \Pi_s u = u \quad \text{and} \quad \int_o^s \Pi_s \, dt = 0 \, . \tag{12}$$

Lemma 1. *Π_s is a compact operator from $L_o^2(0,s)$ into itself. We have $\Pi_s^* = -\Pi_s$ and:*

$$\|\Pi_s\| \leq \frac{s}{2\pi} \; . \tag{13}$$

Proof. Π_s sends $L_o^2(0,s)$ into $W^{1,2}(0,s)$ and the identity map from $W^{1,2}(0,s)$ to $L_o^2(0,s)$ is compact by the Rellich-Kondrachov theorem.

To check that it is antisymmetric, we just integrate by parts:

$$\int_o^s (\Pi_s u, v)\, dt = -\int_o^s (u, \Pi_s v)\, dt + (\Pi_s u, \Pi_s v)\,\Big|_o^s \tag{14}$$

and the last term vanishes since:

$$\Pi_s u(T) - \Pi_s u(0) = \int_o^s u(t) dt = 0 \; . \tag{15}$$

Estimate (13) is the so-called Wirtinger inequality, which we now prove. Take any u in $L_o^2(0,s)$. Expand u and $\Pi_s u$ in Fourier series, remembering that there is no constant term since their mean value in zero:

$$u(t) = \sum_{n\neq 0} u_n e^{2i\pi nt/s} \; , \qquad \text{with} \quad u_{-n} = \overline{u}_n \tag{16}$$

$$\Pi_s u(t) = \sum_{n\neq 0} \frac{s}{2i\pi n} u_n e^{2i\pi nt/s} \; , \qquad \text{with} \quad u_{-n} = \overline{u}_n \; . \tag{17}$$

Using Parseval's equality twice:

$$\begin{aligned} \|\Pi_s u\|^2 &= \sum_{n\neq 0} \frac{s^2}{4n^2\pi^2} |u_n|^2 s \\ &\leq \sum_{n\neq 0} \frac{s^2}{4\pi^2} |u_n|^2 s = \frac{s^2}{4\pi^2} \|u\|^2 \; . \end{aligned} \tag{18}$$

□

So q_s, as defined by formula (11), can be looked upon as a compact perturbation of a positive definite quadratic form. As such, it need no longer be positive definite or even nondegenerate, but the next result will show that it is positive definite on a subspace of finite codimension.

Proposition 2. *There is a splitting:*

$$L_o^2(0,s) = E_+(0,s) \bigoplus E_o(0,s) \bigoplus E_-(0,s) \tag{19}$$

such that:

(a) *$E_+(0,s)$, $E_o(0,s)$ and $E_-(0,s)$ are q_s-orthogonal*

(b) $q_s(u,u) > 0 \quad \forall u \in E_+(0,s) \setminus \{0\}$

(c) $q_s(u,u) = 0 \quad \forall u \in E_o(0,s)$

(d) $q_s(u,u) < 0 \quad \forall u \in E_-(0,s) \setminus \{0\}$

(e) $E_o(0,s)$ *and* $E_-(0,s)$ *are finite-dimensional.*

Proof. Define a self-adjoint operator $\overline{B}_s$ from $L_o^2(0,s)$ into itself by:

$$(\overline{B}_s u, u) = \int_o^s (B(t)Ju, Ju)\, dt\,. \tag{20}$$

By formula (4), we have

$$sb\, \|u\|^2 \geq (\overline{B}_s u, u) \geq sa^{-1}\, \|u\|^2\,. \tag{21}$$

So $\overline{B}_s$ is an isomorphism, and $(\overline{B}_s u, u)$ defines a Hilbert space structure on $L_o^2(0,s)$ which is equivalent to the standard one. Endowing $L_o^2(0,s)$ with the interior product $(\overline{B}_s u, u)$, and applying to $J\Pi_s$ the spectral theory of compact selfadjoint operators on Hilbert space, we see that there is a basis e_n, $n \in \mathbb{N}$, of $L_o^2(0,s)$, and a sequence $\lambda_n \to 0$ in $\mathbb{R}$ such that:

$$(\overline{B}_s e_i, e_j) = \delta_{ij} \tag{22}$$

$$(J\Pi_s e_n, u) = (\overline{B}_s \lambda_n e_n, u) \quad \forall u \in L_o^2(0,s)\,. \tag{23}$$

Hence, expressing any vector $u \in L_o^2(0,s)$ as $u = \Sigma \xi_n e_n$, the expansion:

$$\begin{aligned} q_s(u,u) &= \frac{1}{2}(-J\Pi_s u, u) + \frac{1}{2}(\overline{B}_s u, u) \\ &= -\frac{1}{2}\sum_{n=1}^{\infty} \lambda_n \xi_n^2 + \frac{1}{2}\sum_{n=1}^{\infty} \xi_n^2 \\ &= \frac{1}{2}\sum_{n=1}^{\infty} (1-\lambda_n)\, \xi_n^2\,. \end{aligned} \tag{24}$$

Since $\lambda_n \to 0$ when $n \to \infty$, all the coefficients $(1-\lambda_n)$ are positive except a finite number. Hence the result, with:

$$E_+(0,s) = \left\{ \sum \xi_n e_n | \xi_n = 0 \text{ if } 1-\lambda_n \leq 0 \right\} \tag{25}$$

$$E_o(0,s) = \left\{ \sum \xi_n e_n | \xi_n = 0 \text{ if } 1-\lambda_n \neq 0 \right\} \tag{26}$$

$$E_-(0,s) = \left\{ \sum \xi_n e_n | \xi_n = 0 \text{ if } 1-\lambda_n \geq 0 \right\} \tag{27}$$

□

Definition 3. The *index* of system (1) on the interval $(0,s)$, denoted by i_s, is the dimension of $E_-(0,s)$. The *nullity*, denoted by ν_s, is the dimension of $E_o(0,s)$. □

The rest of this section will develop the meaning of these definitions. The nullity, to begin with, has a ready interpretation. The subspace $E_o(0,s)$ is simply the kernel of q_s; if it is non-trivial, that is $E_o(0,s) \neq \{0\}$, we say that q_s is *degenerate*, and we have:

Theorem 4. *The nullity ν_s is the number of linearly independent solutions of the boundary-value problem*

$$(\mathcal{P}_s)\begin{cases} \dot{x} = JA(t)x \\ x(0) = x(s) \end{cases} . \qquad \square$$

Proof. Call $\mathbb{R}^{2n}$ the subspace of constant functions in $L^2\left(0, s; \mathbb{R}^{2n}\right)$: it is the orthogonal complement of $L^2_o(0, s)$. We have:

$$q_s(u, v) = \frac{1}{2}\int_o^s (-J\Pi_s u - JB(t)Ju, v)\, dt . \tag{28}$$

The kernel of q_s consists of all $u \in L^2_o(0, s)$ such that this interior product vanishes for all $v \in L^2_o(0, s)$. This means that there is some constant $\xi \in \mathbb{R}^{2n}$ such that:

$$-J\Pi_s u - JB(t)Ju = \xi . \tag{29}$$

This we rewrite as:

$$\Pi_s u - J\xi = -B(t)Ju \tag{30}$$

and since $B(t)$ is the inverse of $A(t)$:

$$-Ju = A(t)\left(\Pi_s u - J\xi\right) . \tag{31}$$

Now define x to be $\Pi_s u - J\xi$. Then $x(0) = x(s)$, and Eq. (30) reads as $-J\dot{x} = A(t)x$, which is precisely system (1). □

In other words, ν_s is the dimension of Ker $(R(s) - I)$, where $R(t)$ is the matrizant of system (1). To say that q_s is degenerate means that 1 is an eigenvalue of $R(s)$. By Proposition 3.7, this can happen only at isolated points.

Definition 5. If $\nu_s \neq 0$, we say that s is *conjugate to* 0 with *multiplicity* ν_s. □

We can also define conjugacy between any two points: s_1 and s_2 are conjugate if the equation $\dot{x} = JA(t)x$ has a non-zero solution with $x(s_1) = x(s_2)$. It should be noted that this relation is not transitive: if $(0, s_1)$ and (s_1, s_2) are conjugate times, then s_2 need not be conjugate to 0. It should also be noted that the fact that s is conjugate to 0 does not imply that $(s + T)$ has the same property, even though $A(t)$ is T-periodic.

We are now ready to interpret the index in geometrical terms:

Theorem 6. *The index i_S of system* (1) *on the interval* $(0, S)$ *is equal to the number of $s \in (0, S)$ which are conjugate to* 0, *each counted with multiplicity:*

$$i_s = \sum_{0<s<S} \nu_s . \tag{32}$$

□

The proof proceeds through a sequence of lemmas.

Lemma 7. *When* $0 < s < 2\pi/a$, *we have* $i_s = 0$.

Proof. By formula (4), we have:

$$\int_o^s (B(t)Ju, Ju)\, dt \geq a^{-1} \|u\|^2 \ . \tag{33}$$

Applying Cauchy-Schwarz to formula (11), we get

$$q_s(u,u) \geq \frac{1}{2}\left(- \|u\| \, \|\Pi_s u\| + a^{-1} \|u\|^2\right) \ . \tag{34}$$

We use Wirtinger's inequality of Lemma 1:

$$q_s(u,u) \geq \frac{1}{2}\left(-\frac{s}{2\pi} + a^{-1}\right) \|u\|^2 \ . \tag{35}$$

So q_s is positive definite for $s < 2\pi/a$. □

Lemma 8. *If* $s_1 \leq s_2$ *then*

$$i_{s_1} \leq i_{s_2} \tag{36}$$

$$i_{s_1} + \nu_{s_1} \leq i_{s_2} + \nu_{s_2} \ . \tag{37}$$

□

Proof. Define a map $p : L_o^2(0, s_1) \to L_o^2(0, s_2)$ by:

$$(pu)(t) = \begin{cases} u(t) & \text{if} \quad 0 \leq t \leq s_1 \\ 0 & \text{if} \quad s_1 < t \leq s_2 \end{cases} . \tag{38}$$

Clearly, $q_{s_1}(u,u) = q_{s_2}(pu, pu)$ for every $u \in L_o^2(0, s_1)$. If $E_1 \subset L^2(0, s_1)$ is a negative definite subspace for q_{s_1}, then $E_2 = pE_1$ will be a negative definite subspace for q_{s_2} of the same dimension. Formula (36) follows by taking E_1 to be maximal with that property, $\dim E_1 = i_{s_1}$.

Similarly, if $E_1' \subset L^2(0, s_1)$ is a subspace on which q_{s_1} is negative semi-definite, then q_{s_2} will be negative semi-definite on $E_2' = pE_1'$, and formula (37) follows. □

Lemma 9. *We have:*

$$i_{s+0} \geq i_s + \nu_s \ . \tag{39}$$

Proof. When $\nu_s = 0$, this follows from formula (36). Assume $\nu_\sigma \neq 0$, that is, σ is conjugate to 0. By proposition 3.7, we know that σ is isolated, that is, there are no other points conjugate to 0 in some neighbourhood $\mathcal{N}$ of σ. Taking any point $s \in \mathcal{N}$, with $s > \sigma$, we have $\nu_s = 0$, and hence by Lemma 8:

$$i_s = i_s + \nu_s \geq i_\sigma + \nu_\sigma \ . \tag{40}$$

□

Lemma 10. *There are only finitely many points conjugate to* 0 *in any bounded interval* $(0, s)$.

Proof. It follows from Lemma 9 that each conjugate point σ in $(0, s)$ contributes at least ν_σ to i_s. Since $i_s < \infty$, there must be finitely many σ. □

The next – and final – lemma studies more closely the discontinuities of the function $s \to i_s$.

Lemma 11. *The function $s \to i_s$ is left continuous, and*

$$i_{s+0} - i_{s-0} = \nu_s \ . \tag{41}$$

Proof. We are dealing with a variable quadratic form q_s on a variable space $L^2_o(0, s)$. The first thing to do is to rescale everything to the fixed time interval $(0, 1)$.

Define a map $p : L^2_o(0, s) \to L^2_o(0, 1)$ by $(pu)(t) = u(st)$. Then $q_s(u, u) = sq^1_s(pu, pu)$, where q^1_s is the quadratic form on $L^2_o(0, 1)$ defined by:

$$q^1_s(v) = \frac{1}{2} \int_o^1 [s\,(Jv, \Pi_1 v) + (B(st)Jv, Jv)]\, dt \ . \tag{42}$$

So i_s is also the index of q^1_s. Arguing as in Proposition 2, we can find a basis e^s_i, $i \in \mathbb{N}$, of $L^2_o(0, 1)$ and a sequence λ^s_i in $\mathbb{R}$, with $\lambda^s_i \to 0$ when $i \to \infty$, such that:

$$\left(B^1(s)e^s_i, e^s_j\right) = \delta_{ij} \tag{43}$$

$$sJ\Pi_1 e^s_i = \lambda^s_i(s) B^1(s) e^s_i \ . \tag{44}$$

Here $B^1(s)$ is the self-adjoint isomorphism from $L^2_o(0, 1)$ into itself defined by:

$$\left(B^1(s)u, v\right) = \int_o^1 (B(st)Ju, Jv)\; dt \ . \tag{45}$$

Clearly B^1 depends continuously on s in the operator norm. Writing $u = \Sigma \xi_i e^s_i$, for any $u \in L^2_o(0, 1)$, we get

$$q^1_s(u, u) = \frac{1}{2} \sum_{i=1}^{\infty} (1 - s\lambda^s_i)\, \xi_i^2 \ . \tag{46}$$

Fix some point $\sigma > 0$.

For the sake of convenience, we set $i_\sigma = K$. This means that there is some $\sigma' > \sigma$ such that $i_s = K$ for all s in the interval $I := (\sigma, \sigma')$. By Lemma 10, we may assume that I contains no point conjugate to 0, so that for every $s \in I$, we have $\nu_s = 0$, and the K first eigenvalues of q_s are negative:

$$1 - s\lambda^s_i < 0 \qquad \text{for } 1 \le i \le K \ . \tag{47}$$

Fix $i \le K$. By the preceding inequality, $\lambda^s_i \ge 1/s > 1/\sigma'$ so that the λ^s_i are bounded away from zero for $s \in I$. Because of the orthonormality relation

(43), the e_i^s are bounded in $L_o^2(0,1)$. So the set $\{se_i^s/\lambda_i^s | s \in I\}$ is bounded, and its image by the compact map Π_1 (see Lemma 1) must be precompact. In other words, there is a sequence $s(n) \to \sigma$ with $s(n) > \sigma$, and a vector $w \in L_o^2(0,1)$ such that:

(48) $$s(n)\left(\lambda_i^{s(n)}\right)^{-1} J\Pi_1 e_i^{s(n)} \to w \, .$$

Rewrite Eq. (44):

(49) $$s(n)\left(\lambda_i^{s(n)}\right)^{-1} \Pi_1 e_i^{s(n)} = B^1\left(s(n)\right) e_i^{s(n)} \, .$$

Since the left-hand side converges to w, so does the right-hand side. But $B^1\left(s(n)\right) \to B^1(\sigma)$ by continuity, and so:

(50) $$e_i^{s(n)} \to B^1(\sigma)^{-1} w := e_i \, .$$

Since $\left(B^1\left(s(n)\right) e_i^{s(n)}, e_i^{s(n)}\right) = 1$, we get by continuity $\left(B^1(\sigma)e_i, e_i\right) = 1$. It follows that $w = B^1(\sigma)e_i$ is non-zero, and hence, by formula (49), that $\lambda_i^{s(n)}$ converges to some λ_i. Taking limits in Eq. (49), we get:

(51) $$\sigma J \Pi_1 e_i = \lambda_i B^1(\sigma) e_i$$

with:

(52) $$(1 - \sigma\lambda_i) = \lim_n \left(1 - s(n)\lambda_i^{s(n)}\right) \leq 0 \, .$$

We proceed likewise for $i = 1, \ldots, K$. We get in this way K eigenvectors e_i, associated with K eigenvalues λ_i, satisfying (51) and (52). Taking limits in formula (43), when $s \to \sigma$, we see that $(B(\sigma)e_i, e_j) = \delta_{ij}$; as a consequence, the e_i are distinct and non-zero. It follows from inequality (52) that q_σ^1 is negative semi-definite on the K-dimensional subspace generated by $e_i, \ldots, e_K$. Hence:

(53) $$i_\sigma + \nu_\sigma \geq K = i_{\sigma+0} \, .$$

Lemma 9 provides us with the reverse inequality, so that

(54) $$i_{\sigma+0} = i_\sigma + \nu_\sigma \, .$$

On the other hand, $L_o^2(0,1)$ contains a subspace E_σ of dimension i_σ on which q_σ is negative definite. Since q_s depends continuously on s, the restriction of q_s to E_σ will still be negative definite when s is close to σ. Hence $i_\sigma \leq i_{\sigma-0}$. The converse inequality holds by Lemma 8, so

(55) $$i_\sigma = i_{\sigma-0}$$

and i_s is left continuous. Replacing i_σ by $i_{\sigma-0}$ in formula (54), we get the result.

Proof of Theorem 6. The function $s \to i_s$ is integer-valued, left continuous and non-decreasing on $(0, +\infty)$. Its value at any point S must therefore be equal to the sum of the jumps it incurred in $(0, S)$. By Lemma 11, this is precisely the sum of the ν_s, for $0 < s < S$.

Theorem 4 and 6 provide us with a satisfactory interpretation of the nullity and the index. The next question is whether these numbers are actually non-zero, that is, whether points conjugate to zero exist in general.

We begin by a remark about ordering. If A_1 and A_2 are two symmetric matrices, $A_1 \leq A_2$ means that $A_2 - A_1$ is positive semi-definite:

(56) $$(A_2 x, x) \geq (A_1 x, x) \qquad \text{for all } x .$$

Proposition 12. *Consider two linear systems $\dot{x} = JA_1(t)x$ and $\dot{x} = JA_2(t)x$, with $A_1(t)$ and $A_2(t)$ continuous, symmetric and positive definite. Assume:*

(57) $$A_1(t) \leq A_2(t) \quad \forall t .$$

Then their indices i_s^1 and i_s^2 satisfy:

(58) $$i_s^i \leq i_s^2 \quad \forall s .$$

Proof. Clearly $B_1(t) \geq B_2(t)$ for all t, so that the associated quadratic forms q_s^1 and q_s^2 satisfy $q_s^1(u,u) \geq q_s^2(u,u)$ for all u in $L_o^2(0,s)$. So, q_s^2 must be negative definite on any subspace where q_s^1 is negative definite, and formula (58) follows from the definition of the index. □

This will enable us to compare the system we are investigating, $\dot{x} = JA(t)x$, with constant-coefficients systems, using estimate (3), which states in effect that: there are constants $a \geq b^{-1} > 0$ with

(59) $$aI \geq A(t) \geq b^{-1} I \qquad \text{for all } t .$$

We denote by $E[\alpha]$ the integer part of the real number α, defined as follows:

(60) $$E[\alpha] = k \Leftrightarrow k < \alpha \leq k + 1 .$$

Note that this is not the standard definition: we get $E[a] = a - 1$ for all integers a.

Lemma 13. *Fix $c > 0$. The index of the linear Hamiltonian system with constant coefficients:*

(61) $$\dot{x} = cJx \qquad \text{in } \mathbb{R}^{2n}$$

on the time interval $(0, s)$ is given by:

(62) $$i_s = 2nE\left[\frac{cs}{2\pi}\right] .$$

Proof. The solution of Eq. (61) is $x = e^{cJt}x(0)$. The only times conjugate to 0 are the $s_k = 2k\pi/c$, $k \in \mathbb{Z}$, each one with multiplicity $2n$. The result follows from Theorem 6. □

We can now estimate the distance between two conjugate points, and the index over an interval.

Proposition 14. *Consider the Hamiltonian system*

$$\dot{x} = JA(t)x \tag{63}$$

which satisfies condition (3).

(a) *Take any s_1, and let $s_2 > s_1$ be the first time which is conjugate to s_1. In other words, there is no s in the time interval (s_1, s_2) which is conjugate to s_1. Then:*

$$2\pi a^{-1} \leq s_2 - s_1 \leq 2\pi b \ . \tag{64}$$

(b) *For any $s > 0$, we have the estimate:*

$$2nE\left[\frac{s}{2\pi b}\right] \leq i_s \leq 2nE\left[\frac{as}{2\pi}\right] \ . \tag{65}$$

Proof. (a) There is no loss of generality in taking $s_1 = 0$. Consider the three systems:

$$\dot{x} = aJx \tag{66}$$

$$\dot{x} = JA(t)x \tag{67}$$

$$\dot{x} = b^{-1}Jx \tag{68}$$

and denote by $\bar{i}_s, i_s$ and $\underline{i}_s$ their indices on the interval $(0, s)$. By Proposition 12, we have:

$$\underline{i}_s \leq i_s \leq \bar{i}_s \ . \tag{69}$$

First take $s = s_2$. Then $i_s = 0$ by Theorem 6, so that $\underline{i}_s = 0$ and Lemma 13 gives:

$$\frac{1}{2\pi b} s_2 \leq 1 \ . \tag{70}$$

Then take $s > s_2$. Then $i_s \geq 1$ by Theorem 6 again, so that $\bar{i}_s \geq 1$ and Lemma 13 gives:

$$\frac{a}{2\pi} s > 1 \ . \tag{71}$$

Letting $s \to s_2$, we get formula (64).

(b) Follows from formula (69) and Lemma 13. □

We will show in the next section that $s^{-1}i_s$ has a limit when $s \to \infty$. From formula (65) it will then follow that:

$$\frac{n}{\pi b} \le \lim_{s\to\infty} s^{-1} i_s \le \frac{an}{\pi} \ . \tag{72}$$

Notes and Comments. There are several ways to define the index of a linear Hamiltonian system. The index was first discovered and studied within the Russian school of ordinary differential equations, mostly for the needs of stability theory. After the pionneering work of Krein ([Kre1] to [Kre4]), there was the fundamental paper by Gelfand and Lidsky [GelL]. Again we refer to the treatise [YakS]; a concise account appears in [Lev].

In this approach, the index is understood as a winding number: the group of symplectic matrices M is shown to be homotopic to S^1, and the index of the system $\dot{x} = JA(t)x$ over the interval $[0, s]$ just counts the number of times $R(t)$ (or rather, its image in the homotopy) runs along the circle when moving continuously from $R(0) = I$ to $R(s)$. The direction of rotation has to be taken into account, and therefore this index is a relative integer. Call it the standard index: $i_{\mathbf{standard}} \in \mathbb{Z}$. It is the one that Amann and Zehnder [AmaZ1], [AmaZ2] and then Conley and Zehnder [ConZ1], [ConZ2], [ConZ3] have been using in their celebrated work.

It is of course related to the work of Maslov, which was going on at the same time [Mas], and in fact the standard index can be understood as a Maslov index. This aspect of things was discovered independently by Bott [Bot1], and investigated by Duistermaat [Dui]. The idea is to double the dimension. Consider the system:

$$(\dot{x}_1, \dot{x}_2) = (A(t)x_1, 0)$$

in $\mathbb{R}^{2n} \times \mathbb{R}^{2n}$, endowed with the symplectic structures $[\xi, \eta] = (J\xi_1, \eta_1) - (J\xi_2, \eta_2)$. The diagonal $\xi = \eta$ then is a Lagrangian $2n$-plane, Π_o say, and so is its image Π_t by the matrizant $R(t) \times I$ of this system. Now there is an intersection number for Lagrangian $2n$-planes in symplectic $4n$-space, which is zero if the two planes are in general position, and a relative integer otherwise. The index of the system $\dot{x} = JA(t)x$ on the interval $(0, s)$ is the (algebraic) number of times Π_t intersects Π_o with $0 < t < s$. It is the same as the standard index.

The index of a positive definite Hamiltonian system, as defined in this section, was introduced by the author in the papers [Eke8] and [Eke9]. In contrast with the standard index, it is a Morse index, and therefore a non-negative integer. It was proved by Brousseau [Bro] that

$$\text{Morse index} \ = \ \text{standard index} - n \ .$$

Theorem 6 was proved in [Eke8] (see also [EkeH]). It is of course inspired by Morse's celebrated index theorem in Riemannian geometry (see [Mil]). The other results in this section come from [Eke9].

5. The Iteration Formula

We are dealing with the linear Hamiltonian system

$$\dot{x} = JA(t)x \quad \text{in } \mathbb{R}^{2n} \tag{1}$$

where $A^*(t) = A(t)$ is a continuous T-periodic positive definite matrix:

$$a\|x\|^2 \geq (A(t)x, x) \geq b^{-1}\|x^2\| \quad \text{with } a \geq b^{-1} > 0 \tag{2}$$

$$A(t+T) = A(t)\ . \tag{3}$$

It is natural to single out for consideration the indices $i_T, i_{2T}, \ldots$ corresponding to intervals which are multiples of the basic period T.

In this section, we shall compute i_{kT} in terms of i_T and the Floquet multipliers of system (1).

We begin by a general inequality relating i_{kT} to i_T and ν_T. Recall that the matrizant of system (1) is denoted by $R(t)$, and that:

$$\nu_T = \dim \operatorname{Ker}\,(R(T) - I)\ . \tag{4}$$

Theorem 1. *For all $k \in \mathbb{N}$, we have:*

$$i_{kT} \geq k\,(i_T + \nu_T) - \nu_T\ . \tag{5}$$

Proof. Recall Proposition 4.2 and Definition 4.3 for the index and the nullity. Using the same notations, we see that there is a q_T-orthogonal splitting:

$$L_o^2(0,T) = E_+(0,T) \bigoplus E_o(0,T) \bigoplus E_-(0,T) \tag{6}$$

with:

$$\dim E_-(0,T) = i_1 \quad \text{and } q_1(u) < 0 \quad \text{for } 0 \neq u \in E_-(0,T) \tag{7}$$

$$\dim E_o(0,T) = \nu_1 \quad \text{and} \quad E_o(0,T) = \operatorname{Ker} q_1\ . \tag{8}$$

Denote by $p : L_o^2(0,T) \to L_o^2(0,kT)$ the extension operator (see Lemma 4.8):

$$(pu)(t) = \begin{cases} u(t) & \text{if } 0 \leq t \leq T \\ 0 & \text{if } T < t \leq kT \end{cases} \tag{9}$$

and by $\tau : L^2(0,T) \to L^2(0,kT)$ the translation by T:

$$(\tau w)(t) := pw(t-T)\ . \tag{10}$$

Now define in $L_o^2(0,kT)$ subspaces M_i and N_j by (note the different range of the index)

$$N_j = \tau^j pE_o(0,T) \quad \text{for } 0 \leq j \leq k-2 \tag{11}$$

$$M_i = \tau^i pE_-(0,T) \quad \text{for } 0 \leq i \leq k-1\ . \tag{12}$$

The subspaces are mutually orthogonal. Setting

$$M = \bigoplus_i M_i \quad \text{and} \quad N = \bigoplus_j N_j \tag{13}$$

we check easily that q_{kT} is negative definite on M, and vanishes on N. In other words, N is a q_{kT}-isotropic subspace, and any q_{kT}-isotropic vector in $M \oplus N$ must belong to N. Counting dimensions, we find:

$$\dim M \bigoplus N = (k-1)\nu_T + k i_T \ . \tag{14}$$

We claim that $N \cap \operatorname{Ker} q_{kT} = \{0\}$. Indeed, by Theorem 4.4, the kernel of q_{kT} consists of all $w = \dot{x}$, where x is a kT-periodic solution of $\dot{x} = JA(t)x$. It follows from this equation that w is continuous and $w(t) \neq 0$ for all t, whereas all functions in N must vanish on the interval $((k-1)T, kT)$ by formula (11).

Set $K = (k-1)\nu_T + k i_T$. We have found a K-dimensional subspace $M \oplus N$, on which q_T is negative semi-definite, and which contains no zero-eigenvector. The rest of the proof consists in finding a subspace with the same dimension, on which q_{kT} is negative definite.

We proceed as in Proposition 3.2, that is, we endow $L^2_o(0, kT)$ with the Hilbert structure defined by $\overline{B}_{kT}$. Define a linear selfadjoint operator $\mathcal{L}$ by:

$$\mathcal{L} := I - \overline{B}_{kT}^{-1} J \Pi_{kT} \ . \tag{15}$$

$\mathcal{L}$ is just the gradient of q_{kT} for the inner product $(\overline{B}_{kT} u, v)$. That is, we have:

$$q_{kT}(u, v) = \frac{1}{2} \left(\overline{B}_{kT} \mathcal{L} u, v\right) \ . \tag{16}$$

Now consider the map $I + h\mathcal{L}$. For $|h|$ sufficiently small, it is an automorphism. Denote the linear subspace $(I + h\mathcal{L})\ (M \oplus N)$ by $(M \oplus N)_h$, and the unit sphere of $L^2_o(0, kT)$ by S

$$S := \left\{ w \in L^2(0, kt) \middle| \left(\overline{B}_{kT} w, w\right) = 1 \right\} \ . \tag{17}$$

We have

$$S \bigcap \left(M \bigoplus N\right)_h = \left\{ \varphi_h(w) | w \in S \bigcap \left(M \bigoplus N\right) \right\} \tag{18}$$

$$\varphi_h(w) := (w + h\mathcal{L}w) \left(\overline{B}_{kT}(w + h\mathcal{L}w), w + h\mathcal{L}w\right)^{-1/2} \ . \tag{19}$$

For every $w \in S \cap (M \oplus N)$, we have

$$\frac{d}{dh} q_{kT}\left(\varphi_h(w), \varphi_h(w)\right)\Big|_{h=0} = \left(\overline{B}_{kT} \mathcal{L} w, \frac{d}{dh} \varphi_h(w)\Big|_{h=0}\right) \ . \tag{20}$$

Differentiating formula (19) with $(\overline{B}_{kT} w, w) = 1$, we get:

$$\frac{d}{dh}\varphi_h(w)\bigg|_{h=0} = \boldsymbol{\mathcal{L}}w - q_{kT}(w,w)w\,. \tag{21}$$

Writing this into formula (20) yields:

$$\frac{d}{dh}q_{kT}\left(\varphi_h(w),\varphi_h(w)\right)\bigg|_{h=0} = \left(\overline{B}_{kT}\boldsymbol{\mathcal{L}}w,\boldsymbol{\mathcal{L}}w\right) - q_{kT}(w,w)^2\,. \tag{22}$$

If $q_{kT}(w,w)=0$, that is, if $w\in N\cap S$, we have $\boldsymbol{\mathcal{L}}w\neq 0$ and hence:

$$\frac{d}{dh}q_{kT}\left(\varphi_h(w),\varphi_h(w)\right)\bigg|_{h=0} > 0\,. \tag{23}$$

So $q_{kT}(w,w)\le 0$ on $(M\oplus N)\cap S$, and if $q_{kT}(w,w)=0$ the above inequality holds. Since $(M\oplus N)\cap S$ is compact, it follows that there is an $\epsilon>0$ such that, whenever $-\epsilon<h,0$, we have:

$$\forall w\in\left(M\bigoplus N\right)\cap S\,,\quad q_{kT}\left(\varphi_h(w),\varphi_h(w)\right)<0\,. \tag{24}$$

By formula (18), this means that:

$$\forall w\in\left(M\bigoplus N\right)_h\cap S\,,\quad q_{kT}(w,w)<0\,. \tag{25}$$

So q_{kT} is negative definite on $(M\oplus N)_h$. Its index must therefore be at least $K=(k-1)\nu_T+ki_T$. □

Inequality (5) will be extremely useful in the sequel. For the time being, we seek something more precise, that is, we want a formula giving i_{kT} in terms of i_T.

Recall that i_{kT} is defined to be the index of the quadratic form

$$q_{kT}(u,u)=\frac{1}{2}\int_o^{kT}\left[(Ju,\Pi u)+(B(t)Ju,Ju)\right]dt \tag{26}$$

on the Hilbert space:

$$L_o^2(0,kT)=\left\{u\in L_o^2\left(0,kT;\mathbb{R}^{2n}\right)\Big|\int_o^{kT}u(t)\,dt=0\right\}\,. \tag{27}$$

Alternatively, i_{kT} is also the index of the quadratic form:

$$Q_{kT}(x,x)=\frac{1}{2}\int_o^{kT}\left[(J\dot{x},x)+(B(t)J\dot{x},J\dot{x})\right]dt \tag{28}$$

on the Hilbert space

$$E_{kT}=\left(x\in W^{1,2}\left(0,kT;\mathbb{R}^{2n}\right)\big|\,x(0)=x(T)\right\}\,. \tag{29}$$

Surprisingly enough, to compute i_{kT} in terms of i_T, we have to complexify the situation. With every $\omega\in\boldsymbol{\mathcal{U}}$ (unit circle in $\mathbb{C}$) we associate the Hermitian form:

$$Q^{\omega}_{kt}(x,x) = \frac{1}{2}\int_o^{kT} [(\,J\dot{x},x) + (B(t)J\dot{x},J\dot{x})]\,dt \tag{30}$$

on the complex Hilbert space:

$$E^{\omega}_{kT} = \left\{x \in W^{1,2}\left(0,kT;\mathbb{C}^{2n}\right) \middle|\, x(kT) = \omega x(0)\right\} . \tag{31}$$

In formula (30), we denote by (ξ,η) the standard Hermitian structure in $\mathbb{C}^{2n}$:

$$(\xi,\eta) := \sum_{i=1} \xi_i\overline{\eta}_i = \overline{(\eta,\xi)} . \tag{32}$$

Clearly, the quadratic form Q_{kT} on the real Hilbert space E_{kT} and the Hermitian form Q^1_{kT} on the complex space E^1_{kT} have the same index i_{kT}.

The advantage of complexifying the situation lies in the fact that we now have a splitting lemma. Before we state it, let us identify E^{ω}_T, when ω is a k-th root of unity, with a subspace of E^1_{kT}:

$$E^{\omega}_{T} = \left\{x \in W^{1,2}\left(0,kT;\mathbb{C}^{2n}\right) \middle|\, x(t+T) = \omega x(t)\right\} . \tag{33}$$

If $\omega^k = 1$, this is coherent with definition (31): we have extended the functions of E^{ω}_T from $(0,T)$ to $(0,kT)$ in the obvious way. Note that if $x \in E^{\omega}_T$:

$$\begin{aligned} Q^1_{kT}(x,x) &= \frac{1}{2}\int_o^{kT} [(J\dot{x},x) + (B(t)J\dot{x},J\dot{x})]\,dt \\ &= \frac{1}{2}\sum_{m=0}^{k-1}(\omega\overline{\omega})^m\int_o^{T} [(J\dot{x},x) + (B(t)J\dot{x},J\dot{x})]\,dt \\ &= kQ^{\omega}_T(x,x) \end{aligned} \tag{34}$$

so that we may use Q^1_{kT} or Q^{ω}_T indifferently on E^{ω}_T.

Lemma 2. *Let $k \geq 1$ be given. Then the E^{ω}_T, for $\omega^k = 1$, are orthogonal subspace of E^1_{kT}, both for the standard Hilbert structure and for Q^1_{kT}, and E^1_{kT} splits into a direct sum:*

$$E^1_{kT} = \bigoplus_{\omega^k=1} E^{\omega}_T . \tag{35}$$

Proof. Any $x \in E^1_{kT}$ can be written:

$$x(t) = \sum_{p\in\mathbb{Z}} \xi_p \exp\left(\frac{2i\pi}{kT}p\right) . \tag{36}$$

For $q = 0,1,\ldots,k-1$, denote by $C(q)$ the set of all p such that $p-q \in k\mathbb{Z}$. We may write

$$x(t) = \sum_{q=0}^{k-1} x_q(t) , \qquad \text{with } x_q(t) = \sum_{C(q)} \xi_p \exp\left(\frac{2i\pi}{kT}pt\right) . \tag{37}$$

We then check that:

$$\begin{aligned} x_q(t+T) &= \sum_{C(q)} \xi_p \exp\left(\frac{2i\pi}{kT}pt + 2i\pi\frac{p}{k}\right) \\ &= \exp\left(2i\pi\frac{q}{k}\right) x_q(t) . \end{aligned} \tag{38}$$

So $x_q \in E_T^\omega$ with $\omega = \exp\left(2i\pi\frac{q}{k}\right)$. When q runs from 0 to $k-1$, then ω runs through the k-th roots of unity. Hence the splitting (35).

The rest is easy enough, taking into account the fact that $B(t)$ is T-periodic. Let us check, for instance, that the E_T^ω are Q_{kT}^1-orthogonal. Take $x \in E_T^\omega$ and $y \in E_T^\lambda$, where ω and λ are k-th roots of unity. Then:

$$\begin{aligned} Q_{kT}^1(x,y) &= \frac{1}{2}\int_o^{kT} [(J\dot{x},y) + (B(t)J\dot{x},J\dot{y})]\, dt \\ &= \frac{1}{2}\sum_{m=0}^{k-1} (\omega\overline{\lambda})^m \int_o^T [(J\dot{x},y) + (B(t)J\dot{x},J\dot{y})]\, dt \end{aligned} \tag{39}$$

which is zero unless $\omega = \lambda$. □

As a special and well-known instance of this splitting, take $k = 2$. We then get $\omega = \pm 1$, and Lemma 2 tells us that every $2T$-periodic function ($x(t+2T) = x(t)$) splits into a T-periodic component ($x_1(t+T) = x_1(t)$) and a T-antiperiodic component ($x_2(t+T) = -x_2(t)$).

That this natural splitting answers our questions was first noted by Bott. Consider the index and nullity of Q_{kT}^ω as integer-valued functions of ω:

Definition 3. We define the *Bott maps* j_T and n_T from $\mathcal{U}$ to $\mathbb{N}$ by:

$$j_T(\omega) = \operatorname{index} Q_T^\omega \tag{40}$$

$$n_T(\omega) = \begin{cases} \text{nullity } Q_T^\omega & \text{if } \omega \neq 1 \\ \text{nullity } Q_T^\omega - 2n & \text{if } \omega = 1 \end{cases} . \tag{41}$$

Corollary 4. *For any integer $k \geq 1$, we have:*

$$i_{kT} = \sum_{\omega^k=1} j_T(\omega) \tag{42}$$

$$\nu_{kT} = \sum_{\omega^k=1} n_T(\omega) . \tag{43}$$

Proof. Because of the splitting introduced in Lemma 2, we clearly have:

(44)
$$\operatorname{index} Q^1_{kT} = \sum_{\omega^k=1} \operatorname{index} Q^\omega_T$$

(45)
$$\operatorname{nullity} Q^1_{kT} = \sum_{\omega^k=1} \operatorname{nullity} Q^\omega_T \ .$$

Now index $Q^1_{kT} = i_{kT}$ and formula (42) then follows from (44).

On the other hand, the nullity of Q^1_{kT} is the nullity of Q_{kT}, while ν_{kT} was defined to be the nullity of q_{kT} (see Definition 4.3). But Q_{kT} is translation-invariant – see formula 4.7 – so $\mathbb{R}^{2n} \subset \operatorname{Ker} Q_{kT}$. We can therefore factor out $\mathbb{R}^{2n}$: that is, there is a well-defined quadratic form $\widetilde{Q}_{kT}$ on the quotient space $\widetilde{E}_{kT} := E_{kT}/\mathbb{R}^{2n}$ such that $\widetilde{Q}_{kT}(\widetilde{x}, \widetilde{y}) = Q_{kT}(x, y)$ whenever $x \in \widetilde{x}$ and $y \in \widetilde{y}$. We have:

(46)
$$\operatorname{index} \widetilde{Q}_{kT} = \operatorname{index} Q_{kT}$$

(47)
$$\operatorname{nullity} \widetilde{Q}_{kT} = \operatorname{nullity} Q_{kT} - 2n \ .$$

But the map $x \to \dot{x}$ is an isomorphism of $\widetilde{E}_{kT}$ onto $L^2_o(0, kT)$, and $\widetilde{Q}_{kT}(\widetilde{u}, \widetilde{v}) = q_{kT}(\dot{x}, \dot{y})$. Hence

(48)
$$\operatorname{nullity} q_{kT} = \operatorname{nullity} Q_{kT} - 2n \ .$$

Formula (43) then follows from (45). □

Taking into account the fact that $j_T(\overline{\omega}) = j_T(\omega)$, we can rewrite Bott's formula (42) as follows:

(42)
$$\begin{cases} i_{(2p+1)T} = i_T + 2\displaystyle\sum_{k=1}^{p} j_T\left(e^{2ik\pi/(2p+1)T}\right) \\ i_{2pT} = i_T + j_T(-1) + 2\displaystyle\sum_{k=1}^{p-1} j_T\left(e^{ik\pi/pT}\right) \end{cases} .$$

So all the odd terms in the sequence i_{kT}, $k \in \mathbb{N}$, have the parity of i_T, and all the even terms have the parity of $i_T + j_T(-1)$. The sequence i_{kT} itself falls into one of the three types:

(even)	i_{kT} even	$\forall k$
(odd)	i_{kT} odd	$\forall k$
(alternating)	i_{kT} and	$i_{(k+1)T}$ have different parities.

It follows from the results in Section 4 that the sequence i_{kT} is non-decreasing.

To make use of Bott's formulas, we have to know the functions j_T and $n_T : \boldsymbol{\mathcal{U}} \to \mathbb{N}$ more explicitly. We begin by n_T.

Proposition 5. *For all $\omega \in \mathcal{U}$, we have:*

$$n_T(\omega) = \dim \operatorname{Ker}\,(R(T) - \omega I) \ . \tag{49}$$

So $n_T(\omega) \neq 0$ if and only if ω is a Floquet multiplier of system (1). □

Proof. If $\omega = 1$, this is precisely Theorem 4.4. The case $\omega \neq 1$, $|\omega| = 1$, goes through with minor modifications.

By definition, $n_T(\omega)$ is the nullity of Q_T^ω, that is, the dimension of its kernel. Say $x \in \operatorname{Ker} Q_T^\omega$. This means that, for all $y \in E_T^\omega$:

$$\int_o^T \left[\frac{1}{2}(J\dot{x}, y) + \frac{1}{2}(J\dot{y}, x) + Re\,(B(t)J\dot{x}, J\dot{y})\right] dt = 0 \ . \tag{50}$$

Integrating by parts, we get:

$$\frac{1}{2}\,(Jx(t), y(t))\Big|_o^T + Re \int_o^T (x + B(t)J\dot{x}, J\dot{y})\ dt = 0 \ . \tag{51}$$

The first term is $\frac{1}{2}\,(\omega\overline{\omega} - 1)$, which vanishes since $|\omega| = 1$. Since $\omega \neq 1$, $\dot{y}$ spans $L_o^2\left(O, T; \mathbb{C}^{2n}\right)$ when y spans E_T^ω, so that Eq. (51) boils down to:

$$x + B(t)J\dot{x} = 0 \ . \tag{52}$$

Hence $\dot{x} = JA(t)x$, and $x(T) = \omega x(0)$ since $x \in E_T^\omega$. In other words, the kernel of Q_T^ω consists of all solutions of system (1) such that $x(T) = \omega x(0)$. This is the desired result. □

We now study $j_T : \mathcal{U} \to \mathbb{N}$. We shall determine its points of discontinuities, the amount by which j_T jumps at these points, and the value at a particular point. The function then will be completely determined.

Note first two obvious facts:

$$j_T(1) = i_T \tag{53}$$

$$j_T(\overline{\omega}) = j_T(\omega) \ . \tag{54}$$

For the rest, $j_T(\omega)$ is defined as the index of Q_T^ω on E_T^ω. Both the Hermitian form Q_T^ω and the Hilbert space E_T^ω depend on the parameter $\omega \in \mathcal{U}$. The first thing to do is to fix the underlying space by a suitable change of variable.

Recall that:

$$E_T^\omega = \left\{x \in W^{1,2}\left(0, T; \mathbb{C}^{2n}\right) \middle|\ x(T) = \omega x(0)\right\} \ . \tag{55}$$

If $\omega \neq 1$, we can express $x(0)$ in terms of $\dot{x}$:

$$x(0) + \int_0^T \dot{x}\, dt = \omega x(0) \tag{56}$$

$$x(0) = (\omega - 1)^{-1} \int_0^T \dot{x}\, dt \tag{57}$$

and $\dot{x}$ can be anything in $L^2\left(0, T; \mathbb{C}^{2n}\right)$.

Writing this into Q_T^ω, we get:

$$
\begin{aligned}
Q_T^\omega(x,x) &= \frac{1}{2}\int_o^t \left[\left(J\dot{x}\ ,\ x(0)+\int_o^t \dot{x}(s)\,ds\right) + (B(t)J\dot{x}\ ,\ J\dot{x})\right] dt \\
&= \frac{1}{2}\int_o^T \left[\left(J\dot{x}\ ,\ \int_o^t x(s)\,ds\right) + (B(t)J\dot{x}\ ,\ J\dot{x})\right] dt \\
&\quad + \frac{1}{2}(\overline{\omega}-1)^{-1}\left(J\int_o^T \dot{x}ds\ ,\ \int_o^T \dot{x}ds\right) .
\end{aligned} \tag{58}
$$

Since $|\omega| = 1$, we can write:

$$
(\overline{\omega}-1)^{-1} = -\frac{1}{2} - ih\ , \quad \text{with } h \in \mathbb{R}\ . \tag{59}
$$

This defines a bijection from $\mathcal{U}\setminus\{1\}$ onto $\mathbb{R}$. It will be more convenient to handle the real parameter h than the complex parameter $\omega \in \mathcal{U}$.

We summarize this discussion:

Lemma 6. *For $\omega \in \mathcal{U}$, and $\omega \neq 1$, $j_T(\omega)$ and $\nu_T(\omega)$ are the index and nullity of the Hermitian form q_h defined on $L^2\left(0,T;\mathbb{C}^{2n}\right)$ by:*

$$
\begin{aligned}
q_h(u,u) &= \frac{1}{2}\int_o^T \left[\left(Ju\ ,\ \int_o^t u(s)\,ds\right) + (B(t)Ju, Ju)\right] dt \\
&\quad -\frac{1}{4}\left(J\int_o^T u\ ,\ \int_o^T u\right) - \frac{h}{2}\left(iJ\int_o^T u\ ,\ \int_o^T u\right)
\end{aligned} \tag{60}
$$

where $h \in \mathbb{R}$ and ω are related by formula (59).

Proof. We have:

$$
Q_T^\omega(x,x) = q_h(\dot{x},\dot{x}) \tag{61}
$$

and the result follows immediately. □

We may now proceed.

Proposition 7. *If $\omega_o \neq 1$ is a point of discontinuity of j_T on $\mathcal{U}$, then ω_o must be a Floquet multiplier of system* (1).

Proof. We write:

$$
q_h(u,u) = \frac{1}{2}(K_h u, u) + \frac{1}{2}\left(\widehat{B}u, u\right) \tag{62}
$$

where K_h and $\widehat{B}$ are bounded operators from $L^2\left(0,T;\mathbb{C}^{2n}\right)$ into itself, defined by:

(63)
$$K_h u = -J\int_o^t u(s)\,ds + \frac{1}{2}J\int_o^T u + kiJ\int_o^T u$$

(64)
$$\left(\widehat{B}u\right)(t) = -JB(t)Ju(t)$$

(we have integrated formula (60) by parts).

Since the injection $E_T^{\frac{1}{2}} \hookrightarrow L^2\left(0,T;\mathbb{C}^{2n}\right)$ is compact, K_h is a compact operator, depending continuously on h. Both K_h and $\widehat{B}$ are self-adjoint, and $\widehat{B}$ is positive definite.

We are in the same situation as in Propositions 4.2 and 4.11, so we just sketch the argument. By putting on $L^2\left(0,T;\mathbb{C}^{2n}\right)$ the Hilbertian structure associated with $\widehat{B}$, and applying to K_h the spectral theory of compact self-adjoint operators, we find that q_h has finite index (a fact we already knew) and that if its index (considered as a function of h) changes at the point h_o, then q_h must be degenerate. □

Setting $\left(\overline{\omega}_o - 1\right)^{-1} = -\frac{1}{2} - ih_o$, we find by Lemma 6 that $\nu_T\left(\omega_o\right) \geq 1$. By Proposition 5, this means precisely that ω_o is an eigenvalue of $R(T)$, that is, a Floquet multiplier of system (1).

To prove the converse, we need a perturbation lemma. Endow as above $L^2\left(0,T;\mathbb{C}^{2n}\right)$ with the Hilbert structure defined by $\widehat{B}$, and write:

(65)
$$L(h) = K_h + \widehat{B}$$

so that $q_h(u,v) = \frac{1}{2}\left(L(h)u,u\right)$. Denote by S the unit sphere in $L^2\left(0,T;\mathbb{C}^{2n}\right)$.

Lemma 8. *Assume that the kernel of $L\left(h_o\right)$ is one-dimensional, and spanned by u_o. Then there is some $\epsilon > 0$ and C^∞ maps λ and u from $\left(h_o - \epsilon, h_o + \epsilon\right)$ into $\mathbb{R}$ and S such that:*

(a) $\lambda\left(h_o\right) = 0$ and $u\left(h_o\right) = u_o$

(b) $L(h)u(h) = \lambda(h)u(h)$ for all h

(c) $\frac{\partial\lambda}{\partial h}\left(h_o\right) = \left(\frac{\partial L}{\partial h}\left(h_o\right)u_o, u_o\right)$

(d) if $\frac{\partial\lambda}{\partial h}\left(h_o\right) = 0$, then

$$\frac{\partial^2\lambda}{\partial h^2}\left(h_o\right) = \left(\frac{\partial^2 L}{\partial h^2}\left(h_o\right)u_o, u_o\right) - 2\left(L\left(h_o\right)\frac{\partial u}{\partial h}\left(h_o\right), \frac{\partial u}{\partial h}\left(h_o\right)\right).$$

Proof. Apply the implicit function theorem to the map

(66)
$$(h,\lambda,u) \to L(h)u - \lambda u$$

from $\mathbb{R}\times\mathbb{R}\times S$ into $L^2\left(0,T;\mathbb{C}^{2n}\right)$, near $\left(h_o,0,u_o\right)$. Its derivative is the map:

(67)
$$(\lambda,u) \to L\left(h_o\right)u - \lambda u_o$$

which is clearly invertible (with u in the tangent space to S at u_o, that is, u orthogonal to u_o). It follows that (a) and (b) define $u(h)$ and $\lambda(h)$ uniquely, and they are C^∞ functions of h.

We write the Taylor expansions:

(68) $$L(h) = L_o + (h - h_o) L_1 + \frac{1}{2} (h - h_o)^2 L_2 + \dots$$

(69) $$u(h) = u_o + (h - h_o) u_1 + \frac{1}{2} (h - h_o)^2 u_2 + \dots$$

(70) $$\lambda(h) = (h - h_o) \lambda_1 + \frac{1}{2} (h - h_o)^2 \lambda_2 + \dots$$

Write $L(h)u(h) = \lambda(h)u(h)$, and identify terms of the same order. The terms of order 0 give $L_o u_o = 0$, which we already knew since $L_o = L(h_o)$. The terms of order 1 give:

(71) $$L_o u_1 + L_1 u_o = \lambda_1 u_o \ .$$

Take the inner product of both sides with u_o. The first term on the left vanishes, since:

(72) $$(L_o u_1, u_o) = (u_1, L_o u_o) = (u_1, 0) = 0$$

and we are left with:

(73) $$\lambda_1 = (L_1 u_o, u_o)$$

which is precisely formula (c).

If $\lambda_1 = 0$, we go to order 2. We get:

(74) $$\frac{1}{2} L_o u_2 + L_1 u_1 + \frac{1}{2} L_2 u_o = \frac{1}{2} \lambda_2 u_o \ .$$

Again we take the inner product with u_o. Again the first term on the left vanishes, and we are left with:

(75) $$\lambda_2 = 2 (L_1 u_1, u_o) + (L_2 u_o, u_o) \ .$$

But formula (71) with $\lambda_1 = 0$ gives $L_1 u_o = -L_o u_1$. Writing this into the above equation yields:

(76) $$\lambda_2 = (L_2 u_o, u_o) - 2 (L_o u_1, u_1)$$

which is precisely formula (a). □

If $\frac{\partial}{\partial h} q_h (u_o, u_o)$ is non-zero, its sign is the sign of $\frac{\partial \lambda}{\partial h} (h_o)$, and determines how the index of q_h will change when h crosses the value h_o: if for instance it is positive, $\lambda(h)$ goes from negative to positive values as h increases through h_o, and the index of q_h decreases by 1.

If $\frac{\partial}{\partial h} q_h (u_o, u_o)$ vanishes at h_o, the variation of the index is not determined by the second derivative $\frac{\partial^2}{\partial h^2} q_h (u_o, u_o)$. There is an additional term, which takes into account the change in the eigenvector, and which has to be computed if any conclusion is to be reached. This is the content of formula (d).

Proposition 9. *Let $\omega_o \neq 1$ be a Floquet multiplier of system (1), with $|\omega_o| = 1$. Let (p_o, q_o) be its Krein type. Then ω_o is a point of discontinuity of j_T, and*

$$j_T\left(\omega_o e^{i0}\right) - j_T\left(\omega_o e^{-i0}\right) = q_o - p_o \ . \tag{77}$$

Proof. Assume first that ω_o is a simple eigenvalue, and let u_o, with $\|u_o\| = 1$, span the corresponding eigenspace. By Proposition 5, we must have $u_o = \dot{x}_o$, where x_o satisfies

$$\dot{x}_o = JA(t)x_o \quad \text{and} \quad x_o(T) = \omega_o x_o(0) \ . \tag{78}$$

Differentiating formula (60) with respect to h, we get:

$$\frac{\partial}{\partial h} q_h\left(u_o, u_o\right)\Big|_{h=h_o} = -\frac{1}{2}\left(iJ\int_o^T u_o \ , \ \int_o^T u_o\right) \ . \tag{79}$$

Using formula (57) with $u_o = \dot{x}_o$, this becomes

$$\begin{aligned} \frac{\partial}{\partial h} q_h\left(u_o, u_o\right)\Big|_{h=h_o} &= -\frac{|1-\omega_o|^2}{2}\left(iJx_o(0), x_o(0)\right) \\ &= \frac{|1-\omega_o|^2}{2}\left(Gx_o(0), x_o(0)\right) \ . \end{aligned} \tag{80}$$

But $x_o(0)$ is an eigenvector of $R(T)$ associated with the eigenvalue ω_o, and we recognize on the right-hand side the Krein-sign of ω_o. If for instance ω_o is Krein-positive, the above derivative is positive. By Lemma 8, the index decreases by 1 as h increases through h_o, that is, $j_T(\omega)$ decreases by 1 as ω moves through ω_o in the positive direction. This is formula (77) with $p_o = 1$ and $q_o = 0$. The case $p_o = 0$ and $q_o = 1$ is proved in the same way.

Assume next that ω_o is a Krein-definite eigenvalue of $R(T)$, of type $(p_o, 0)$ say. Choose $\epsilon > 0$ so small that

$$j_T\left(\omega_o e^{i0}\right) = j_T\left(\omega_o e^{i\epsilon}\right) \tag{81}$$

$$j_T\left(\omega_o e^{-i0}\right) = j_T\left(\omega_o e^{-i\epsilon}\right) \tag{82}$$

and there are no eigenvalues of $R(T)$ of the form $\omega_o e^{i\theta}$, with $0 < \theta \leq \epsilon$.

Now $R(T)$ is the limit of a sequence M_n of symplectic matrices with simple eigenvalues. By Proposition 1.4, there is a sequence $A_n(t)$ converging uniformly to $A(t)$ such that $R_n(T) = M_n \to R(T)$, where $R_n(t)$ is the matrizant of the system $\dot{x} = JA_n(t)x$. Each of these systems defines a map $j_T^n : \mathcal{U} \to \mathbb{N}$.

Clearly n can be chosen so large that:

$$j_T^n\left(\omega_o e^{i\epsilon}\right) = j_T\left(\omega_o e^{i\epsilon}\right) \tag{83}$$

$$j_T^n\left(\omega_o e^{-i\epsilon}\right) = j_T\left(\omega_o e^{-i\epsilon}\right) \tag{84}$$

and that there are exactly p_o eigenvalues of $R_n(T)$ on the arc $\omega_o e^{i\theta}$, $0 \leq \theta \leq \epsilon$, each of them simple. Applying the preceding result to the system $\dot{x} = JA_n(t)x$, we get:

$$j_T^n\left(\omega_o e^{i\epsilon}\right) - j_T^n\left(\omega_o e^{-i\epsilon}\right) = -p_o \ . \tag{85}$$

Using the Eqs. (81), (82) and (83), (84), this gives:

$$j_T\left(\omega_o e^{i0}\right) - j_T\left(\omega_o e^{-i0}\right) = -p_o \tag{86}$$

which is precisely formula (77) for the Krein type $(p_o, 0)$.

Lastly, we deal with the case of Krein-indefinite eigenvalues. Again, choose $\epsilon > 0$ so small that (81) and (82) hold, and there are no eigenvalues of $R(T)$ of the form $\omega_o e^{i\theta}$, with $0 < \theta \leq \epsilon$.

By Corollary 3.5, if $t \neq T$ but $|T - t|$ is small enough, $R(t)$ will have only Krein-definite eigenvalues on the unit circle. Take $t < T$ for instance. Then there will be exactly p_o^- Krein-positive and q_o^- Krein-negative eigenvalues on the arc $e^{i\theta}$, $-\epsilon < \theta < \epsilon$ (counting multiplicities). We have the relation:

$$p_o^- - q_o^- = p_o - q_o \ . \tag{87}$$

On the other hand, since $\nu_T\left(\omega_o e^{i\theta}\right)$ and $\nu_T\left(\omega_o e^{-i\theta}\right)$ are zero, we have, always provided $|T - t|$ is small enough:

$$j_t\left(\omega_o e^{i\epsilon}\right) = j_T\left(\omega_o e^{i\epsilon}\right) \tag{88}$$

$$j_t\left(\omega_o e^{-i\epsilon}\right) = j_T\left(\omega_o e^{-i\epsilon}\right) \tag{89}$$

The result obtained in the Krein-definite case applies to j_t. We get:

$$j_t\left(\omega_o e^{i\epsilon}\right) - j_t\left(\omega_o e^{-i\epsilon}\right) = q_o^- - p_o^- \ . \tag{90}$$

Comparing (81), (82) with (88), (89), this becomes:

$$j_T\left(\omega_o e^{i0}\right) - j_T\left(\omega_o e^{-i0}\right) = q_o^- - p_o^- \ . \tag{91}$$

which is the desired result, since $q_o^- - p_o^- = q_o - p_o$. □

It turns out that 1 is also a point of discontinuity of j_T on $\mathcal{U}$, even if it is not a Floquet multiplier. This is due to the fact that, in the splitting (35), the subspace $\mathbb{C}^{2n}$, which is contained in the kernel of Q_{kT}^1, lies in E_T^1: none of the E_T^ω, for $\omega \neq 1$, contains constant functions.

Proposition 10. *Assume* 1 *is not a Floquet multiplier of system* (1). *Then:*

$$j_T\left(e^{0i}\right) = j_T\left(e^{-0i}\right) = j_T(1) + n \ . \tag{92}$$

Proof. We separate in q_h a fixed and a variable part:

$$q_h(u, u) = q_o(u, u) - \frac{h}{2}\left(iJ\int_o^T u \ , \ \int_o^T u\right) \tag{93}$$

where q_o is obtained by setting $h = 0$ in formula (60).

It will be convenient to write:

$$q_o(u,u) = \frac{1}{2}(Lu,u) \tag{94}$$

where L is a bounded self-adjoint operator in $L^2(0,T;\mathbb{C}^{2n})$ defined by

$$(Lu)(t) = -J\int_o^t u\,ds + \frac{1}{2}J\int_o^T u\,dt - JB(t)Ju(t)\ . \tag{95}$$

Split $L^2(0,T;\mathbb{C}^{2n})$ into $L_o^2 \oplus \mathbb{C}^{2n}$, where $\mathbb{C}^{2n}$ is the subspace of constants and L_o^2 the subspace of functions with mean value zero. Write $u = u_o + \xi$ the corresponding decomposition of a vector u in $L^2(0,T;\mathbb{C}^{2n})$. We shall try to write $q_h(u,u)$ as a sum of squares:

$$q_h(u_o+\xi, u_o+\xi) = \frac{1}{2}(Lu_o,u_o) + Re\left(\int_o^T Lu_o dt, \xi\right) + \frac{1}{2}(M_h\xi,\xi) \tag{96}$$

$$M_h := -\int_o^T JB(t)J\,dt - hT^2 iJ\ . \tag{97}$$

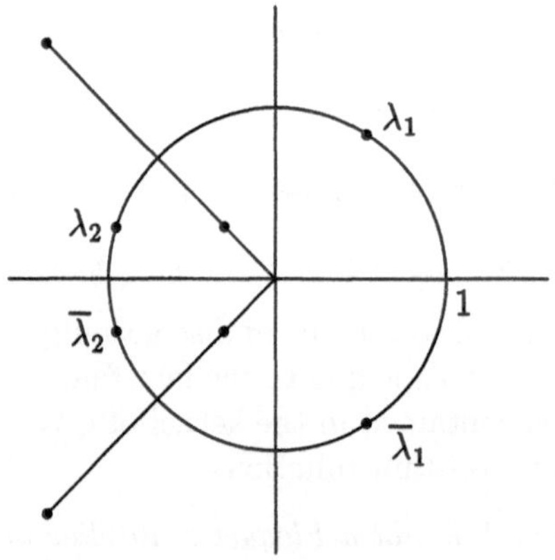

Fig. 4. The Bott map j when 1 is not a Floquet multiplier. Then j is constant on the arcs $(1,\lambda_1)$, (λ_1,λ_2), $(\lambda_2,\overline{\lambda}_2)$, $(\overline{\lambda}_2,\overline{\lambda}_1)$ and $(\overline{\lambda}_1,1)$; the Floquet multipliers which are off the unit circle play no role. Note the discontinuity which occurs at 1, even though it is not a Floquet multiplier: $j(1) = i_T$ but $j\left(e^{i0}\right) = i_T + n$.

The matrix M_h defines a Hermitian operator in $\mathbb{C}^{2n}$. It is clearly invertible for $|h|$ large. Let us proceed with the calculations:

$$q_h(u,u) = \frac{1}{2}(Lu_o, u_o) - \frac{1}{2}Re\left(\int_o^T Lu_o\,dt\ ,\ M_h^{-1}\int_o^T Lu_o\,dt\right) \tag{98}$$
$$+\frac{1}{2}(M_h\xi,\xi)$$

$$\zeta := \xi + M_h^{-1}\int_o^T Lu_o\,dt\ . \tag{99}$$

The map $(u_o,\xi) \to (u_o,\zeta)$ clearly is a linear isomorphism of $L_o^2 \times \mathbb{C}^{2n}$ into itself, and we may therefore adopt (u_o,ζ) as a new coordinate system in $L^2(0,T;\mathbb{C}^{2n})$. It follows from formula (98) that the subspaces $u_o = 0$ and $\zeta = 0$ are q_h-orthogonal, so the index of q_h must be the sum of the indices of its restrictions to these subspaces.

Setting $u_o = 0$, we get the Hermitian form $\frac{1}{2}(M_h\zeta,\zeta)$ on $\mathbb{C}^{2n}$. For $|h|$ large, it will behave like $-hT^2iJ$, which is nondegenerate and has index n.

Setting $\zeta = 0$, we get the Hermitian form

$$q_h(u_o,u_o) = \frac{1}{2}(Lu_o,u_o) - \frac{1}{2}Re\left(\int_o^T Lu_o\,dt\ ,\ M_h^{-1}\int_o^T Lu_o\,dt\right)\ . \tag{100}$$

When $|h| \to \infty$, $M_h^{-1} \to 0$. In the limit, we get $\frac{1}{2}(Lu_o,u_o)$, which is the restriction of the Hermitian form q_o to L_o^2. Writing $\int_o^T u_o = 0$ in formula (95), we get:

$$q_o(u_o,u_o) = \frac{1}{2}\int_o^T\left[\left(Ju\ ,\ \int_o^t u(s)\,ds\right) + (B(t)Ju,Ju)\right]dt \tag{101}$$

and we recognize on the right-hand side the Hermitian form q_T of formula (4.11). Its index is $i_T = j_T(1)$ by definition. Since 1 is not a Floquet mutliplier, q_T is nondegenerate (Theorem 4.4), and so the restriction of q_h to L_o^2 will also have index i_T for large $|h|$. □

All that remains to do is to compute $j(\omega_o)$ when ω_o is a Floquet multiplier. Once this is done, the function $j_T : \mathcal{U} \to \mathbb{N}$ is completely determined.

Proposition 11. *Let $\omega_o \in \mathcal{U}$ be a Floquet multiplier of system (1), with Krein type (p_o,q_o). Let $2r_o^-$ be the number of Floquet multipliers which arrive on the unit circle at ω_o*

$$j_T\left(\omega_o e^{i0}\right) = j_T(\omega_o) + q_o - r_o^- \tag{102}$$

$$j_T\left(\omega_o e^{-i0}\right) = j_T(\omega_o) + p_o - r_o^-\ . \tag{103}$$

If $\omega_o = 1$, with multiplicity m_o, we have $p_o = q_o = \frac{1}{2}m_o$, and

$$j_T\left(e^{i0}\right) = j_T(1) + n + \frac{1}{2}m_o - r_o^- = j_T\left(e^{-i0}\right)\ . \tag{104}$$

Proof. Choose $\epsilon > 0$ so small that there is no Floquet multiplier of the form $\omega_o e^{i\theta}$ with $0 < \theta \leq \epsilon$. Then

$$j_T\left(\omega_o e^{i\epsilon}\right) = j_T\left(\omega_o e^{i0}\right) \tag{105}$$

$$j_T\left(\omega_o e^{-i\epsilon}\right) = j_T\left(\omega_o e^{-i0}\right) \ . \tag{106}$$

Now apply Corollary 3.5 with $\lambda = \omega_o$. For $t \in S$ and $t < T$, there are p_o^- Krein-positive eigenvalues on the arc $e^{i\theta}\omega_o$, $-\epsilon < \theta < 0$, and q_o^- Krein-negative eigenvalues on the arc $e^{i\theta}\omega_o$, $0 < \theta < \epsilon$. We have by Definition 3.6:

$$r_o^- = p_o - p_o^- = q_o - q_o^- \ . \tag{107}$$

We now pause to note that the theory developped in Section 4, can easily be extended to other eigenvalues than 1. We define an instant s to be ω_o-conjugate to 0 if the boundary-value problem in $\mathbb{C}^{2n}$

$$\dot{x} = JA(t)x \tag{108}$$

$$x(s) = \omega_o x(0) \tag{109}$$

has a non-trivial solution. The multiplicity of s is the number of $\mathbb{C}$-linearly independent solutions. There is then a theorem according to which $j_T(\omega_o)$ is the number of points $s \in (0,t)$ which are ω_o-conjugate to 0.

The upshot is that, if $t < T$ and $|t - T|$ is small enough, there are no points $s \in [t, T)$ which are ω_o-conjugate to 0, and hence:

$$j_t(\omega_o) = j_T(\omega_o) \ . \tag{110}$$

By Proposition 3.5, ω_o is not an eigenvalue of $R(t)$, so $j_t(\omega_o)$ can be computed readily. If $\omega_o \neq 1$, Proposition 9 gives:

$$j_t\left(\omega_o e^{i\epsilon}\right) - j_t(\omega_o) = q_o^- \ . \tag{111}$$

Now if $|T - t|$ is small enough: $j_t\left(\omega_o e^{i\epsilon}\right) = j_T\left(\omega_o e^{i\epsilon}\right)$. Writing this equality into the preceding one, together with (110) and (105), we get:

$$j_T\left(\omega_o e^{i0}\right) - j_T(\omega_o) = q_o^- = q_o - r_o^- \ . \tag{112}$$

Formula (103) is proved in the same way.

If $\omega_o = 1$, we apply Proposition 10 instead of Proposition 9, and Eq. (111) becomes:

$$j_t\left(e^{i\epsilon}\right) - j_t(1) = n + q_o^- \tag{113}$$

yielding :

$$j_T\left(e^{i0}\right) - j_T(1) = n + q_o^- = n + q_o - r_o^- \ . \tag{114}$$

□

Corollary 12. *If ω_o is Krein-positive, we have:*

$$j_T\left(\omega_o e^{i0}\right) = j_T(\omega_o) \tag{115}$$

$$j_T\left(\omega_o e^{-i0}\right) = j_T(\omega_o) + p_o \tag{116}$$

If ω_o is Krein-negative, we have:

(117) $$j_T\left(\omega_o e^{-i0}\right) = j_T\left(\omega_o\right) + q_o$$

(118) $$j_T\left(\omega_o e^{-i0}\right) = j_T\left(\omega_o\right) .$$

Proof. We have $\omega_o \neq 1$ and $r_o^- = 0$ since ω_o is Krein-definite. Just apply formula (102) and (103). □

We can use the theory of ω_o-conjugate points to relate r_o^+ (the number of pairs of Floquet multipliers which leave the unit circle at ω_o), r_o^- and the dimension of the eigenspace of ω_o:

Proposition 13. *Let ω_o be a Floquet multiplier of system* (1), *with Krein type (p_o, q_o). Set $m_o = p_o + q_o$ and $d_o = \dim \mathrm{Ker}\,(R(T) - \omega_o I)$. Then:*

(119) $$r_o^+ + r_o^- = m_o - d_o .$$ □

Proof. Choose $\epsilon > 0$ so small that $R(t)$ has no eigenvalue of the form $e^{i\theta}\omega_o$, with $0 < \theta \leq \epsilon$. Then choose $t > T$ so small that

(a) we are in the situation of Corollary 3.5, that is the arc $e^{i\theta}\omega_o$, with $0 < \theta < \epsilon$, carries p_o^+ eigenvalues of $R(t)$,

(b) the interval (T, t) contains no point which is ω_o-conjugate to 0,

(c) $j_t\left(\omega_o e^{i\epsilon}\right) = j_T\left(\omega_o e^{i\epsilon}\right)$.

Since T is ω_o-conjugate to 0 with multiplicity d_o, we have, by a suitable variant of Theorem 4.6:

(120) $$j_t\left(\omega_o\right) - j_T\left(\omega_o\right) = d_o .$$

On the other hand, since $R(t)$ has Krein-definite eigenvalues only, we may apply Proposition 9, and we get:

(121) $$j_t\left(\omega_o e^{i\epsilon}\right) - j_t\left(\omega_o\right) = -p_o^+ .$$

Replacing the two terms on the left by their values, given by (c) and (120), we get:

(122) $$j_T\left(\omega_o e^{i0}\right) - j_T\left(\omega_o\right) = d_o - p_o^+ .$$

Comparing with the value given for the left-hand side by Proposition 11, we get

(123) $$d_o - p_o^+ = q_o - r_o^- .$$

Now $p_o^+ = p_o - r_o^+$ by definition. The result follows. □

Corollary 14. *Assume ω_o is semi-simple: $m_o = d_o$. Then, if $\omega_o \neq 1$:*

(124) $$j_T\left(\omega_o e^{i0}\right) = j_T\left(\omega_o\right) + q_o$$

(125) $$j_T\left(\omega_o e^{-i0}\right) = j_T\left(\omega_o\right) + p_o$$

and if $\omega_o = 1$:

$$j_T\left(e^{i0}\right) = j_T(1) + \frac{m_o}{2} \tag{126}$$

$$j_T\left(e^{-i0}\right) = j_T(1) + \frac{m_o}{2}\,. \tag{127}$$ □

Proof. If follows from (119) that $r_o^+ = r_o^- = 0$. Write this into Proposition 11. □

Corollary 15. *For* $\omega_o \neq 1$, *we have:*

$$\operatorname{Max}\left\{-p_o + d_o, q_o - \frac{m_o}{2}\right\} \leq j_T\left(\omega_o e^{i0}\right) - j_T\left(\omega_o\right) \leq q_o \tag{128}$$

$$\operatorname{Max}\left\{-q_o + d_o, p_o - \frac{m_o}{2}\right\} \leq j_T\left(\omega_o e^{-i0}\right) - j_T\left(\omega_o\right) \leq p_o\,. \tag{129}$$

For $\omega_o = 1$, *we have*

$$\operatorname{Max}\left\{n - \frac{m_o}{2} + d_o, n\right\} \leq j_T\left(e^{\pm i0}\right) - j_T(1) \leq n + \frac{m_o}{2}\,. \tag{130}$$ □

Proof. By Proposition 13, we have $r_o^- \leq m_o - d_o$. By formula (3.34) we have $2r_o^- \leq m_o$. We have $r_o^- \geq 0$. Write all this information into Proposition 11. □

We have now completely determined j_T: we know its value at $\omega = 1$, namely i_T, and the rules we have given determine its values for all $\omega \in \mathcal{U}$. Let us do the calculations in two important cases.

Proposition 16 (Hyperbolic case). *Assume there is no Floquet multiplier on the unit circle. Then:*

$$i_{kT} = k\left(i_T + n\right) - n\,. \tag{132}$$ □

Proof. The function $j_T : \mathcal{U} \to \mathbb{N}$ has no point of discontinuity except $\omega = 1$. We have $j_T(1) = i_T$, and $j_T\left(e^{i0}\right) = i_T + n$ by Proposition 10. Hence:

$$j_T(\omega) = i_T + n \quad \forall \omega \neq 1 \tag{133}$$

and Eq. (132) follows from Bott's formula (42). □

Proposition 17 (Simple eigenvalues on the unit circle). *Assume that there are* $2N$ *Floquet multipliers on* $\mathcal{U}$, *all of which are simple, and none of which is a root of unity:*

$$\omega_\nu = e^{\pm 2i\pi\alpha_\nu} \quad 0 < \alpha_\nu < \frac{1}{2} \tag{134}$$

$$\alpha_\nu \notin \mathbb{Q} \quad 1 \leq \nu \leq N\,. \tag{135}$$

Set $\alpha_o = 0$ *and* $\alpha_{N+1} = \frac{1}{2}$. *Define* $j^\nu \in \mathbb{N}$ *to be the value of* j_T *between* $\omega_{\nu-1}$ *and* ω_ν:

$$j^\nu := j_T\left(e^{2i\pi\alpha}\right)\,, \quad \alpha_{\nu-1} < \alpha < \alpha_\nu\,. \tag{136}$$

Then:

$$(137)\qquad \begin{aligned} i_{kT} = i_T + 2j^1 E[k\alpha_1] + 2\sum_{\nu=2}^{N} j^\nu \left(E[k\alpha_\nu] - E[k\alpha_{\nu-1}]\right) \\ + j^{N+1}\left(k - 1 - 2E[k\alpha_\nu]\right) . \end{aligned}$$

Alternatively:

$$(138)\qquad i_{kT} = i_T + 2\sum_{\nu=1}^{N} \left(j^\nu - j^{\nu+1}\right) E[k\alpha_\nu] + (k-1)j^{N+1} .$$

Proof. The number $N_{k,\nu}$ of k-th roots of unity which lie between $\omega_{\nu-1}$ and ω_ν, for $2 \leq \nu \leq N$, is:

$$(139)\qquad \begin{aligned} N_{k,\nu} &= \#\{q \in \mathbb{N} | \alpha_{\nu-1} < q/k < \alpha_\nu\} \\ &= E[k\alpha_\nu] - E[k\alpha_{\nu-1}] . \end{aligned}$$

For $\nu = 1$ and $\nu = N+1$, we have to take into account $+1$ (which is always a root of unity) and -1 (which is a k-th root of unity if k is even).

Using Bott's formula, and the symmetry of j_T with respect to the real axis we get:

$$(140)\qquad \begin{aligned} i_{kT} = \sum_{\omega^k=1} j_T(\omega) = i_1 + 2j^1 E[k\alpha_1] \\ + 2\sum_{\nu=2}^{N} j^\nu \left(E[k\alpha_\nu] - E[k\alpha_{\nu-1}]\right) + j^{N+1}F . \end{aligned}$$

The last term F is equal to:

$$(141)\qquad 1 + 2\left(k' - 1 - E[k\alpha_N]\right] \quad \text{if } \; k = 2k' \text{ is even}$$

$$(142)\qquad 2\left(k' - E[k\alpha_N]\right)\text{if } \; k = 2k'+1 \text{ is odd} .$$

It turns out to be equal to $k - 1 - 2E[k\alpha_n]$ in both cases. Hence formula (137). Formula (138) follows by regrouping terms differently. □

There are other, more general, consequences of Bott's formulas. The most important is the following:

Theorem 18. *We have:*

$$(143)\qquad \lim_{k\to\infty} \frac{1}{k} i_{kT} = \frac{1}{2\pi} \int_{\mathcal{U}} j_T(\omega) d\omega .$$

Proof. We have:

$$(144)\qquad \frac{2\pi}{k} i_{kT} = \frac{2\pi}{k} \sum_{\omega^k=1} j_T(\omega) .$$

The right-hand side is a Riemann sum, which converges to the corresponding integral when $k \to \infty$. □

Definition 19. The limit of $k^{-1} i_{kT}$ is denoted by $\hat{\imath}$, and is called the *mean index per period* of system (1).

The mean index per period is an important characteristic of system (1). It follows immediately from the definitions that if is non-negative. We can do better, and show it is always positive:

Proposition 20. *We have $\hat{\imath} > 0$ always. We also have:*

$$\widehat{\imath} \geq i_T + \nu_T \; . \tag{145}$$
□

Proof. Formula (145) follows immediately from Theorem 1, and it yields $\widehat{\imath} > 0$ unless $i_T = \nu_T = 0$.

If $i_T = \nu_T = 0$, we can apply Proposition 10, and we get:

$$j_T\left(e^{0i}\right) = j_T\left(e^{-0i}\right) = n \; . \tag{146}$$

So j_T is positive on a small arc on each side of 1, and the integral on the right-hand side of (143) cannot vanish. □

We conclude this section by a slightly refined version of Theorem 18:

Proposition 21. *There exists a constant $c \leq 2(n+1)(i_T + 2n + 1)$ such that:*

$$\left| i_{kT} - k\hat{\imath} \right| \leq c \quad \forall k \in \mathbb{N} \; . \tag{147}$$

Proof. Write Bott's formula again:

$$i_{kT} = \sum_{\omega^k = 1} j_T(\omega) \; . \tag{148}$$

It follows from Proposition 9 and Corollary 15 that j_T is bounded, namely $|j_T(\omega) - i_T| \leq 2n$ for every $\omega \in \mathcal{U}$.

Denote by Λ the set of Floquet multipliers on $\mathcal{U}$, and set $\Lambda' = \Lambda \cup \{+1\}$. We write

$$A_k = \left\{\omega | \omega^k = 1 \; , \; \omega^k \in \Lambda'\right\} \tag{149}$$

$$B_k = \left\{\omega | \omega^k = 1 \; , \; \omega^k \notin \Lambda'\right\} \tag{150}$$

and:

$$\sum_{\omega^k = 1} j_T(\omega) = \sum_{A_k} j_T(\omega) + \sum_{B_k} j_T(\omega) \; . \tag{151}$$

There are at most $2(n+1)$ terms in A_k each of which is bounded by $i_T + 2n$, so the first term on the right is bounded by $2(n+1)(i_T + 2n)$.

To compute the second term, we denote by $e^{\pm 2i\pi\alpha_\nu}$, $0 \leq \nu \leq N$, $0 \leq \alpha_\nu \leq \frac{1}{2}$, the points in Λ'. Clearly $N \leq (n+1)$. Write

$$N_{k,\nu} = \#\left\{q \middle| \alpha_{\nu-1} < \frac{q}{k} < \alpha_\nu\right\} . \tag{152}$$

On the other hand, we have from the definition of B_k

$$\sum_{B_k} j_T(\omega) = 2\sum_{\nu=1}^{N} N_{k,\nu} j^\nu \tag{153}$$

where j^ν denotes the value of j_T on the arc $e^{2i\pi\alpha}$, $\alpha_{\nu-1} < \alpha < \alpha_\nu$.

Formula (152) yields:

$$k(\alpha_\nu - \alpha_{\nu-1}) - 1 < N_{k,\nu} < k(\alpha_\nu - \alpha_{\nu-1}) + 1 . \tag{154}$$

Multiplying both sides by $2j^\nu$, adding up and taking Eq. (153) into account, we get:

$$2k\sum_\nu j_\nu(a_\nu - \alpha_{\nu-1}) - 2N < \sum_{B_k} j_T(\omega) < 2k\sum_\nu j_\nu(a_\nu - \alpha_{\nu-1}) + 2N . \tag{155}$$

We recognize on both sides the integral of j_T:

$$2\sum_\nu j_\nu(\alpha_\nu - \alpha_{\nu-1}) = \frac{1}{2\pi}\int j_T(\omega)d\omega = \hat{\imath} \tag{156}$$

and hence:

$$\left|\sum_{B_k} j_T(\omega) - k\hat{\imath}\right| \leq 2N \leq 2(n+1) . \tag{157}$$

The result follows by formula (151). □

Note for future reference the following estimate:

$$\left|i_{kT} - k\hat{\imath}\right| \leq 2nk \tag{158}$$

which follows immediately from Bott's formula and the bound on j_T.

Notes and Comments. The splitting in Lemma 2 and the iteration formulas of Corollary 4 are due to Bott [Bot1] and have become classical in the theory of closed geodesics on Riemannian surfaces (see Klingenberg's book [Kli]). Cushman and Duistermaat [CusD] have extended Bott's formulas to the case when the Hamiltonian is indefinite.

This theory has to be adapted to our definition of the index, which relies on the introduction of the quadratic form Q_s. This was done in [Eke9], which contains the main results of this section, except for Theorem 1, which appeared in [Eke11], and Propositions 11 to 15. In fact, the treatment of $j_T(e^{i0})$ in [Eke9] was erroneous. The formula $j_T(e^{i0}) = j_T(1) + n + \left[\frac{d}{2}\right]$, given in that paper, is wrong (except when $d_o = 1$) and must be replaced by formula (104).

6. The Index of a Periodic Solution to a Nonlinear Hamiltonian System

We introduce nonautonomous systems first. Let $H : \mathbb{R} \times \mathbb{R}^{2n} \to \mathbb{R}$ be a C^2 function, T-periodic with respect to time:

$$H(t+T,x) = H(t,x) \quad \forall (t,x) \ . \tag{1}$$

We shall refer to H as the *Hamiltonian*, and we shall use repeatedly the following notations:

$$H'(t,x) = (\partial H/\partial x_i)_{1\leq i\leq n} \in \mathbb{R}^{2n} \tag{2}$$

$$H''(t,x) = \left((\partial^2 H/\partial x_i \partial x_j)\right)_{1\leq i,j\leq n} \in \mathcal{L}\left(\mathbb{R}^{2n}\right) \ . \tag{3}$$

We associate with H the $2n$-dimensional system of ordinary differential equations:

$$\dot{x} = JH'(t,x) \tag{4}$$

and we assume that we have found a T-periodic solution

$$x(t+T) = x(t) \quad \forall t \ . \tag{5}$$

Clearly, under the periodicity assumption (1), this condition is equivalent to the simpler one:

$$x(0) = x(T) \ . \tag{6}$$

The *linearized system* around the solution x is:

$$\dot{y} = JH''(t,x(t))\, y \ . \tag{7}$$

This is a linear Hamiltonian system: it has the standard form we have studied in the preceding section, with $A(t) = H''(t,x(t))$. From now on, we shall require that it is positive definite:

Definition 1. Let x be a T-periodic solution of $\dot{x} = JH'(t,x)$. We shall say that x is *admissible* if $H''(t,x(t))$ is positive definite for all t. □

Definition 2. Let x be an admissible T-periodic solution of $\dot{x} = JH'(t,x)$.

(a) The *Floquet multipliers* of (T,x) are the Floquet multipliers of the linearized system (7).

(b) The *index* $i_s(x)$ and *nullity* $n_s(x)$ of x over the interval $(0,s)$ are the index and nullity of the linearized system over the same interval.

(c) The *Bott functions* $j_{(T,x)}$ and $n_{(T,x)}$ from $\mathcal{U}$ to $\mathbb{N}$ are the Bott maps of the linearized system.

(d) The *tower* $\mathcal{T}(T,x)$ is the infinite sequence:

$$\mathcal{T}(T,x) = \{i_{kT}(x) | k \in \mathbb{N}_+\} \ . \tag{8}$$

(e) The *mean index per period* $\hat{\imath}(T,x)$ is the positive number:

$$\hat{\imath}(T,x) := \lim_{k\to\infty} i_{kT}(x)/k = \int_{\mathcal{U}} \jmath_{(T,x)} d\omega \ . \tag{9}$$ □

Note that if x is T-periodic, it is also kT-periodic for all $k \geq 1$. There will also be Floquet multipliers for (kT,x), and a tower $\boldsymbol{T}(kT,x)$, with obvious relations: the Floquet multipliers of (kT,x) are the k-th powers of the Floquet multipliers of (T,x);

$$\boldsymbol{T}(kT,x) \subset \boldsymbol{T}(T,x) \quad \forall k \geq 1 \ ; \tag{10}$$

$$\hat{\imath}(kT,x) = k\hat{\imath}(T,x) \ . \tag{11}$$

The theory developed for linear Hamiltonian systems thus extends to periodic solutions of nonlinear systems, provided the admissibility condition is met. For a general, nonautonomous, admissible Hamiltonian $H(t,x)$, there is nothing special about the linearized system, which can be just any positive definite linear Hamiltonian systems.

On the other hand, if the system is *autonomous*, that is, if the Hamiltonian H does not depend on the time variable t, the linearized system (7) exhibits very particular features, which have to be taken into account when applying the results in the preceding sections. In the rest of this section, since there is nothing further to be said about nonautonoamous Hamiltonian systems, we shall confine our investigations to autonomous ones.

From now on, $H : \mathbb{R}^{2n} \to \mathbb{R}$ is a C^2 function, and we are dealing with a solution (T,x) of the boundary-value problem:

$$\dot{x} = JH'(x) \tag{12}$$

$$x(0) = x(T) \ . \tag{13}$$

It should be noted that, in contrast with the non-autonomous case, there is now no natural period for the nonlinear system (12), that is, no a priori restriction on T. For instance, if (T,x) solves problem (12), (13), it is clear that x is T-periodic; but it may in fact have a smaller period, that is, x might be T/k-periodic for some integer $k \geq 2$. If x is not constant, there is a smallest positive number $\overline{T}$ such that x is $\overline{T}$-periodic; we then have $\overline{T} = T/\overline{k}$ for some integer $\overline{k} \geq 1$, and $\overline{T}$ is called the *minimal period* of x.

It is assumed that x is admissible: $H''(x(t))$ is positive definite for all t. If the period T is not specified, it will be assumed that it is the minimal period $\overline{T}$. Thus, the Floquet multipliers of x are the Floquet multipliers of $(x,\overline{T})$, the tower of x is the tower of $(x,\overline{T})$, and the Bott functions of x are the Bott functions of $(\overline{T},x)$:

$$\boldsymbol{T}(x) := \boldsymbol{T}\left(\overline{T},x\right) \tag{14}$$

$$\hat{\imath}(x) := \hat{\imath}\left(\overline{T},x\right) \tag{15}$$

$$\jmath_x(\omega) := \jmath_{\left(\overline{T},x\right)}(\omega) \ . \tag{16}$$

Proposition 4. *Any non-constant solution* (T,x) *of problem* (12), (13) *has the Floquet multiplier* 1:

$$\nu_T \geq 1 . \tag{17}$$

Proof. Differentiate Eq. (12) with respect to time:

$$\ddot{x} = JH''(x(t))\,\dot{x} . \tag{18}$$

This means that $y := \dot{x}$ is a solution or the linearized system (7).

We know that $x(t+T) = x(t)$ for all t. Differentiate this relation with respect to time:

$$\dot{x}(t+T) = \dot{x}(t) \tag{19}$$

This means that $y = \dot{x}$ ist T-periodic.

So the linearized system (7) has a non-trivial T-periodic solution. In other words, 1 is a Floquet multiplier, and $\dot{x}(0) \neq 0$ is the corresponding eigenvector. □

Corollary 5. *For all integers* $k \geq 1$, *we have:*

$$i_{kT} \geq k i_T + (k-1) \geq k-1 \tag{20}$$ □

Proof. Apply Theorem 5.1 with $\nu_T \geq 1$. □

Corollary 6. *We have* $j_x(\omega) \geq 1$ *for all* $\omega \in \boldsymbol{\mathcal{U}}$. □

Proof. By Bott's formula, we have $i_{2T} = i_T + j(-1)$, and inequality (20) for $k=2$ then yields $j(-1) \geq 1$. Having $j(\omega) = 0$ for some $\omega \neq -1$ would require, by Corollary 5.15, that all $2n$ Floquet multipliers should lie on the unit circle, on the arc between ω and $\overline{\omega}$ containing 1, and that $j\left(\omega e^{-i0}\right) = 0$. It would then follow that $j(-1) = 0$. □

By Corollary 1.6, the multiplicity m_o of the Floquet multiplier 1 must be even, so it is at least 2. The computation of the Bott function $j_{(T,x)}$ then gives rise to certain difficulties: by Proposition 5.11, to compute its value at $e^{\pm 0i}$, we have to know the number of eigenvalues which arrive on the unit circle at 1. This kind of information is not readily available, except in some special cases, for instance if 1 is semi-simple. In that case, by Corollary 5.14, we have:

$$j_x\left(e^{\pm 0i}\right) = j_x(1) + \frac{1}{2}m_o . \tag{21}$$

In more general case, we are left with the estimates provided by Corollary 5.15. If for instance 1 is a double eigenvalue, we have:

$$n \leq j_x\left(e^{\pm 0i}\right) - j_x(1) \leq n+1 . \tag{22}$$

As for the eigenspace associated with the Floquet multiplier 1, it is generated by $\dot{x}(0)$. Its dimension is at least 1; it need not be higher, even though the multiplicity m_o is at least two. If this dimension actually is 2 or more, we shall say that (T,x) is *degenerate*. By our previous convention, to say that x is degenerate means that $(\overline{T},x)$ is degenerate, where $\overline{T}$ is the minimal period of x.

Definition 7. A solution (T,x) of problem $(\mathcal{H})$ is called *nondegenerate* if

$$\dim \operatorname{Ker}\,(R(T)-I)=1\ . \tag{23}$$

□

Degenerate periodic solutions provide us with a litmus test for the presence of additional integrals of the motion. An *integral of the motion* for the autonomous system (24) is a function $G:\mathbb{R}^{2n}\to\mathbb{R}$ such that $G\,(x(t))$ is constant along all trajectories of the system. If G is at least C^1, this boils down to the identity:

$$(G'(x),JH'(x))=0\quad \forall x\ , \tag{24}$$

an immediate consequence of which is that the Hamiltonian $H(x)$ itself is an integral of the motion. Autonomous Hamiltonians serve to model conservative systems, the energy of which is constant throughout time.

Proposition 8. *If $G:\mathbb{R}^{2n}\to\mathbb{R}$ is an integral of the motion, then*

$$JG'\,(x(0))\in \operatorname{Ker}\,(R(T)-I) \tag{25}$$

Proof. Let (T,x) be a solution of problem (12), (13). The flow $\varphi^t(\xi)$ is well-defined in a neighbourhood Ω of $(T,x(0))$, and we have the identity:

$$G\,(\varphi^\tau(\xi))=G(\xi)\qquad \text{for all } (\tau,\xi)\in\Omega\ . \tag{26}$$

Differentiating with respect to ξ, at $\tau=T$ and $\xi=x(0)$, we get:

$$(G'\,(x(T))\,,R(T)\zeta)=(G'\,(x(0))\,,\zeta)\quad \forall\zeta\in\mathbb{R}^{2n}\ . \tag{27}$$

This boils down to:

$$R^*(T)G'\,(x(0))=G'\,(x(0))\ . \tag{28}$$

Now $R^*(T)=-JR^{-1}(T)J$; the above equation thus becomes:

$$R^{-1}(T)JG'\,(x(0))=JG'\,(x(0))\ , \tag{29}$$

which is the result we announced. □

Corollary 9. *Let x be a T-periodic solution of $\dot{x}=JH'(x)$. If there is an integral of the motion G such that $G'\,(x(0))$ and $H'\,(x(0))$ are not collinear, then (T,x) is a degenerate solution.*

Proof. By the preceding results, both $H'\,(x(0))=\dot{x}(0)$ and $G'\,(x(0))$ belong to Ker $(R(T)-I)$, which must therefore be at least two-dimensional. □

Corollary 10. *Let x be a T-periodic solution of $\dot{x} = JH'(x)$, and ξ an eigenvector of $R(T)$ associated with a Floquet multiplier $\lambda \neq 1$. Then ξ is tangent to $\sum$ at $x(0)$.*

Proof. We have $R(T)\xi = \lambda\xi$ by definition. Taking the interior product of both sides with $H'(x(0))$, and using formula (27), we get $\lambda = 1$ or $(\xi, H'(x(0))) = 0$. Hence the result. □

We now turn to fixed-energy problems. Start from a C^2 compact hypersurface $\sum \subset \mathbb{R}^{2n}$; assume that $\sum$ bounds a convex set with non-empty interior. Consider a C^2 function $H : \mathbb{R}^{2n} \to \mathbb{R}$, such that:

$$\forall x \ , \ H(x) \geq 0 \tag{30}$$

$$\sum = (x | H(x) = 1\} \tag{31}$$

$$\forall x \in \sum \ , \ H'(x) \neq 0 \tag{32}$$

$$\forall x \in \sum \ , \ H''(x) \ \text{ is positive definite} \ . \tag{33}$$

Consider the Hamiltonian flow $\dot{x} = JH'(x)$ associated with H. Since H is an integral of the motion, $\sum$ is invariant under the flow. Assume we are given a periodic solution $z(t)$ on $\sum$:

$$\dot{z} = JH'(z) \tag{34}$$

$$H(z) = 1 \tag{35}$$

$$z(0) = z(T) \ . \tag{36}$$

Denote by $i_T(y; H)$ the index of this periodic solution on the interval $[0, T]$ and by $\hat{\imath}\,(T, z; H) := \lim i_{kT}\,(z; H)\,/k$ its mean index per period.

As usual, we denote by $T_x \sum$ the tangent hyperplane to $\sum$ at $x \in \sum$; it is defined by the equation:

$$(\xi, H'(x)) = 0 \quad \forall \xi \in \sum \ . \tag{37}$$

Denote by $R_H(t)$ the matrizant of the linearized system around $z(t)$, and note for future reference the followings:

Lemma 11. $T_{z(0)} \sum$ *is invariant by* $R_H(T)$*; more generally:*

$$\forall t \ , \ R_H(t) T_{z(0)} \sum \subset T_{z(t)} \sum \ . \tag{38}$$

Proof. Consider the flow φ^t associated with the equation $\dot{x} = JH'(x)$. We have $H\left(\varphi^t(x)\right) = H(x)\ \forall t$. Differentiating with respect to x at $x = z(0)$, we get:

$$\left(H'\left(z(t)\right), R_H(t)\xi\right) = \left(R_H^*(t) H'\left(z(t)\right), \xi\right) = \left(H'\left(z(0)\right), \xi\right) \ . \tag{39}$$

If $\xi \in T_{z(0)}\Sigma$, the right-hand side vanishes by Eq. (37), and it then follows from the same equation that $R_H(t)\xi \in T_{z(t)}\Sigma$. □

The conditions (30) to (33) do not determine H uniquely. If we choose another analytic representation for $\sum$, that is, another function G satisfying the same conditions, z will still be, after a suitable reparametrization, a periodic solution for the new Hamiltonian system $\dot{x} = JG'(x)$, with a new period S:

Proposition 12. *Let $G : \mathbb{R}^{2n} \to \mathbb{R}$ be a C^2 function satisfying conditions* (30) *to* (33). *Then there is an increasing C^1 diffeomorphism σ of $[0, T]$ onto an interval $[0, S]$ such that $z_\sigma(s) := z\left(\sigma^{-1}(s)\right)$ is a S-periodic solution of $\dot{x} = JG'(x)$.*

Proof. At every point $x \in \Sigma$, the vectors $H'(x)$ and $G'(x)$ must be collinear, since both carry the normal to $\sum$. It follows from condition (33) that they are both oriented outwards with respect to Σ. This implies that there must be some non-negative scalar $\lambda(x)$ such that:

$$H'(x) = \lambda(x)G'(x) \quad \forall x \in \sum . \tag{40}$$

Since neither H' nor G' vanish on Σ, and both are C^1 functions of x, the number $\lambda(x)$ must be positive, and a C^1 function of x. Define the reparametriyation σ of z by:

$$\frac{d}{dt}\sigma = \lambda\left(z(t)\right) . \tag{41}$$

We get:

$$\begin{aligned} \dot{z}_\sigma(s) &= \dot{z}(t)\left(dt/ds\right) = \dot{z}(t)(d\sigma/dt)^{-1} = \dot{z}(t)\lambda\left(z(t)\right)^{-1} \\ &= JH'\left(z(t)\right)\lambda(t)^{-1} = JG'\left(z_\sigma(s)\right) . \end{aligned} \tag{42}$$

□

We have associated with z various numbers: its Floquet multipliers, its index and its mean index per period. Because of Proposition 12, we have to ask which of these numbers depend on the particular representation H we have chosen for Σ, and which ones do not. In other words, if we replace H by G and z by z_σ, are of these numbers unchanged? It turns out that the Floquet multipliers and the mean index per period are unaffected, while the index changes; that is, $i_T(z; H)$ and $i_S(z_\sigma; G)$ are different in general.

Proposition 13. *Let $G : \mathbb{R}^{2n} \to \mathbb{R}$ be a C^2 function satisfying conditions* (30) *to* (33), *an let σ be defined as above. Then (T, z) and (S, z_σ) have the same Floquet multipliers with the same multiplicity and the same Krein sign.*

Proof. Denote by φ_H and φ_G the Hamiltonian flow associated with H and G. The argument in Proposition 12 tells us that for every $x \in \sum$ and every $t \in \mathbb{R}$ there is some $s \in \mathbb{R}$ such that

$$\varphi_H^t(x) = \varphi_G^s(x) , \tag{43}$$

namely:

$$s(t,x) := \int_0^t \lambda\left(\varphi_H^u(x)\right)\,du\ . \tag{44}$$

This formula clearly defines s as a C^2 function of (x,t) on $\sum$. Note that, by the periodicity of z and z_σ, we have:

$$z(t+T) = z(t) = z_\sigma(s) = z_\sigma(s+S) \quad \forall t \in \mathbb{R}\ . \tag{45}$$

This means that $s\,(z(t),t) = S$ for all t, and hence:

$$\left(s'\,(z(t),t)\,,\dot{z}(t)\right) = 0 \tag{46}$$

where s' denotes, as usual, the derivative with respect to x.

Differentiating formula (43) with respect to x in $\sum$, at $x = z(0)$ and $t = T$, we get:

$$\forall \xi \in T_{z(0)} \sum\ ,\ R_H(T)\xi = R_G(S)\xi + \left(s'\,(z(0),T)\,,\xi\right) JG'\,(z(0))\ . \tag{47}$$

Denote by E the one-dimensional subspace spanned by the collinear vectors $\dot{z}(0) = JH'\,(z(0))$ and $\dot{z}_\sigma(0) = JG'\,(z(0))$. Writing formula (46) into (47), we find that $R_H(T)$ and $R_G(S)$ coincide on E; in fact, we know that $\dot{z}(0)$ and $\dot{z}_\sigma(0)$ are eigenvectors associated with the eigenvalue 1, so that the restriction of $R_H(T)$ and $R_G(S)$ to E is the identity.

Now consider the quotient space $T_{z(0)} \sum /E$. Since E is an eigenspace, the maps $R_H(T)$ and $R_G(S)$ from $T_{z(0)} \sum$ into itself give rise to quotient maps $\overline{R}_H(T)$ and $\overline{R}_G(S)$ from $T_{z(0)} \sum /E$ into itself. Formula (47) now tells us that these quotient maps coincide: $\overline{R}_H(T) = \overline{R}_G(S)$. It is a standard fact from linear algebra that the eigenvalues of $\overline{R}_H(T)$ are also eigenvalues of $R_H(T)$; similarly for $\overline{R}_G(T)$ and $R_G(T)$.

We have thus shown that the restrictions of $R_H(T)$ and $R_G(S)$ to the invariant hyperplane $T_{y(0)} \sum$ in $\mathbb{R}^{2n}$ have the same spectrum. There is one missing eigenvalue. It is found to be 1 in both cases by the following argument. Write the orthogonal decomposition $\mathbb{R}^{2n} = T_{z(0)} \sum \oplus N$, where N is the normal to $\sum$ at $z(0)$, which is spanned by $H'\,(z(0))$. If $\xi \in N$, we get $R_H(T)\xi = \lambda\xi + \zeta$, for some $\zeta \in T_{z(0)} \sum$ and $\lambda \in \mathbb{R}$. The missing eigenvalue is λ, and by relation (27), taking $\xi = H'\,(z(0))$, we get:

$$\begin{aligned} \lambda \left\|\xi\right\|^2 &= \left(R_H(T)\xi, \xi\right) = \left(R_H(T)H'\,(z(0))\,, H'\,(z(0))\right) \\ &= \left(H'\,(z(0))\,, H'\,(z(0))\right) = \left\|\xi\right\|^2\ . \end{aligned} \tag{48}$$

The rest of the proposition concerning the Krein sign we leave to the reader. We note that for the eigenvalue 1 there is nothing to be proved, since it is always of Krein-type (p,q) with $p = q$, by Lemma 2.9. □

Theorem 14. *Let $G : \mathbb{R}^{2n} \to \mathbb{R}$ be a C^2 function satisfying conditions* (30) *to* (33), *and let S and σ be defined as above. Then:*

$$\hat{\imath}\,(T, z; H) = \hat{\imath}(S, z_\sigma; G)\ . \tag{49}$$

□

To prove Theorem 14, we have to imbed H into a one-parameter family of Hamiltonians, namely H^θ, for $\theta \geq 1$. Clearly $H^1 = H$, and each of the H^θ satisfies conditions (30) to (33).

For $x \in \sum$, all the equations:

$$\dot{x} = J\left(H^\theta\right)'(x) = J\theta H(x)^{\theta-1} H'(x) = J\theta H'(x) \tag{50}$$

differ by a constant scalar factor on the right-hand side. It follows that they have the same trajectories. If we denote by φ_θ the flow associated with Eq. (50), we have:

$$\varphi_\theta^\tau(x) = \varphi_1^t(x) \tag{51}$$

$$t = \theta H(x)^{\theta-1}\tau \ . \tag{52}$$

Set $z_\theta(t) := z(t\theta)$. The curve z_θ is a T/θ-periodic solution of Eq. (50); for $\theta = 1$, we get $z_1 = z$. All these curves are reparametrizations of the same closed trajectory on $\sum$.

Lemma 15. *$i_{t/\theta}\left(z_\theta; H^\theta\right)$ is a non-decreasing function of θ on the interval $[0, \infty]$.*

Proof. Write down the linearized system around z_θ:

$$\dot{y} = J\theta H''\left(z_\theta(t)\right)y + J\theta(\theta-1)\left(H'\left(z_\theta(t), y\right)\right)H'\left(z_\theta(t)\right). \tag{53}$$

Rescale time by setting $s = \theta t$. We are then dealing with a linear Hamiltonian system, with T-periodic coefficients, depending on the parameter θ:

$$\dot{y} = JH''\left(z(s)\right)y + J\left(\theta-1\right)\left(H'\left(z(s), y\right)\right)H'\left(z(s)\right) \ . \tag{54}$$

Denote by $JA_\theta(s)$ the right-hand side, with $A_\theta = A_\theta^*$. The quadratic Hamiltonian for this system is:

$$\frac{1}{2}\left(A_\theta(s)y, y\right) := \frac{1}{2}\left(H''\left(z(s)\right)y, y\right) + \frac{1}{2}\left(\theta-1\right)\left(H'\left(z(s)\right), y\right)^2 \ . \tag{55}$$

It is clearly increasing with θ, so that its index on any interval is a non-increasing function of θ by Proposition 4.12. □

Lemma 16. We have:

$$i_T\left(z; H\right) \leq i_{T/\theta}\left(z_\theta; H^\theta\right) \leq i_T(z; H) + 2n \ . \tag{56}$$

Proof. The first inequality is a direct consequence of Lemma 15. To prove the second one, we argue as in Section 4.

We are dealing with the linear Hamiltonian system $\dot{y} = JA_\theta(s)y$. The right-hand side is given by formulas (54) and (55); it depends linearly on θ. Proposition 4 tells us that the nullity $\nu_{T/\theta}\left(z_\theta; H^\theta\right)$ is non-zero for all values of θ.

Fix some $\overline{\theta} > 1$. Choose some $t < T$ such that:

$$\nu_t(z;H) = 0 \text{ and } i_t(z;H) = i_T(z;H) \tag{57}$$

$$\nu_{t/\overline{\theta}}(z;H) = 0 \text{ and } i_{t/\overline{\theta}}(z;H) = i_{T/\overline{\theta}}(z;H)\ . \tag{58}$$

This is possible if t is close enough to T (see formula 4.55).

The quadratic form on $L^2_o\left(\mathbb{R}/t\mathbb{Z};\mathbb{R}^{2n}\right)$ given by:

$$q_\theta(u,u) := \int \left[(Ju, \Pi u) + \left(A_\theta(s)^{-1}Ju, Ju\right)\right] ds \tag{59}$$

is the sum of a compact term, independent of θ, and a positive definite term depending continuously on θ. It follows that $i_{t/\theta}\left(z_\theta; H^\theta\right)$ is piecewise constant as a function of θ, and the discontinuity points must be values of θ where the quadratic form q_θ degenerates. Such parameter values will be called *singular* and their *multiplicity* is the dimension of the corresponding kernel. The jump in the index at a singular value is at most equal to its multiplicity. Since $i_{t/\theta}\left(z_\theta; H^\theta\right)$ is a non-decreasing function of θ, we end up with the inequality:

$$i_{t/\overline{\theta}}\left(z;H^{\overline{\theta}}\right) - i_t(z;H) \le N\left(\overline{\theta},0\right) \tag{60}$$

where $N\left(\overline{\theta},0\right)$ is the number of singular values of θ in the interval $\left(0,\overline{\theta}\right)$, each counted with multiplicity. One can in fact show that equality holds in formula (60).

To say that θ is a singular value means that the matrizant R_θ of Eq. (54) has the eigenvalue 1 at t; the multiplicity of θ is the dimension of Ker $(A(t/\theta) - I)$. Differentiating formula (51) with respect to x at $x = z(0)$, we get:

$$R_\theta(t/\theta)\xi = R(t)\xi + (\theta - 1)\,t\left(H'(z(0)), \xi\right) JH'(z(t))\ . \tag{61}$$

The right-hand side depends linearly on θ. It follows that Det $(R_\theta(t/\theta)$ $-I)$ is a polynomial in θ of degree $2n$. It is not identically zero, since it does not vanish for $\theta = 1$. It can have at most $2n$ roots in the interval $[0,\infty]$, each counted with its algebraic multiplicity.

Let τ be such an m-uple root. Let d be the dimension of Ker $(R_\tau(t/\tau) - I)$. Choose a basis $\xi_1, \dots, \xi_{2n}$ such that $\xi_1, \dots, \xi_d$ are eigenvectors associated with the eigenvalue 1, and write the matrix $((m_{ij}(\theta)))$ for the operator $R_\theta(t/\theta) - I$ in that particular basis. We have $m_{ij}(\theta) = a_{ij}\theta + b_{ij}$ for all (i,j), and all the coefficients in the first d rows and columns vanish for $\theta = \tau$. It follow that $(\theta - \tau)$ factors in these terms, and $(\theta - \tau)^d$ factors in the determinant. Hence $d \le m$: the multiplicity of τ as a root of the equation Det $(R(T/\tau) -) = 0$ is not less than its nultiplicity as a singular value of the parameter θ.

We have found at most $2n$ singular values of the parameter, each counted with multiplicity. Hence:

$$i_{t/\overline{\theta}}\left(z;H^{\overline{\theta}}\right) - i_t(z;H) \le 2n \tag{62}$$

and the result follows from formulas (57) and (58). □

Because of Lemma 15 and 16, we may define a (finite) integer $g_T(z;H)$ by:

$$g_T(z;H) := \lim_{\theta\to\infty} i_{T/\theta}\left(z_\theta; H^\theta\right) \tag{63}$$

and we have:

$$i_T(z;H) \le g_T(z;H) \le i_T(z;H) + 2n \ . \tag{64}$$

The last step in the proof of Theorem 14 consits in showing that $g_T\,(z;H)$ is the same for all the representations H satisfying (30) to (33). This requires a preliminary computation:

Lemma 17. *Let G be another Hamitonian satisfying conditions* (30) *to* (33), *and define $\lambda(x)$ as above:*

$$\forall x \in \sum, \quad H'(x) = \lambda(x)G'(x) \ . \tag{65}$$

Then:

$$\forall \xi \in T_x \sum, \quad \lambda(x)\left(G''(x)\xi,\xi\right) = \left(H''(x)\xi,\xi\right) \ . \tag{66}$$

Proof. This equality is well-known in differential geometry, where the quadratic form defined by (66) is known as the first fundamental form of the hypersurface $\sum$.

We have seen in Proposition 13 that $\lambda(x)$ is smooth and positive everywhere. Differentiate the identity (65) on $\sum$:

$$\forall \xi \in T\sum, \quad H''(x) = \left(\lambda'(x),\xi\right) G'(x) + \lambda(x)G''(x)\xi \ . \tag{67}$$

Taking interior product with ξ again, the term $(\xi, G'(x))$ vanishes because ξ is a tangent vector, and we are left with formula (66). □

Lemma 18. *Let G be another Hamiltonian satisfying conditions* (30) *to* (33). *Define σ and S as above. We have:*

$$g_T(z;H) = g_S\left(z_\sigma;G\right) \ . \tag{68}$$

Proof. We have to compare the linear Hamiltonian systems

$$\dot{y} = JH''\left(z(t)\right)y + (\theta-1)\left(H'\left(z(t),y\right)JH'\left(z(t)\right)\right) \tag{69}$$

and

$$\dot{y} = JG''\left(z_\sigma(s)\right)y + (\gamma-1)\left(G'\left(z_\sigma(s),y\right)JG'\left(z_\sigma(s)\right)\right) \tag{70}$$

for large value of θ and γ. To get the same time interval $[0,T]$, we rescale time in the second equation by Proposition 12. It becomes:

$$\dot{y} = \lambda\left(z(t)\right)\left[JG''\left(z(t)\right)y + J(\gamma-1)\left(G'\left(z(t),y\right)G'\left(z(t)\right)\right)\right] \ . \tag{71}$$

We have to compare the indices of the linear Hamiltonian systems (69)a and (71) on the time interval $[0,T]$. The corresponding Hamiltonians are:

$$\frac{1}{2}\left(A_\theta(t)y,y\right) := \frac{1}{2}\left(H''\left(z(t)\right)y,y\right) + \frac{1}{2}(\theta-1)\left(H'\left(z(t)\right),y\right)^2 \tag{72}$$

$$\begin{aligned}\frac{1}{2}\left(B_\gamma(t)y,y\right) := &\frac{1}{2}\lambda\left(z(t)\right)\left(G''\left(z(t)\right)y,y\right)\\ &+\frac{1}{2}(\gamma-1)\lambda\left(z(t)\right)\left(G'\left(z(t)\right),y\right)^2 .\end{aligned} \tag{73}$$

We may write $\mathbb{R}^{2n} = T_{z(t)}\sum \oplus \mathbb{R}H'\left(z(t)\right)$. On the hyperplane $T_{z(t)}\sum$, A_θ and B_γ coincide by Lemma 17. On the normal, that is, when y is collinear to $H'\left(z(t)\right)$ and $G'\left(z(t)\right)$, we can make $\frac{1}{2}\left(B_\gamma(t)y,y\right)$ as large as we want by suitably increasing γ; for instance, we can make it bigger than $\frac{1}{2}\left(A_\theta(t)y,y\right)$ for any fixed value of θ. It follows that:

$$\forall\theta\ , \quad \exists\gamma_\theta : \forall\gamma \geq \gamma_\theta\ , \quad B_\gamma \geq A_\theta\ . \tag{74}$$

Similarly:

$$\forall\gamma\ , \quad \exists\theta_\gamma : \forall\theta \geq \theta_\gamma\ , \quad A_\theta \geq B_\gamma\ . \tag{75}$$

By formula (63), we know that there are some $\overline{\theta}$ and $\overline{\gamma}$ such that, for all $\theta \geq \overline{\theta}$, the system (69) has index $g_T(z;H)$, while for all $\gamma \geq \overline{\gamma}$ the system (71) has index $g_S(z_\sigma;G)$. By formula (74), with $\theta \geq \overline{\theta}$, and Proposition 4.12, it follows that $g_T(z;H) \geq g_S(z_\sigma;G)$, while the converse inequality follows from formula (75). □

Proof of Theorem 14. From (64) and (68), we deduce that:

$$\left|i_T(z;H) - i_S(z_\sigma;G)\right| \leq 2n\ . \tag{76}$$

The same inequality holds for any period of z, for instance kT, with $k \in \mathbb{N}$:

$$\left|i_{kT}(z;H) - i_{kS}(z_\sigma;G)\right| \leq 2n\ . \tag{77}$$

Dividing by k and letting $k \to \infty$, we get formula (49). □

To conclude, let us note an important fact:

$$\hat{\imath}(T,z;H) > 2\ . \tag{78}$$

This will be proved in the next section (Theorem 7.7).

Notes and Comments. Proposition 8 and the resulting corollary go back to Poincaré. We refer to the survey paper by Kozlov [Koz1] for more results on non-integrable systems. We do not go into the theory of near-integrable systems, which is dominated by the celebrated Kolmogorov-Arnold-Moser theorem; see the forthcoming book by Zehnder and Moser [ZehM].

Theorem 14, asserting that the mean index per period is a geometric concept, depending only on the trajectory and the energy hypersurface it lies on, appears here for the first time. It may well be that the constant $2n$ in the estimates (66) and (76) can be improved.

7. Examples

The time has now come to illustrate all these concepts on examples. Unfortunately, there are not too many of them: the equations $\dot{x} = JH'(x)$ can be solved analytically only if the system is completely integrable. But there are very few completely integrable Hamiltonian systems around, and even fewer if one adds the requirement of convexity.

Example 1: The Harmonic Oscillator

We have $\mathbb{R}^{2n} = \mathbb{R}^n \times \mathbb{R}^n$, so that $x = (p, q)$, and J is defined by the $2n \times 2n$ matrix:

$$J = \begin{pmatrix} 0 & I_n \\ -I_n & 0 \end{pmatrix}$$

We are given n positive numbers:

$$0 < \alpha_1 \leq \alpha_2 \leq \ldots \leq \alpha_n \tag{1}$$

and we consider the quadratic Hamiltonian:

$$H(x) = \sum_{i=1}^{n} \frac{1}{2} \alpha_i \left(p_i^2 + q_i^2\right) . \tag{2}$$

The corresponding Hamiltonian system is linear. In fact, this $2n$-dimensional system decouples into n independent linear systems of dimension two:

$$\dot{q}_i = \alpha_i p_i \tag{3}$$

$$\dot{p}_i = -\alpha_i q_i \tag{4}$$

the solution of which is:

$$q_i(t) = A_i \sin\left[2\pi\left(t - t_0\right)/T_i\right] \tag{5}$$

$$p_i(t) = A_i \cos\left[2\pi\left(t - t_0\right)/T_i\right] \tag{6}$$

$$T_i = \frac{2\pi}{\alpha_i} . \tag{7}$$

T_i is the period of the motion of the ith projection. The solution also depends on $(n+1)$ additional parameters, which can be chosen arbitrarily: the amplitudes $A_i > 0$, $1 \leq i \leq n$, and the phase t_0.

It should finally be noted that there are n integrals of the motion, namely the functions G_i, $1 \leq i \leq n$, defined by:

$$G_i(x) = \frac{1}{2}\left(p_i^2 + q_i^2\right) \tag{8}$$

Let I be a subset of $\{1, \ldots, n\}$. The $G_i'(x)$, for $i \in I$, are linearly independent at all points $x = (x_1, \ldots, x_n\}$ such that $x_i \neq 0$ whenever $i \in I$. Note also that the G_i are in involution:

$$(JG_i'(x), G_j(x)) = 0 \quad \forall x \ . \tag{9}$$

In the study of periodic solutions, there are two case to consider.

Definition 1. The linear system $\dot{x} = JH'(x)$ will be called *weakly nonresonant* if all the α_i/α_j are irrational. If this is not the case, the system will be called *strongly resonant.* □

1.1. Weakly Nonresonant Systems. If A_i and A_j are non-zero, consider the point $M(t) = (q_i(t), q_j(t))$ in $\mathbb{R}^2$ defined by the equations:

$$q_i(t) = A_i \sin\left[2\pi\left(t - t_0\right)/T_i\right] \tag{10}$$

$$q_j(t) = A_j \sin\left[2\pi\left(t - t_0\right)/T_j\right] \ . \tag{11}$$

It is well-known that the curve $M(t)$ is either dense in the rectangle $-A_i \leq q_i \leq A_i$, $-A_j \leq q_j \leq A_j$ or periodic. The first case happens when T_i/T_j is irrational: we get a so-called *Lissajous curve.* The second case happens when T_i/T_j is rational. Say $T_i/T_j = a/b$; then the minimal period of the motions is $T = mT_i/a$, where m is the smallest common multiple of a and b.

Clearly, if $x(t)$ is a periodic solution, all its two-dimensional projections (q_i, q_j) must be periodic. This is impossible if T_i/T_j is irrational, unless A_i or A_j vanishes. It follows that if the system is weakly nonresonant, there are exactly n families of periodic solutions, given by

$$q_i(t) = A_i \sin\left[2\pi\left(t - t_0\right)/T_i\right] \tag{12}$$

$$p_i(t) = A_i \cos\left[2\pi\left(t - t_0\right)/T_i\right] \tag{13}$$

$$p_j(t) = 0 = q_j(t) \quad \forall j \neq 1 \ . \tag{14}$$

Each of these families is a two-dimensional vector space, on which $\mathbb{S}^1 = \mathbb{R}/\mathbb{Z}$ acts by the phase shift: if $x(t)$ is a T-priodic solution, so is $x(t - \tau T)$ for all $\tau \in \mathbb{S}^1$.

Let x be a solution belonging to the i-th family; that is, the equations for x are given by (12), (13), (14).

The linearized system around x is the original system (13), (14) itself, since it is already linear. It will be convenient to write it in the following form.

Set $x_i = (q_i, p_i)$, so that $x = (x_1, \ldots, x_n)$. In this way, we have split $\mathbb{R}^{2n}$ into two-dimensional subspaces which are stable by J. The equations becomces:

$$\dot{x}_i = J\alpha_i x_i \ , \qquad 1 \le i \le n \tag{15}$$

the matrizant of which is the diagonal operator:

$$R(t) = \left(e^{J\alpha_i t}, \ldots, e^{J\alpha_n t}\right) \ . \tag{16}$$

We know that $e^{2i\pi J} = I$. It follows that t is conjugate to 0 if and only if there is some i such that $\alpha_1 t \in 2\pi\mathbb{Z}$, and that its multiplicity is two.

Let x be a T-periodic solution. We must have $T = kT_i$ for some $i \in \{1, \ldots, n\}$ and some $k \in \mathbb{N}$. Its index is twice the number of points conjugate to 0 in the interval $(0, T)$, that is:

$$\begin{aligned} i(T,x) &= 2\sum_{j=1}^{n} E\left[T/T_j\right] = 2\sum_{j=1}^{n} E\left[T\alpha_j/2\pi\right] \\ &= 2(k-1) + 2\sum_{j \neq i} E\left[k\alpha_j/\alpha_i\right] \ . \end{aligned} \tag{17}$$

The function E has been defined in Sect. 4. Recall that $E[\alpha]$ is the integer $a \in \mathbb{Z}$ such that $a < \alpha \le a+1$, so that $E[a] = a - 1$ for all integers a. The mean index per period follows immediately:

$$\widehat{i}(T,x) := \lim_{N\to\infty} N^{-1} i(NT,x) = 2k + 2\sum_{j\neq i} k\alpha_j/\alpha_i = 2k\sum_{j=1}^{n} \alpha_j/\alpha_i \ . \tag{18}$$

The Floquet multipliers of (T,x) are the eigenvalues of the $e^{JT\alpha_j}$; we get $\lambda_j = e^{\pm 2ik\pi\alpha_j/\alpha_i}$. If the system satisfies a slightly stronger condition than weak nonresonance, namely if:

$$k\alpha_j + k\alpha_{j'} \neq k'\alpha_i \quad \forall (j,j') \neq (i,i) \quad \forall k' \in \mathbb{N} \tag{19}$$

then the $e^{2ik\pi\alpha_j/\alpha_i}$, for $j \neq i$, are all Krein-positive (see Proposition 2.11). To determine the Bott function $j_{(T,x)}$ on $\mathcal{U}$, all we need to know is its value near $\omega = 1$, which is given by Corollary 5.14:

$$j_{(T,x)}\left(e^{0i}\right) = i_{(T,x)} + n + 1 \ . \tag{20}$$

If for instance we are dealing with the minimal period $\overline{T} = T_i$, so that $k = 1$, we have, counting conjugate points and using the inequalities (1) (which are now strict, since the system is weakly nonresonant):

$$i(x) = 2(n-1) \tag{21}$$

$$\widehat{i}(x) = 2\sum_{j} \alpha_j/\alpha_i \tag{22}$$

$$j_x\left(e^{0i}\right) = 2(n-1) + n + 1 = 3n - 2i + 1 \ . \tag{23}$$

We can now check the iteration formula of Sect. 5 by setting $T = T_i$, or $T = kT_i$, in Eq. (17). Suppose for instance that we have

$$\alpha_n > 2\alpha_{n-1}$$

and take $\overline{T} = T_n$. Equation (17) gives:

$$i(\overline{T}, x) = 0 \tag{24}$$

$$i(2\overline{T}, x) = 2\ . \tag{25}$$

On the other hand, formula (23) gives $j_{(\overline{T},x)}\left(e^{i0}\right) = n+1$, and the Floquet multipliers $e^{2i\pi\alpha_j/\alpha_i}$, $1 \leq j \leq n = 1$, belong to the upper half-circle, and are all Krein-positive. This means that the value of $j_{(\overline{T},x)}$ must drop by $(n-1)$ between $\omega = e^{i0}$ and $\omega = -1$, and hence $j(-1) = 2$. By Bott's formula, we find:

$$i(2\overline{T}, x) = j_{(\overline{T},x)}(1) + j_{(\overline{T},x)}(-1) = 0 + 2 = 2\ . \tag{26}$$

A final remark: all periodic solutions are degenerate, since the eigenspaces of $R(kT_i)$ are two-dimensional. However, the degeneracy is not as bad as one would expect from the number of first integrals. There are n of them, and one would think from Proposition 6.8 that the dimension of $\operatorname{Ker}[R(T) - 1]$ must be at least n, and even $2n$ from the theory of completely integrable systems. This is not so, because along every periodic solution, all the integrals of the motion degenerate, except for a single one: if $\overline{T} = T_i$, we have $x_j(t) = 0$ for $j \neq 1$, and hence $G'_j(x(t)) = 0$. So Proposition 6.8 gives only two non-trivial vectors in the kernel, $H'(x(0))$ and $G'_i(x(0))$.

1.2. Strongly Resonant Systems. Let us deal immediately with the worst case: all periods T_i are the same. That is, we have $\alpha_1 = \ldots = \alpha_n = 1$ say, and the Hamiltonian then is:

$$H(x) = \frac{1}{2}\|x\|^2\ . \tag{27}$$

The equation becomes:

$$\dot{x} = Jx \tag{28}$$

and all its solutions are 2π-periodic. All the points in $2\pi\mathbb{Z}$ are conjugate to 0 with multiplicity $2n$.

If x is a T-periodic solution, we must have $T = 2k\pi$ for some $k \in \mathbb{N}$. We have:

$$i(T, x) = 2n(k-1) \tag{29}$$

$$\widehat{i}(T, x) = 2nk \tag{30}$$

$$j_{(T,x)}\left(e^{i0}\right) = i(T, x) + 2n = 2nk\ .$$

It should be noted that in this case, there are periodic solutions with $x_i(t) \neq 0$ for all i, so $G_i'(x(0)) \neq 0$, and Proposition 6.7 gives us n linearly independent vectors in the kernel of $R(T) - I$. In fact, we find that this kernel is the whole space, so that the periodic solutions are totally degenerate.

In the general case of a strongly resonant system, with $T = kT_i$, the formulas are:

$$i(T,x) = 2\sum E[T/T_j]\, d_j = 2(k-1)d_1 + 2\sum_{j\neq i} E[k\alpha_j/\alpha_i]\, d_j \tag{31}$$

$$\widehat{i}(T,x) = 2k\sum_j d_j\alpha_j/\alpha_i\ , \tag{32}$$

where d_j is the dimension of Ker $[R(T_j) - I]$, and

$$j_{(T,x)}\left(e^{i0}\right) = i(T,x) + n + \frac{1}{2}d_0\ , \tag{33}$$

where d_0 is the dimension of Ker $[R(T) - I]$, that is, the number of linearly independent T-periodic solutions. Note that $d_0 = \sum d_j$, where the summation is extended to the j such that T_j divides T. □

Example 2: Positively Homogeneous Hamiltonians

We start from a compact convex set C containing 0 in its interior. We define the *gauge* of C to be the function $J_C : \mathbb{R}^{2n} \to \mathbb{R}$ given by the formula:

$$j_C(x) := \text{Min}\left\{\lambda | \lambda^{-1}x \in C\right\} \qquad \text{for } x \neq 0 \tag{34}$$

and $j_C(0) = O$. It is a convex function, positively homogeneous of degree one:

$$\forall\lambda \geq 0\ , \quad j_C(\lambda x) = \lambda j_C(x)\ . \tag{35}$$

Assume now that the boundary Σ of C is a C^2 hypersurface. Then j_C is C^2 at every point $x \neq 0$ in $\mathbb{R}^{2n}$. We easily check that:

$$j_C'(x) = x\,\|x\|^{-1} \qquad \text{for } x \neq 0\ . \tag{36}$$

For each $\alpha > 1$, we introduce the function H_α defined by:

$$H_\alpha(x) = j_C(x)^\alpha\ . \tag{37}$$

H_α is convex and C^1 on $\mathbb{R}^{2n}$; in fact:

$$H_\alpha'(x) = x\,\|x\|^{\alpha-1} \tag{38}$$

and we see that H_α is even C^2 if $\alpha > 2$.

H_α is also positively homogeneous of degree α:

$$H_\alpha(\lambda x) = \lambda^\alpha H_\alpha(x) \quad \forall\lambda \geq 0\ , \quad \forall x \in \mathbb{R}^{2n}\ . \tag{39}$$

Differentiating this relation with respect to λ at $\lambda = 1$, we get the well-known Euler identities:

(40) $$\alpha H_\alpha(x) = (H'_\alpha(x), x)$$

(41) $$\alpha(\alpha - 1)H_\alpha(x) = (H''_\alpha(x)x, x) \ .$$

Differentiating with respect to x, we find that H'_α is positively homogeneous of degree $(\alpha - 1)$:

(42) $$H'_\alpha(\lambda x) = \lambda^{\alpha-1} H'_\alpha(x) \quad \forall \lambda > 0$$

(43) $$(\alpha - 1)H'_\alpha(x) = (H''_\alpha(x), x) \ .$$

Also:

(44) $$H''_\alpha(\lambda x) = \lambda^{\alpha-2} H''_\alpha(x) \quad \forall \lambda > 0 \ .$$

For $\alpha > 0$, the Hamiltonian H_α satisfies conditions (6.30) to (6.33) with respect to Σ. We may write $H_\alpha = (H_2)^\theta$, with $\theta = \alpha/2$. This is the situation we have just analyzed, and Lemmas 6.14 to 6.16 apply in this particular case. Because of the homogeneity however, we will be able to carry the calculations a little further.

Fix $\alpha > 1$. Consider a non-zero T-periodic solution z of the system $\dot{x} = JH'_\alpha(x)$, with $z(t) \in \Sigma$ for every t; denote by $R(t)$ the matrizant of the linearized system. It is easily seen that z belongs to a one parameter familiy of periodic solutions, indexed by the energy level h:

Lemma 2. *The curve:*

(45) $$z_h(t) := h^{1/\alpha} z\left(t h^{1-2/\alpha}\right)$$

is a periodic solution of $\dot{x} = JH'_\alpha(x)$; *we have:*

(46) $$H_\alpha\left(z_h(t)\right) = h$$

(47) $$z_h\left(T_h\right) = z_h(0)$$

(48) $$T_h = T h^{2/\alpha - 1} \ .$$

□

Lemma 3. $\mathbb{R}^{2n}$ *splits into* $T_{z(0)}\Sigma \oplus \{z(0)\}$; *the first subspace is invariant by* $R(T)$, *and the second one belongs to a two-dimensional invariant subspace associated with the eigenvalue* 1 :

(49) $$R(T)T_{z(0)}\Sigma \subset T_{z(0)}\Sigma$$

(50) $$R(T)\dot{z}(0) = \dot{z}(0)$$

(51) $$R(T)z(0) = T(\alpha - 2)\dot{z}(0) + z(0) \ .$$

Proof. Formulas (49) and (50) have been proved earlier (see Lemma 6.11).

Differentiating Eq. (47) with respect to h, we get:

(52) $$z(T)\frac{d}{dh}T_h + \frac{d}{dh}z_h(T) = \frac{d}{dh}z_h(0) \ .$$

Replacing z_h and T_h by their values, given by formulas (45) and (48), and setting $h = 1$, we get:

$$(2/\alpha - 1)T\dot{z}(T) + R(T)\alpha^{-1}z(0) = \alpha^{-1}z(0) \tag{53}$$

and hence:

$$R(T)z(0) = z(0) - (2-\alpha)T\dot{z}(0) \ . \tag{54}$$

The result follows immediately. □

We shall now consider the dependence on α. We write T_α, z_α, R_α instead of T, z, R; the (T_α, z_α) are related by the reparametrization formulas (51) and (52); for instance, $T_\alpha = 2\alpha^{-1}T_2$. Formula (6.61) gives us $R_\alpha(T_\alpha)$:

$$R_\alpha(T_\alpha)\zeta = R_2(T_2)\zeta + (\alpha/2 - 1)T_2\left(H_2'(z(0)), \zeta\right) JH_2'(z(0)) \ . \tag{55}$$

It follows that all the $R_\alpha(T_\alpha)$ coincide on $T_{z(0)}\Sigma$. Write $\mathbb{R}^{2n} = T_{z(0)}\Sigma \oplus \{z(0)\}$ as in Lemma 3, and $T_{z(0)}\Sigma = E \oplus \{\dot{z}(0)\}$. Write the matrix of the operator $R_\alpha(T_\alpha)$ in the decomposition $\{z(0)\} \oplus \{\dot{z}(0)\} \oplus E$. It consists of two blocks, a two-dimensional one in the upper left corner which is given by (50) and (51), and a $(n-2)$-dimensional one in the lower right corner which is independent of α. Hence:

Lemma 4. *Set* $K := \mathrm{Ker}\,(R_2(T_2) - I) \cap T_{z(0)}\Sigma$. *We have:*

$$\forall \alpha \neq 2\ , \quad \mathrm{Ker}\,(R_\alpha(T_\alpha) - I) = K \tag{56}$$

$$\mathrm{Ker}\,(R_2(T_2) - I) = K \oplus \{z(0)\} \ . \tag{57}$$

So $\mathrm{Ker}\,(R_2(T_2) - I)$ *is two-dimensional at least.* □

It follows that $\theta = 1$ is the only possible singular value of the parameter, and its multiplicity is one. In other words, $\alpha = 2$ is the only possible point of discontinuity of $i_T(z_\alpha, H^\alpha)$, and the jump is at most one. In fact, it is exactly one, as we now check.

After suitable time rescaling, as in Lemmas 6.14 and 6.15, we find that $i_T(z_\alpha, H^\alpha)$ is the index of the quadratic form q_α on $L^2_O\left(\mathbb{R}/T_2\mathbb{Z}; \mathbb{R}^{2n}\right)$ given by:

$$q_\alpha(u,u) := \int_0^{T_2} \left[(Ju, \Pi u) + \left(A_\alpha(s)^{-1}Ju, Ju\right)\right] ds \ , \tag{58}$$

where

$$\frac{1}{2}\left(A_\alpha(s)y, y\right) := \frac{1}{2}\left(H_2''(z(s))\,y, y\right) + \frac{1}{2}(\alpha/2 - 1)\left(H'(z(s)), y\right)^2 \ . \tag{59}$$

Now apply Lemma 5.8: the change in the index of q_α when α crosses the singular value will be given by the first-order term $\partial q_\alpha(u_2, u_2)/\partial\alpha$ at $\alpha = 2$, provided it does not vanish. Substituting, we get:

$$\partial\left(A_\alpha(s)^{-1}y, y\right)/\partial\alpha = -\frac{1}{2}\left(H_2'\left(z(s)\right), H_2''\left(z(s)\right)^{-1}y\right)^2 \tag{60}$$

and hence:

$$\begin{aligned} \partial q_\alpha\left(u_2, u_2\right)/\partial\alpha &= \int -\frac{1}{2}\left(H_2', \left(H_2''\right)^{-1} J\dot{z}_2\right)^2 dt \\ &= -\int \frac{1}{2}\left(H_2', \left(H_2''\right)^{-1} H_2'\right)^2 dt \end{aligned} \tag{61}$$

which is clearly negative. So the index decreases by 1 as α increases through 2. We have proved:

Proposition 5. *Consider a compact C^2 hypersurface Σ which bounds a convex set C containing the origin in its interior, and define a family H^α of Hamiltonian by:*

$$H_\alpha(x) = j_C(x)^\alpha \tag{62}$$

where j_C is the gauge of C. Let z be a closed Hamiltonian trajectory on Σ, and denote by z_α the corresponding periodic solutions of $\dot{x} = JH_\alpha'(x)$. Then $i\left(z_\alpha; H_\alpha\right)$ is constant on the intervals $(1,2]$ and $(2,\infty)$; we have:

$$i\left(z_\alpha; H_\alpha\right) = i\left(z_\beta; H_\beta\right) + 1 \qquad \text{for } \alpha > 2 \geq \beta\,. \tag{63}$$

□

In this statement, T_α is taken to be the minimal period of z_α, but the same result holds of course for any period.

Denote by $j_\alpha(\omega)$ the Bott function of (T_α, z_α). We know from Proposition 6.13 that the z_α, for $\alpha \geq 1$, have the same Floquet multipliers with the same multiplicity and Krein sign. So the Bott functions coincide on the intervals of continuity, up to an additive constant which is determined by $j_\alpha\left(e^{\pm 0}\right)$. We are now able to determine this value, at least in the generic case:

Proposition 6. *Assume that 1 is a double Floquet multiplier. Then:*

$$j_\beta\left(e^{\pm i0}\right) = j_\beta(1) + n + 1 \qquad \text{if } 1 < \beta \leq 2 \tag{64}$$

$$j_\alpha\left(e^{\pm i0}\right) = j_\alpha(1) + n \qquad \text{if } \alpha > 2\,. \tag{65}$$

Proof. As we just mentioned, for all α and β, and for all $\omega \in \mathcal{U}$ which is not a Floquet multiplier, we have:

$$j_\alpha(\omega) = j_\beta(\omega) + c(\alpha, \beta)\,, \tag{66}$$

where $c(\alpha,\beta)$ is a constant depending on α and β. On the other hand, we know that the mean index per period $\widehat{i}_\alpha$ does not depend on α (see Theorem 6.14). But Theorem 5.18 tells us that $\widehat{i}_\alpha$ is the mean value of the function j_α on the unit circle; it follows that $c(\alpha,\beta) \equiv 0$. It then follows from formula (66) that, for all α and β:

$$j_\alpha\left(e^{\pm i0}\right) = j_\beta\left(e^{\pm i0}\right) . \tag{67}$$

On the other hand, if we take into account the identity $j_\alpha(1) = i_T(z_\alpha; H_\alpha)$, and the similar one for β, Proposition 5 gives:

$$j_\alpha(1) = j_\beta(1) + 1 \qquad \text{for } \alpha > 2 \geq \beta \tag{68}$$

and formula (5.130), which is applicable since 1 is a double Floquet multiplier, yields:

$$n \leq j_\alpha\left(e^{\pm i0}\right) - j_\alpha(1) \leq n + 1 \tag{69}$$

$$n \leq j_\beta\left(e^{\pm i0}\right) - j_\beta(1) \leq n + 1 . \tag{70}$$

Putting the inequalities (67) to (70) together, we get (64) and (65) as the only possibility. Note that, for $\alpha = 2$, this agrees with Corollary 5.14. □

A General Result. We go back to the assumptions of Section 6. Start from a C^2 compact hypersurface $\Sigma \subset \mathbb{R}^{2n}$; assume that Σ bounds a convex set with non-empty interior. Consider a C^2 function $H : \mathbb{R}^{2n} \to \mathbb{R}$, such that:

$$\forall x , \quad H(x) \geq 0 \tag{71}$$

$$\Sigma = \{x | H(x) = 1\} \tag{72}$$

$$\forall x \in \Sigma , \quad H'(x) \neq 0 \tag{73}$$

$$\forall x \in \Sigma , \quad H''(x) \ \text{ is positive definite.} \tag{74}$$

Consider the Hamiltonian flow $\dot{x} = JH'(x)$ associated with H and a periodic solution $z(t)$ on Σ:

$$\dot{z} = JH'(z) \tag{75}$$

$$H(z) = 1 \tag{76}$$

$$z(0) = z(T) . \tag{77}$$

Denote by $i_T(z; H)$ the index of this periodic solution on the interval $(0, T]$, and by $\hat{i}(T, z; H) := \lim i_{kT}(z; H)/k$ its mean index per period.

Theorem 7. *If $n \geq 2$, we have:*

$$\hat{i}(T, z; H) > 2 . \tag{78}$$

Proof. By Theorem 6.14, we may assume that H is positively homogeneous of degree 2.

By Lemma 3, Ker $(R_2(T_2) - I)$ contains at least two vectors, $x(0)$ and $\dot{x}(0)$. So T_2 is conjugate to 0 with multiplicity 2, and Theorem 5.1 implies that $\hat{i}(T, z; H) \geq 2$. We want a strict inequality.

If 1 is a double Floquet multiplier, Proposition 6 gives:

$$j_2\left(e^{\pm i0}\right) = j_2(1) + n + 1 \geq 3 \tag{79}$$

There are at most $2(n-1)$ more Floquet multipliers on the unit circle, to be placed symmetrically with respect to the real axis, and each of them can decrease j_2 by its multiplicity at most (see Proposition 5.9). It follows that, for any $\omega \in \mathcal{U}$, we have:

$$j_2(\omega) \geq j_2(1) + n + 1 - (n-1) = j_2(1) + 2\ . \tag{80}$$

So j_2 is a step function, which is bounded below by 2, and which takes a value strictly greater than 2 on a neighbourhood of 1 (except at 1 itself). Its mean value must therefore be strictly greater than 2, and formula (78) follows from Theorem 5.18:

$$\widehat{\imath}(T, z; H) = (2\pi)^{-1} \int_{\mathcal{U}} j_2(\omega) d\omega > 2\ . \tag{81}$$

If 1 has higher multiplicity, some even number $m \geq 4$ say, Proposition 5.11 gives

$$j_2\left(e^{\pm i0}\right) = j_2(1) + n + m/2 - r_0^- \tag{82}$$

with $2r_0^-$ denoting the number of Floquet multipliers which arrive on the unit circle at 1, so that $2r_0^- \leq m$. There are at most $(2n-m)$ more Floquet multipliers on the unit circle. As before, it follows that:

$$j_2(\omega) \geq j_2(1) + n + m/2 - r_0^- - (2n-m)/2 = j_2(1) + m - r_0^-\ . \tag{83}$$

Hence:

$$j_2(\omega) \geq j_2(1) + m/2 \geq j_2(1) + 2\ . \tag{84}$$

So $j_2 \geq 2$ everywhere, and formula (82) yields $j_2 \geq n$ near 1. We are in the same situation as when $m = 2$. □

Notes and Comments. The strict inequality $\widehat{\imath} > 2$ will be crucial far the multiplicity result of the last chapter (Corollary V.3.16). It was first proved in [EkeH3].

8. Non-periodic Solutions: The Mean Index

Let $H : \mathbb{R}^{2n} \to \mathbb{R}$ be a convex Hamiltonian of class C^2. We will be working on a fixed energy level, $H = 1$ say. That is, we consider the set:

$$\Sigma = \left\{x \in \mathbb{R}^{2n} \middle|\ H(x) = 1\right\}$$

and we assume that it is compact, non-empty, with $H'(x)$ non-vanishing and $H''(x)$ positive definite for every $x \in \Sigma$. It follows that Σ is a C^2 hypersurface in $\mathbb{R}^{2n}$.

Denote by φ^t the flow associated with the Hamiltonian system

(1) $$\dot{x} = JH'(x)$$

and choose some $\epsilon > 0$ such that the set

(2) $$\mathcal{U} := \{x | 1 - \epsilon < H(x) < 1 + \epsilon\}$$

is bounded, with $H'(x) \neq 0$ for all $x \in \mathcal{U}$. Then $\varphi^t : \mathcal{U} \to \mathcal{U}$ is well-defined for all t. We have

(3) $$(\varphi^t)^* (dx_1 \wedge \ldots \wedge dx_{2n}) = dx_1 \wedge \ldots \wedge dx_{2n}$$

by Liouville's theorem and

(4) $$(\varphi^t)^* dH = dH$$

since $H \circ \varphi^t = H$.

There is a unique $(2n-1)$-form μ on $\mathbb{R}^{2n}$ such that

(5) $$dx_1 \wedge \ldots \wedge dx_{2n} = dH \wedge \mu \,.$$

Applying $(\varphi^t)^*$ to both sides, we get $dH \wedge \mu = dH \wedge (\varphi^t)^* \mu$. By the uniqueness of the decomposition (5), we conclude that μ itself is preserved by the flow:

(6) $$(\varphi^t)^* \mu = \mu \,.$$

Denote by $i_\Sigma : \Sigma \to \mathbb{R}^{2n}$ the standard injection, and set $\mu_\Sigma := (i_\Sigma)^* \mu$.

Lemma 1. *μ_Σ is a nondegenerate $(2n-1)$-form on Σ which is preserved by the Hamiltonian flow φ^t.*

Proof. All we have to do is to prove that μ_Σ is nondegenerate. This follows immediately from Eq. (5) and the fact that dH vanishes on $T\Sigma$. Indeed, if we could find independent vectors $\xi_1, \ldots, \xi_{2n-1}$ in $T_x\Sigma$ such that $\mu_\Sigma(\xi_1, \ldots, \xi_{2n-1}) = 0$, we would have

(7) $$(dx_1 \wedge \ldots \wedge dx_{2n}) (H'(x), \xi_1, \ldots, \xi_{2n}) = 0$$

which is clearly impossible. □

From now on, we will consider μ_Σ as a finite positive measure on Σ, invariant by φ^t. This is the setting for ergodic theory. From now on, a.e. will mean "almost everywhere with respect to μ_Σ".

Take some $\xi \in \Sigma$, and consider the (not necessarily periodic) solution $x(t) = \varphi^t(\xi)$ of the Cauchy problem. For each T, the index $i_T(\xi)$ of the linearized system

$$\dot{y} = JH''(x(t))\, y$$

on the interval $(0, T)$ is well-defined, but there is no reason why $i_T(\xi)/T$ should converge to some limit – unless $x(t)$ happens to be periodic. What we will show, however, is that this limit exists for a.e. $\xi \in \Sigma$.

Theorem 2. *There is a borelian function $I : \Sigma \to [0, \infty)$ such that*

$$I \circ \varphi^t = I \quad \forall t \tag{8}$$

$$\frac{1}{T} i_T \to I \quad \text{when} \quad T \to \infty \tag{9}$$

the convergence being L^1 and a.e. □

The proof consists in checking the assumption of Kingman's subadditive ergodic theorem.

For each fixed T, consider the function $i_T : \Sigma \to [0, \infty)$. By definition, $i_T(\xi)$ is the index of the quadratic form

$$q_t^\xi(u, u) = \int_0^T \left[(Ju, \Pi u) + \left(H'' \left(\varphi^t(\xi) \right)^{-1} Ju, Ju \right) \right] dt \tag{10}$$

on the space $L_o^2 \left(0, T; \mathbb{R}^{2n} \right)$. It depends continuously on $\xi \in \Sigma$. Arguing by compactness, as in Sect. 4, we find that i_T is a lower semi-continuous function of ξ, and hence that it is borelian.

By Lemma 4.8, we have:

$$T \geq S \Rightarrow i_S(\xi) \leq i_T(\xi) . \tag{11}$$

To show that $T^{-1} i_T$ converges when $T \to \infty$, it is therefore sufficient to show that $N^{-1} i_N$ converges when $N \to \infty$, $N \in \mathbb{N}$. So we restrict our attention to the sequence i_N, $N \in \mathbb{N}$.

Lemma 3. *The sequence i_N is φ-subadditive, that is:*

$$i_{N+K} \leq i_N + i_K \circ \varphi^N \quad \forall (N, K) . \tag{12}$$

Proof. Fix some ξ in Σ. By definition, $L_o^2 \left(0, N; \mathbb{R}^{2n} \right)$ and $L_o^2 \left(0, K; \mathbb{R}^{2n} \right)$ contain subspaces E_N and E_K, with dimension $i_N(\xi)$ and $i_K \left(\varphi^N(\xi) \right)$, such that the restrictions of q_N^ξ and $q_K^{\varphi^N(\xi)}$ to E_N and E_K are negative definite.

Now imbed everything into $L_o^2 \left(0, N + K; \mathbb{R}^{2n} \right)$. More precisely, consider the subspace $\widetilde{E}_N$ and $\widetilde{E}_K$ defined as follows:

$$\widetilde{E}_K = \left\{ u \in L_o^2(0, N + K) |\ u|_{(0,N)} \in E_N \ \text{ and } \ u|_{(N,N+K)} = 0 \right\} \tag{13}$$

$$\widetilde{E}_K = \left\{ u \in L_o^2(0, N + K) |\ u|_{(0,N)} = 0 \ \text{ and } \ u|_{(N,N+K)} \in E_K \right\} . \tag{14}$$

The quadratic form q_{N+K}^ξ restricts to $\widetilde{E}_N$ as q_N^ξ, and to $\widetilde{E}_K$ as $q_K^{\varphi^N(\xi)}$. They are orthogonal subspaces, both for the standard Hilbert structure on

$L^2_o\left(0, N+K; \mathbb{R}^{2n}\right)$ and for q^ξ_{N+K}. If follows that the restriction of q^ξ_{N+K} on $\widetilde{E}_N \oplus \widetilde{E}_N^{N+K}$ is negative definite. Hence formula (12). □

To apply Kingman's theorem, we have to show that the i_N are integrable. In fact, they are uniformly bounded. Introduce the following definition:

Definition 4. Let δ and Δ be two be two positive numbers, with $0 < \delta \leq \Delta$. We say that H'' is (δ, Δ)-*pinched* on Σ if:

$$\delta \|\xi\|^2 \leq \frac{1}{2}\left(H''(x)\xi, \xi\right) \leq \Delta \|\xi\|^2 \quad \forall x \in \Sigma\,, \quad \forall \xi\,. \tag{15}$$ □

If H'' is (δ, Δ)-pinched on Σ, it will be (δ', Δ')-pinched for every $\delta' > \delta$ and $\Delta' < \Delta$. The best values for δ and Δ are given by:

$$2\overline{\delta} = \operatorname*{Min}_{x\in\Sigma} \operatorname{Spec} H''(x) \tag{16}$$

$$2\overline{\Delta} = \operatorname*{Max}_{x\in\Sigma} \operatorname{Spec} H''(x) \tag{17}$$

both numbers being well-defined and positive, since H is C^2, Σ is compact and $H''(x)$ is positive definite for all $x \in \Sigma$.

Lemma 5. *Let H'' be (δ, Δ)-pinched on Σ, and take any $\xi \in \Sigma$. We have:*

$$2nE\left[\frac{T\delta}{\pi}\right] \leq i_T(\xi) \leq 2nE\left[\frac{T\Delta}{\pi}\right]\,. \tag{18}$$

Proof. We simply apply Proposition 4.14 with $a = \Delta$ and $b = \delta^{-1}$. Recall that we have a non-standard definition of the integer part:

$$E[\alpha] = k \Leftrightarrow k < \alpha \leq k+1\,. \tag{19}$$ □

We have thus found uniform bounds for i_T. Theorem 2 now follows from:

Kingman's Subadditive Ergodic Theorem. *Assume $\varphi : \Sigma \to \Sigma$ is measure-preserving and $i_N \in L^1$ is a φ-subadditive sequence such that:*

$$\operatorname*{Inf}_N \frac{1}{N} \int i_N > -\infty\,. \tag{20}$$

Then $N^{-1} i_N$ converges a.e. and L^1 to some φ-invariant function I such that

$$\int I = \operatorname*{Inf}_N \frac{1}{N} \int i_N\,. \tag{21}$$ □

We get from Lemma 5 an additional estimate:

$$2\frac{n\delta}{\pi} \leq I(\xi) \leq 2\frac{n\Delta}{\pi} \quad \forall \xi \in \Sigma\,. \tag{22}$$

Notes and Comments. The mean index of a non-periodic trajectory is introduced here for the first time. It shall not be used in the rest of the book, and the only purpose of this very short section is to show that it exists, and to ask what are its properties. All questions are open: we would particularly like to know

(a) what are the relations, if any, between the mean index $I(\xi)$ and the Liapounov exponents, as defined in the general theory of dynamical systems via Oseledec's noncommutative ergodic theorem

(b) how representative the periodic points are: for instance, is I continuous at periodic points? In the case of an elliptic periodic point, with suitable generic properties, KAM analysis might throw some light on the matter.

I learned Kingman's subadditive ergodic theorem from N. Ghoussoub.

Chapter II. Convex Hamiltonian Systems

1. Fundamentals of Convex Analysis

We start from a *duality pairing* $(X, X^*, \langle \cdot , \cdot \rangle)$, that is, two real vector spaces X and X^*, and a bilinear map $(x, x^*) \to \langle x, x^* \rangle$ into $\mathbb{R}$ which separates points:

$$\forall x \in X, \exists x^* \in X^* : \quad \langle x, x^* \rangle \neq 0$$
$$\forall x^* \in X^*, \exists x \in X : \quad \langle x, x^* \rangle \neq 0 \ .$$

We endow X and X^* with the corresponding weak topologies, $\sigma(X, X^*)$ and $\sigma(X^*, X)$.

A standard example of such a situation, and the only one which is of any use, is the case when X is a Banach space and X^* its topological dual. If we do not take this as the starting situation, it is just to stress the fact that everything in this section depends on the weak topologies only.

We will consider functions with values in $\mathbb{R} \cup \{+\infty\}$. The introduction of $+\infty$ as an admissible value does not stem from a abstract desire for generality, but from very precise needs which will appear presently.

Let $F : X \to \mathbb{R} \cup \{+\infty\}$ be such a function. We define its *domain* $\operatorname{dom} F$ to be the set of points $x \in X$ where $F(x)$ is finite:

$$\operatorname{dom} F := \{x \in X \mid F(x) < +\infty\}$$

and its *epigraph* $\operatorname{epi} F \subset X \times \mathbb{R}$ as follows:

$$\operatorname{epi} F = \{(x, a) \in X \times \mathbb{R} \mid F(x) \geq a\} \ .$$

Note that, if $x \notin \operatorname{dom} F$, there is no point in $\operatorname{epi} F$ which lies above x. We shall say that F is *proper* if it is not identically $+\infty$, that is, if $\operatorname{epi} F$ is non-empty.

Definition 1. $F : X \to \mathbb{R} \cup \{+\infty\}$ is a *convex function* if $\operatorname{epi} F \subset X \times \mathbb{R}$ is a convex set. □

Definition 2. $F : X \to \mathbb{R} \cup (+\infty\}$ is *lower semicontinuous* (l.s.c.) if $\operatorname{epi} F \subset X \times \mathbb{R}$ is a closed set. □

These definitions can immediately be translated into equivalent properties:

Proposition 3. $F : X \to \mathbb{R} \cup \{+\infty\}$ *is a convex function iff the following inequality holds:*

$$F(\alpha x + \beta y) \le \alpha F(x) + \beta F(y)$$

for all x and y in X, and all positive real numbers α and β such that $\alpha+\beta = 1$. □

Proposition 4. $F : X \to \mathbb{R} \cup \{+\infty\}$ *is a l.s.c. function iff one of the following holds:*

(i) for every $a \in \mathbb{R}$, the set $\{x | F(x) \le a\}$ is closed in X.

(ii) $\liminf_{y \to x} F(y) \ge F(x)$. □

As an example, consider the *indicator function* δ_K of a subset $K \subset X$:

$$\delta_K(x) = \begin{cases} 0 & \text{if } x \in K \\ +\infty & \text{otherwise} . \end{cases}$$

The function δ_K is convex iff K is convex, and it is l.s.c. iff K is closed.

If F is convex (l.s.c.) so is λF for all $\lambda > 0$. If F and G are convex (l.s.c.) so is $F + G$. The set of all convex (l.s.c.) functions on X is a convex cone (not a linear space: substracting is forbidden). There is one more operation which can be defined on convex functions: inf-convolution.

Definition 5. Let F_1 and F_2 be convex functions on X. Consider in $X \times \mathbb{R}$ the subset:

$$\operatorname{epi} F_1 + \operatorname{epi} F_2 := \left\{ (x_1, a_1) + (x_2, a_2) \middle| \begin{array}{l} (x_1, a_1) \in \operatorname{epi} F_1 \\ (x_2, a_2) \in \operatorname{epi} F_2 \end{array} \right\}$$

and define $\overline{F} : X \to \mathbb{R} \cup \{-\infty\} \cup \{+\infty\}$ by:

(1) $$\overline{F}(x) := \inf \{a | (x, a) \in \operatorname{epi} F_1 + \operatorname{epi} F_2\} .$$

If $\overline{F}(x) > -\infty$ for every $x \in X$, we say that $\overline{F}$ is the *inf-convolute* of F_1 and F_2, and we write:

$$\overline{F} =: F_1 \Box F_2 .$$ □

We can write formula (1) in another way:

(2) $$F_1 \Box F_2(x) = \inf_{x_1 + x_2 = x} \{F_1(x_1) + F_2(x_2)\} .$$

So the inf-convolute of F_1 and F_2 is well-defined if the right-hand side of this formula is never $-\infty$. It is sufficient, for instance, that F_1 and F_2 both a bounded from below. Then $F_1 \Box F_2$ is convex, and

$$\operatorname{epi} F_1 + \operatorname{epi} F_2 \subset \operatorname{epi} (F_1 \Box F_2) \subset \overline{\operatorname{epi} F_1 + \operatorname{epi} F_2} .$$

Note that, in general, epi $(F_1 \Box F_2)$ will *not* be equal to epi F_1 + epi F_2. This is because the infimum need not be attained on the right-hand side of formula

(2): we construct epi $(F_1 \square F_2)$ from $(\text{epi}\, F_1 + \text{epi}\, F_2)$ by "vertical closure", that is, taking the intersection with every vertical line $\{x\} \times \mathbb{R}$ in $X \times \mathbb{R}$, which is either empty or an unbounded interval, and closing it if it is open.

Note also that $\text{epi}\, F_1 + \text{epi}\, F_2$ need not be closed even if $\text{epi}\, F_1$ and $\text{epi}\, F_2$ both are. It follows that $F_1 \square F_2$ need *not* be l.s.c. even if F_1 and F_2 both are convex l.s.c.

The main theoretical tool in the study of duality pairings $(X, X^*, \langle \cdot\,, \cdot \rangle)$ is the Hahn-Banach theorem, and the separation theorems which follow from it. For instance, we know that every closed convex subset of $X \times \mathbb{R}$ is the intersection of all closed half-spaces which contain it. But epigraphs of convex l.s.c. functions are particular cases of closed convex sets, and every non-vertical closed half-space can be written as the set of (x, a) such that $\langle x, x^* \rangle - a \leq m$, for some $x^* \in X^*$ and $m \in \mathbb{R}$. Note that the separating hyperplane $a = \langle x, x^* \rangle - m$ then is the graph of the continuous affine functions $x \to \langle x, x^* \rangle - m$. It turns out that vertical half-spaces can be neglected, and we get the fundamental result.

Theorem 6. *Let $F : X \to \mathbb{R} \cup \{+\infty\}$ be a convex l.s.c. function. Denote by $M(F)$ the set of its continuous affine minorants:*

$$(3) \qquad (x^*, m) \in M(F) \Leftrightarrow \langle x, x^* \rangle - m \leq F(x) \quad \forall x \in X \; .$$

Then F is the pointwise supremum of all functions in $M(F)$:

$$(4) \qquad F(x) = \sup \{ \langle x, x^* \rangle - m |\, (x^*, m) \in M(F) \} \; . \qquad \square$$

The next thing to do is clearly to gain some understanding of the set $M(F)$. It is defined by formula (3) as a subset of $X^* \times \mathbb{R}$. Let us fix x^* in X, and look at all the $m \in \mathbb{R}$ such that $(x^*, m) \in M(F)$. Formula (3) gives:

$$(5) \qquad \begin{aligned} (x^*, m) \in M(F) &\Longleftrightarrow m \geq \langle x, x^* \rangle - F(x) \quad \forall x \in X \\ &\Longleftrightarrow m \geq \sup_{x \in X} \{ \langle x, x^* \rangle - F(x) \} \; . \end{aligned}$$

If we now define

$$(6) \qquad F^*(x^*) := \sup_{x \in X} \{ \langle x, x^* \rangle - F(x) \}$$

then formula (5) becomes $m \geq F^*(x^*)$, that is, $M(F)$ is simply the epigraph of F^*. Note that, if F is proper, then the right-hand side of formula (6) is not $-\infty$, so that F^* is well-defined as a function from X^* into $\mathbb{R} \cup \{+\infty\}$.

Formula (4) now becomes:

$$\begin{aligned} F(x) &= \sup \{ \langle x, x^* \rangle - m | x^* \in X^*, m \geq F^*(x^*) \} \\ &= \sup_{x^* \in X^*} \{ \langle x, x^* \rangle - F^*(x^*) \} \; . \end{aligned}$$

We now summarize this discussion:

Definition 7. Let $F : X \to \mathbb{R} \cup \{+\infty\}$ be a proper function. The function $F^* : X^* \to \mathbb{R} \cup \{+\infty\}$ defined by

$$F^*(x^*) = \sup_{x \in X} \{\langle x, x^* \rangle - F(x)\} \tag{7}$$

is called the *Fenchel conjugate* or *Legendre transform* of F. □

Proposition 8. *Let F be a proper convex function and F^* its Legendre transform. We have:*

$$F(x) = \sum_{x^* \in X^*} \{\langle x, x^* \rangle - F^*(x^*)\} \ . \tag{8}$$

□

As an easy consequence of formulas (7) or (8), we have *Fenchel's inequality*

$$F(x) + F^*(x^*) \geq \langle x, x^* \rangle \quad \forall x \in X \ \ \forall x^* \in X^* \ . \tag{9}$$

Proposition 9. *If F is proper, F^* is convex l.s.c. If in addition F is convex l.s.c., we have:*

$$(F^*)^* = F \ . \tag{10}$$

□

Proof. Formula (7) defines F^* as the pointwise supremum of a family of affine continuous functions on X^*, namely $x^* \to \langle x, x^* \rangle - F(x)$, indexed by $x \in X$. It follows that F^* must be convex l.s.c.

Replacing F by F^* in Definition 7, we get:

$$F^{**}(x) := \sum_{x^* \in X^*} \{\langle x, x^* \rangle - F^*(x^*)\} \ .$$

If F is convex l.s.c., the right-hand side coincides with $F(x)$ by Proposition 8.

The Fenchel conjugation $F \to F^*$ sends the set of proper convex l.s.c. functions on X onto the set of proper convex l.s.c. functions on X^*; its inverse is itself. Therefore, any property of F must translate into some property of F^*, even though the transformation may not be obvious.

Note the following easy relations:

$$F \leq G \Longleftrightarrow F^* \geq G^* \tag{11}$$

$$(\lambda F)^* (x^*) = \lambda F^* \left(\frac{x^*}{\lambda} \right) \qquad \text{for } \ \lambda > 0 \ . \tag{12}$$

On the other hand, the direct computation of $(F+G)^*$ is surprisingly difficult. We must use a detour.

Proposition 10. *Let F_1 and F_2 be convex l.s.c. functions on X, such that $F_1 \Box F_2$ is well-defined. Then:*

(13)
$$(F_1 \Box F_2)^* = F_1^* + F_2^* \ . \qquad \square$$

Proof. By Definition 7:

$$(F_1 \Box F_2)^* (x^*) = \sup_x \{\langle x, x^* \rangle - (F_1 \Box F_2)(x)\} \ .$$

Plug in formula (2):

$$\begin{aligned}(F_1 \Box F_2)^* (x^*) &= \sup_x \left\{ \langle x, x^* \rangle - \inf_{x_1 + x_2 = x} \{F_1(x_1) + F_2(x_2)\} \right\} \\ &= \sup_{x_1, x_2} \{\langle x_1 + x_2, x^* \rangle - F_1(x_1) - F_2(x_2)\} \\ &= \sup_{x_1} \{\langle x_1, x^* \rangle - F_1(x_1)\} + \sup_{x_2} \{\langle x, x_2 \rangle - F_2(x_2)\} \\ &= F_1^*(x^*) + F_2^*(x^*) \ . \qquad \square\end{aligned}$$

Corollary 11. *Let F_1 and F_2 be convex l.s.c. functions on X, such that $F_1 + F_2$ is proper. Then $F_1^* \Box F_2^*$ is well-defined and:*

(14)
$$(F_1 + F_2)^* = (F_1^* \Box F_2^*)^{**} \ . \qquad \square$$

Proof. Replace F_1 by F_1^* and F_2 by F_2^* in Proposition 10. We get:

$$(F_1^* \Box F_2^*)^* = F_1 + F_2 \ .$$

Taking conjugates on both sides, we get formula (14). $\square$

We already know that $F_1^* \Box F_2^*$ is convex. If it also happened to be l.s.c., we could apply Proposition 9, and formula (14) would reduce to a much more pleasant identity, $(F_1 + F_2)^* = F_1^* \Box F_2^*$. More about this in the next section.

Similarly, there are difficulties in finding $(G \circ A)^*$, when A is a linear map. We shall allow A to be an unbounded operator. To be precise, A will map a linear subspace $\operatorname{dom} A \subset X$ into Y; we define

(15)
$$\operatorname{graph} A : \{(x, Ax) | x \in \operatorname{dom} A\} \ .$$

It is a linear subspace of $X \times Y$. We say that A is *closed* if its graph is closed. The orthogonal of $\operatorname{graph} A$ in $X^* \times Y^*$ is the graph of another closed operator $-A^*$ from $\operatorname{dom} A^* \subset Y^*$ into X^*:

(16)
$$\operatorname{graph} - A^* := (\operatorname{graph} A)^{\perp} \ .$$

In other words:

(17)
$$x^* + A^* y^* = 0 \Longleftrightarrow \langle x, x^* \rangle + \langle Ax, y^* \rangle = 0 \quad \forall x \in \operatorname{dom} A \ .$$

If A is a closed operator from X to Y and $G : Z \to \mathbb{R} \cup \{+\infty\}$ is a function on Y, we define $G \circ A : X \to \mathbb{R} \cup \{+\infty\}$ as follows:

(18)
$$(G \circ A)(x) = \begin{cases} G(Ax) & \text{if } x \in \operatorname{dom} A \\ +\infty & \text{otherwise} \ . \end{cases}$$

Proposition 12. *Let* $A : \operatorname{dom} A \to Y$ *be a closed operator, with* $\operatorname{dom} A \subset X$, *and* $G : Y \to \mathbb{R} \cup \{+\infty\}$ *a convex l.s.c. function. Define a function* $\Phi : X^* \to \mathbb{R} \cup \{+\infty\}$ *by:*

$$\Phi(x^*) = \inf \{G^*(y^*) | A^* y^* = x^*\} \tag{19}$$

the right-hand side being $+\infty$ *if* $x^* \notin A^*(Y^*)$. *Then* Φ *is convex, and we have:*

$$\Phi^* = G \circ A \tag{20}$$

$$(G \circ A)^* = \Phi^{**} . \tag{21}$$

□

Proof. It is easy to check that Φ is convex, and (21) follows from (20) by Fenchel conjugation. All we have to do is to check formula (20).

Use Definition 7 to write:

$$\begin{aligned}
\Phi^*(x) &= \sup_{x^*} \{\langle x, x^* \rangle - \Phi(x^*)\} \\
&= \sup_{x^*} \sup_{A^* y^* = x^*} \{\langle x, x^* \rangle - G^*(y^*)\} \\
&= \sup_{y^*} \{\langle x, A^* y^* \rangle - G^*(y^*)\} \\
&= \sup_{y^*} \{\langle Ax, y^* \rangle - G^*(y^*)\} \\
&= \Phi(Ax) .
\end{aligned}$$

□

By now, we have paid enough attention to the Fenchel conjugation. Let us move on to another important operation in convex analysis: subdifferentiation.

The idea is quite simple. We have seen that any convex l.s.c. function F on X is the pointwise supremum of its continuous affine minorants $M(F)$, a fact which is expressed in Proposition 8:

$$F(x) = \operatorname{Sup}_{x^*} \{\langle x, x^* \rangle - F^*(x^*)\} . \tag{22}$$

We now raise the question: given $x \in X$ such that $F(x) < +\infty$, is the supremum attained on the right-hand side? This would mean that there is some x^* such that:

$$F(x) = \langle x, x^* \rangle - F^*(x^*) . \tag{23}$$

On the other hand, for any $y \in X$, we have the Fenchel inequality:

$$F(y) \geq \langle y, x^* \rangle - F^*(x^*) . \tag{24}$$

Substracting (23) from (24), we see an equivalent condition:

$$F(y) \geq \langle y - x, x^* \rangle + F(x) \tag{25}$$

which means that the continuous affine function $y \to \langle y - x, x^* \rangle + F(x)$, which coincides with F at $y = x$, is everywhere less then F. In other words, we have an affine minorant of slope x^* through $(x, F(x))$.

This is by no means always the case, even for $X = \mathbb{R}$. For instance, the function $F(t)$ defined by

$$F(t) = \begin{cases} -\sqrt{t} & \text{if } t \geq 0 \\ +\infty & \text{otherwise} \end{cases}$$

is finite at $t = 0$, with $F(0) = 0$, convex and l.s.c., but there is no affine minorant of F through $(0,0)$. So we need a definition.

Definition 13. Let $F : X \to \mathbb{R} \cup \{+\infty\}$ be a proper function. We shall say that x^* is a *subgradient* of F at x if one of the equivalent conditions (23) or (25) holds. The set of all subgradients of F at x is denoted by $\partial F(x)$. If $\partial F(x) \neq \emptyset$, we shall say that F is *subdifferentiable* at x. □

Proposition 14. *$\partial F(x)$ is a closed convex subset of X^*.* □

Proof. By definition:

$$\partial F(x) = \{x^* \in X^* | F(x) + F^*(x^*) = \langle x, x^* \rangle\} .$$

By Fenchel's inequality, this amounts to:

$$\begin{aligned} \partial F(x) &= \{x^* \in X^* | F(x) + F^*(x^*) \leq \langle x, x^* \rangle\} \\ &= \{x^* \in X^* | F^*(x^*) - \langle x, x^* \rangle \leq F(x)\} . \end{aligned}$$

The function $x^* \to F^*(x^*) - \langle x, x^* \rangle$ is convex and l.s.c. The set of points where its value is less than $F(x)$ is convex and closed. □

The most remarkable property of subgradients, is the so-called *Legendre reciprocity formula*:

Proposition 15. *The three properties are equivalent:*

(a) $x^* \in \partial F(x)$

(b) $x \in \partial F^*(x^*)$

(c) $F(x) + F^*(x^*) = \langle x, x^* \rangle$. □

Proof. The equivalence of (a) and (c) follows from the definition of the subdifferential. It is remarkable that formula (c) is symmetric. Indeed, if we write $x \in \partial F^*(x^*)$, and seek what it means, we get:

$$F^*(x^*) + F^{**}(x) = \langle x, x^* \rangle$$

which is formula (c) since $F^{**} = F$ (Proposition 9). □

It is now natural to investigate how ∂ operates on convex functions. We easily find that:

(26) $$\forall \lambda > 0 \quad \partial(\lambda F)(x) = \lambda \partial F(x)$$

$$\partial\left(F_1+F_2\right)(x) \subset \partial F_1(x)+\partial F_2(x) \tag{27}$$

$$\partial(F \circ A)(x) \subset A^* \partial F(A x) \tag{28}$$

but we can do better. The two last inclusions will usually be strict, unless additional conditions are met: we tackle this problem in the next section.

Notes and Comments. Convex analysis is a creation of Fenchel, Moreau and Rockafellar. The treatise by Rockafellar [Roc] is limited to finite dimension. The infinite-dimensional case is treated in the books [EkeT] and [AubE].

2. Convex Analysis on Banach Spaces

We now change our framework: X will be a Banach space, and X^* its dual. This means that we have two, and even three, more topologies to deal with: the norm topologies on X and X^*, and the weak topology $\sigma\left(X^*, X^{**}\right)$ on X^*, which is different from the weak-$*$ topology $\sigma\left(X^*, X\right)$ unless X is reflexive. For this reason, we shall concentrate our attention on X; from now on "continuous" or "l.s.c." will mean "norm-continuous" or "norm-l.s.c.".

We begin by a fundamental fact:

Theorem 1. *Let $F: X \rightarrow \mathbb{R} \cup\{+\infty\}$ be convex and l.s.c. Then F is weakly l.s.c.* □

Proof. This is again a consequence of the Hahn-Banach theorem: all closed convex subsets of $X \times \mathbb{R}$ are weakly closed. The converse is of course trivial.□

So the whole theory we described in the preceding section applies to convex l.s.c. functions on X. As an example, let us compute the Fenchel conjugate of $F(x)=\frac{1}{\alpha}\|x\|^\alpha$, with $\alpha \geq 1$:

$$\begin{aligned}
F^*\left(x^*\right) & =\operatorname{Sup}_x\left\{\left\langle x, x^*\right\rangle-\frac{1}{\alpha}\|x\|^\alpha\right\} \\
& =\operatorname{Sup}_{t \geq 0} \operatorname{Sup}_{\|x\|=t}\left\{\left\langle x, x^*\right\rangle-\frac{1}{\alpha}\|x\|^\alpha\right\} \\
& =\operatorname{Sup}_{t \geq 0}\left\{\|x\|^*-\frac{1}{\alpha} t^\alpha\right\} \\
& =\frac{1}{\beta}\left\|x^*\right\|_*^\beta \quad \text { if } \alpha>1, \text { with } \beta=\frac{\alpha}{\alpha-1} \\
& =\delta_I\left(\left\|x^*\right\|_*\right) \quad \text { if } \alpha=1, \text { with } I=[0,1] .
\end{aligned} \tag{1}$$

But there is more information to be derived from the additional topological structure. As an example, let us indicate that:

Proposition 2. *Let $F : X \to \mathbb{R} \cup \{+\infty\}$ be a convex function, finite and continuous at $\overline{x} \in X$. Then F is subdifferentiable at $\overline{x}$:*

$$\partial F(\overline{x}) \neq \emptyset \ .$$

Proof. Since F is continuous at $\overline{x}$, the interior of its epigraph, $\operatorname{int}(\operatorname{epi} F)$, is a non-empty open convex set. It does not contain $(\overline{x}, F(\overline{x}))$, and we just separate this point from $\operatorname{int}(\operatorname{epi} F)$, using a standard version of the Hahn-Banach theorem. This slope of the separating hyperplane turns out to be a gradient of F at $\overline{x}$. □

The greatest advantage of dealing with Banach spaces is the fact that we will be able to resolve all the ambiguities of the preceding section under very reasonable assumptions. We begin by removing the two stars in Formula (1.14).

Theorem 3. *Let F_1 and F_2 be convex, l.s.c. functions on X such that:*

$$0 \in \operatorname{int}(\operatorname{dom} F_1 - \operatorname{dom} F_2) \ . \tag{2}$$

Then $F_1^* \square F_2^*$ is a convex function on X^* which is l.s.c. for the $\sigma(X^*, X)$-topology. In fact:

$$F_1^* \square F_2^* = (F_1 + F_2)^* \ . \tag{3}$$

In addition, for every $x^* \in X^*$ we can find some x_1^* and x_2^*, with $x_1^* + x_2^* = x^*$, such that

$$(F_1^* \square F_2^*)(x^*) = F_1^*(x_1^*) + F_2^*(x_2^*) \ . \tag{4}$$

Proof. Note first that condition (2) implies that $\operatorname{dom} F_1 \cap \operatorname{dom} F_2 \neq \emptyset$, so that $F_1 + F_2$ is proper.

Recall the definition of $F_1^* \square F_2^*$:

$$(F_1^* \square F_2^*)(x^*) = \inf_{y^*} \{F_1^*(y^*) + F_2^*(x^* - y^*)\} \ . \tag{5}$$

Unfortunately, the topology $\sigma(X^*, X)$ is not metrizable, so that we must use filters instead of sequences. Fix some point $\overline{x}^* \in X^*$, and take an ultra-filter $\mathcal{F}$ converging to $\overline{x}^*$ for $\sigma(X^*, X)$, such that:

$$(F_1^* \square F_2^*)(x^*) \xrightarrow{\mathcal{F}} \ell \quad \text{in } \mathbb{R} \ . \tag{6}$$

$$B := \{x^* \mid \|z^*\|_* \leq 1 + \|x^*\|_*\} \in \mathcal{F} \ . \tag{7}$$

With every $\epsilon > 0$ and $A \in \mathcal{F}$, we associate a subset $\widehat{A}(\epsilon)$ of $X^* \times Y^*$;

$$\widehat{A}(\epsilon) = \{(x^*, y^*) | x^* \in A, F_1^*(y^*) + F_2^*(x^* - y^*) \leq \ell + \epsilon\} \ . \tag{8}$$

Because of (5) and (6), $\widehat{A}(\epsilon) \neq \emptyset$. The family $\widehat{\mathcal{F}} := \left\{\widehat{A}(\epsilon) | A \in \mathcal{F}, \epsilon > 0\right\}$ generates a filter on $X^* \times Y^*$. We choose an ultrafilter $\mathcal{U} \supset \widehat{\mathcal{F}}$.

We now take some $A \in \mathcal{F}$, $A \subset B$. For $(x^*, y^*) \in \widehat{A}(1)$ and $x \in X$, we write Fenchel's inequality:

$$\langle x, y^* \rangle \leq G_{x^*}(x) + G_{x^*}^*(y^*) \tag{9}$$

where $G_{x^*}^*$ is defined as follows:

$$G_{x^*}^*(z^*) := F_1^*(z^*) + F_2^*(x^* - z^*) \tag{10}$$

and of course $G_{x^*} = G_{x^*}^{**}$. An explicit computation, using Corollary 1.11 gives:

$$G_{x^*} = \left[F_1 \Box \left(F_2^- - x^*\right)\right]^{**} \tag{11}$$

where $F_2^-(x) := F_2(-x)$. In other words, G_{x^*} is the greatest l.s.c. convex function such that:

$$G_{x^*} \leq \left[F_1 \Box \left(F_2^- - x^*\right)\right] . \tag{12}$$

Now define a new convex l.s.c. function G by:

$$G := \sup_{x^* \in B} G_{x^*} \tag{13}$$

We have, remembering the definition of B:

$$\begin{aligned} G(x) &\leq \sup_{x^* \in B} \left[F_1 \Box \left(F_2^- - x^*\right)\right](x) \\ &= \sup_{x^* \in B} \inf_{x_1 - x_2 = x} \{F_1(x_1) + F_2(x_2) + \langle x^*, x_2 \rangle\} \\ &\leq \inf_{x_1 - x_2 = x} \{F_1(x_1) + F_2(x_2) + \|x_2\| (1 + \|\overline{x}^*\|_*)\} . \end{aligned} \tag{14}$$

By assumption (2), there is some $\alpha > 0$ such that, if $\|x\| \leq \alpha$, we can write $x = x_1 - x_2$ with $F_1(x_1) < \infty$ and $F_2(x_2) < \infty$, so that the right-hand side of inequality (14) is finite. So the function G is finite on the ball $\|x\| \leq \alpha$; since it is convex and l.s.c., it must be continuous on the interior $\|x\| \leq \alpha$. It follows that it is uniformly bounded on some ball around 0:

$$\exists \eta > 0 , \quad \exists c : \quad \|x\| \leq \eta \Rightarrow G(x) \leq c . \tag{15}$$

Writing this into Fenchel's inequality (9), and remembering formula (10) as well, we get:

$$(x^*, y^*) \in \widehat{A} \text{ and } \|x\| \leq \eta \Rightarrow \langle x, y^* \rangle \leq c + \ell + 1 . \tag{16}$$

Replacing x by $-x$, we get the reverse inequality:

$$(x^*, y^*) \in \widehat{A} \text{ and } \|x\| \leq \eta \Rightarrow |\langle x, y^* \rangle| \leq c + \ell + 1 . \tag{17}$$

We now use the ultrafilter:

$$\langle x, y^* \rangle \xrightarrow{\mathcal{U}} \langle x, \overline{y}^* \rangle \quad \text{with } \|\overline{y}^*\| \leq \frac{c + \ell + 1}{\eta} . \tag{18}$$

This means that $y^* \stackrel{\mathcal{U}}{\to} \overline{y}^*$ in the $\sigma(X^*, X)$-topology. Since F_1^* and F_2^* are l.s.c. in this topology, we get from formula (8), since $\epsilon \stackrel{\mathcal{U}}{\to} 0$

(19) $$F_1^*(\overline{y}^*) + F_2^*(\overline{x}^* - \overline{y}^*) \leq \ell \ .$$

But from the definition of the inf-convolution, we have:

(20) $$(F_1^* \Box F_2^*)(\overline{x}^*) \leq F_1^*(\overline{y}^*) + F_2^*(\overline{x}^* - \overline{y}^*) \leq \ell \ .$$

This shows that $F_1^* \Box F_2^*$ is l.s.c. Formula (4) follows from taking for $\mathcal{F}$ the ultrafilter consisting of all $A \subset X^*$ with $\overline{x}^* \in A$ (which amounts to fixing x^* at $\overline{x}^*$ in the preceding argument). □

This immediately settles the question of finding the subdifferential of a sum:

Corollary 4. *With the same assumptions, we have:*

(21) $$\forall x \in X \quad \partial(F_1 + F_2)(x) = \partial F_1(x) + \partial F_2(x) \ .$$

Proof. In the preceding section, we saw that the left-hand side contains the right-hand side. We have to prove the converse, that is, given any $x^* \in \partial(F_1 + F_2)(x)$, to split it into $x_1^* \in \partial F_1(x)$ and $x_2^* \in \partial F_2(x)$.

Start from $x^* \in \partial(F_1 + F_2)(x)$. By definition:

(22) $$(F_1 + F_2)(x) + (F_1 + F_2)^*(x^*) = \langle x, x^* \rangle \ .$$

Replace $(F_1 + F_2)^*(x^*)$ by its value, given by formula (4):

(23) $$F_1(x) + F_2(x) + F_1^*(x_1^*) + F_2^*(x_2^*) = \langle x, x^* \rangle$$

(24) $$x^* = x_1^* + x_2^* \ .$$

Rewrite Eq. (23) as follows:

$$[F_1(x) + F_1^*(x_1^*) - \langle x, x_1^* \rangle] + [F_2(x) + F_2^*(x_2^*) - \langle x, x_2^* \rangle] = 0 \ .$$

Each of the bracket is non-negative, by Fenchel's inequality. Since their sum is zero, both must be zero:

(25) $$F_1(x) + F_1^*(x_1^*) - \langle x, x_1^* \rangle = 0$$

(26) $$F_2(x) + F_2^*(x_2^*) - \langle x, x_2^* \rangle = 0 \ .$$

But this means precisely that $x_1^* \in \partial F_1(x_1)$ and $x_2^* \in \partial F_2(x_2)$, and Eq. (24) is the desired splitting. □

We have similar results for $f \circ A$:

Proposition 5. *Let X and Y be Banach spaces, $A : \operatorname{dom} A \to Y$ a closed linear operator from X to Y, and $G : Y \to \mathbb{R} \cup \{+\infty\}$ a convex l.s.c. function. Assume that:*

$$0 \in \operatorname{int}(\operatorname{dom} G - A(\operatorname{dom} A)) \tag{27}$$

Then:

$$(G \circ A)^* (x^*) = \min \{G^*(y^*) \,|A^* y^* = x^*\} \ . \tag{28}$$

The minimum on the right-hand side is $+\infty$ *if* $(A^*)^{-1}(x^*) = \emptyset$, *and is achieved at some point* $\overline{y}^* \in (A^*)^{-1}(x^*)$ *otherwise.* □

Proof. Define maps G_A and $\overline{G}$ on $X \times Y$ by:

$$G_A(x,y) = \begin{cases} G(y) & \text{if } \ y = Ax \\ +\infty & \text{otherwise} \end{cases} \tag{29}$$

$$\overline{G}(x,y) = G(y) \ . \tag{30}$$

We have $\operatorname{dom} \overline{G} = x \times \operatorname{dom} G$, so that assumption (27) is equivalent to:

$$0 \in \operatorname{int}\left(\operatorname{dom} \overline{G} + \operatorname{graph} A\right) \ . \tag{31}$$

Clearly $G_A = \overline{G} + \delta_{\operatorname{graph} A}$. Applying Theorem 3, we get

$$G_A^* = \overline{G}^* \Box \delta^*_{\operatorname{graph} A} \ . \tag{32}$$

An easy computation gives:

$$\overline{G}^* (x^*, y^*) = G^*(y^*) + \delta_0(x^*) \tag{33}$$

$$\delta^*_{\operatorname{graph} A} = \delta_{(\operatorname{graph} A)^\perp} = \delta_{\operatorname{graph} A^*} \ . \tag{34}$$

We write this information back into formula (32), together with formula (4). We get:

$$G_A^*(x^*, y^*) = \min \{G^*(y_1^*) |\, y_1^* + y_2^* = y^*, 0 + x_2^* = x^*, -A^* y_2^* = x_2^*\} \ .$$

Hence:

$$G_A^*(x^*, 0) = \min \{G^*(y^*) |\, A^* y^* = x^*\} \ . \tag{35}$$

On the other hand, by the definition of Fenchel conjugation, we have:

$$\begin{aligned} G_A^*(x^*, 0) &= \sup_{x,y} \{\langle x, x^* \rangle + \langle y, 0 \rangle - G_A(x,y)\} \\ &= \sup_x \{\langle x, x^* \rangle - G(Ax)\} = (G \circ A)^* (x^*) \ . \end{aligned} \tag{36}$$

Comparing formulas (35) and (36), we get our result. □

Corollary 6. *With the same assumption, we have:*

$$(G \circ A)(x) = A^* \partial G(Ax) \quad \forall x \in X \ . \tag{37}$$

□

Proof. Start from $x^* \in \partial(G \circ A)(x)$. By definition:

$$G(Ax) + (G \circ A)^*(x^*) = \langle x, x^* \rangle \ . \tag{38}$$

By Proposition 5, there is some y^* such that:

$$(G \circ A)^*(x^*) = G^*(y^*) \tag{39}$$

$$A^* y^* = x^* \ . \tag{40}$$

Hence:

$$G(Ax) + G^*(y^*) = \langle x, A^* y^* \rangle = \langle Ax, y^* \rangle \ . \tag{41}$$

This means precisely that $y^* \in \partial G(Ax)$. Equation (40) then yields $x^* \in A^* \partial G(Ax)$, and the result follows. □

If for instance we take $G = \delta_0$, the indicator function of $\{0\}$, we find $G^* \equiv 0$. Condition (27) then means that $A(\operatorname{dom} A) = Y$. Proposition 5 then becomes: if a closed operator A is onto, then A^* has closed range.

Finally, if X is a Banach space, we can talk about differentiable functions on X. There are several possible definitions, the two most important being the following:

Definition 7. A function $F : A \to \mathbb{R} \cup \{+\infty\}$, which is finite in a neighbourhood of x, is *Gâteaux-differentiable* at x if there exists some $x^* \in X^*$ such that:

$$\forall y \in X \ , \quad \lim_{h \to 0} \frac{F(x + hy) - F(x)}{h} = \langle x^*, y \rangle \ . \tag{42}$$

We then write $F'(x) := x^*$. □

Definition 8. A function $F : X \to \mathbb{R} \cup \{+\infty\}$, which is finite in a neighbourhood of x, is *Fréchet-differentiable* at x if there exists some $x^* \in X^*$ such that:

$$\lim_{y \to 0} \frac{F(x + y) - F(x) - \langle x^*, x \rangle}{\|y\|} = 0 \ . \tag{43}$$

We then write $F'(x) := x^*$. □

Clearly Fréchet-differentiability implies Gâteaux-differentiability. Conversely, if F is Gâteaux-differentiable on an open set $\Omega \subset X$ and $F' : \Omega \to X^*$ is continuous, then F is Fréchet-differentiable on Ω: we then say that F is a C^1 function on Ω.

When F is l.s.c., convex and finite on some neighbourhood of $x \in X$, it must be continuous at x, and hence subdifferentiable at x, but it need not be differentiable in any sense. So subdifferentiability is a strict extension of differentiability, as the following result shows:

Proposition 9. *If $F : X \to \mathbb{R} \cup \{+\infty\}$ is convex, l.s.c. and if its subdifferential at x is a singleton, $\partial F(x) = \{x^*\}$, then F is Gâteaux-differentiable at x and $x^* = F'(x)$.*

If $F : X \to \mathbb{R} \cup \{+\infty\}$ is Gâteaux-differentiable and continuous at x, then $\partial F(x) = \{F'(x)\}$. □

The first part of the proposition immediately reduces to a one-dimensional problem, and the second part is another consequence of the Hahn-Banach separation theorems.

If we go one step further and look into C^2 functions, we get the following result:

Proposition 10. *Let X be a Banach space and $F : X \to \mathbb{R}$ a C^2 function. Assume that its Hessian is positive definite everywhere:*

$$\forall x \in X \ , \quad \langle F''(x)y, y\rangle > 0 \quad \forall y \in Y/\{0\} \tag{44}$$

so that F is strictly convex. Then $\operatorname{dom} F^*$ has non-empty interior in X^*, and F^* is C^2 on $\operatorname{int}(\operatorname{dom} F^*)$. We have:

$$F'(x) = x^* \Rightarrow \begin{cases} x = [F^*]'(x^*) \\ I_X = [F^*]''(x^*)F''(x) \end{cases} . \tag{45}$$

□

Proof. Fix $x \in X$, and define $x^* := F'(x)$. Consider the equation:

$$y^* = F'(y) \ . \tag{46}$$

F' is a C^1 map from X to X^*. Since $F''(x) \in \mathcal{L}(X, X^*)$ is positive definite, it is invertible and we may apply the inverse function theorem, to the effect that Eq. (46) can be solved uniquely for y close to x and y^* close to x^*. Every such y^* belongs to $\operatorname{dom} F^*$, which therefore has non-empty interior.

Inverting Eq. (46) by the Legendre reciprocity formula yields $y \in \partial F^*(y^*)$. By local uniqueness, this must coincide with the C^1 solution we found by the inverse function theorem. So $\partial F^*(y^*)$ must be a singleton, $\partial F^*(y^*) = \{[F^*]'(y^*)\}$ where $[F^*]'$ is C^1 on a neighbourhood of x^*. The Legendre reciprocity formula now reads:

$$y^* = F'(y) \Longleftrightarrow y = [F^*]'(y^*) \ .$$

Differentiating the right-hand side with respect to y:

$$I_X = [F^*]''(F'(y)) \circ F''(y) \ .$$

□

3. Integral Functionals on L^α

Let Ω be some borelian subset of $\mathbb{R}^n$, endowed with the Lebesque measure, and $f : \Omega \times \mathbb{R}^N \to \mathbb{R} \cup \{+\infty\}$ some non-negative Borel function:

$$f(\omega, \xi) \geq 0 \quad \forall (\omega, \xi) \in \Omega \times \mathbb{R}^N . \tag{1}$$

Take $1 \leq \alpha \leq \infty$ and consider the space $L^\alpha\left(\Omega; \mathbb{R}^N\right)$, henceforth shortened to L^α. For every $x \in L^\alpha$, define $F(x)$ by:

$$F(x) := \int_\Omega f(\omega, x(\omega))\, d\omega . \tag{2}$$

We have thus defined a function $F : L^\alpha \to \mathbb{R} \cup \{+\infty\}$. In this section, we shall collect its main properties, both to illustrate the preceding sections by examples, and to serve for future reference.

Proposition 1. *Assume that f is non-negative and l.s.c. with respect to ξ.*

$$\forall \omega \in \Omega , \quad f(\omega, \cdot) \quad \text{is l.s.c. on } \mathbb{R}^N . \tag{3}$$

Then F is l.s.c. on L^α.

Proof. If x_n is a sequence converging to x in L^α, we may extract a subsequence $x_{n'}$, such that:

$$\liminf_{n\to\infty} F(x_n) = \lim_{n'\to\infty} F(x_{n'}) \tag{4}$$

and from the $x_{n'}$ we may extract another subsequence $x_{n''}$ which converges almost everywhere. We get:

$$\liminf_{n\to\infty} F(x_n) = \lim_{n''\to\infty} F(x_{n''}) \tag{5}$$

$$= \lim_{n''\to\infty} \int_\Omega f(\omega, x_{n''}(\omega))\, d\omega . \tag{6}$$

Since f is non-negative, we may use Fatou's lemma:

$$\geq \int_\Omega \liminf_{n'\to\infty} f(\omega, x_{n''}(\omega))\, d\omega . \tag{7}$$

We now use the pointwise convergence $x_{n''}(\omega) \to x(\omega)$ and the lower semi-continuity of $f(\omega, \cdot)$:

$$\geq \int_\Omega f(\omega, x(\omega))\, d\omega = F(x) . \tag{8}$$

We have proved that $\liminf_{n\to\infty} F(x_n) \geq F(x)$; that is, F is l.s.c. □

Clearly, if f is convex with respect to ξ:

$$\forall \omega \in \Omega \quad f(\omega, \cdot) \quad \text{is convex} \tag{9}$$

then F will be a convex function on L^α. If (1), (3) and (9) are satisfied, then F will be a l.s.c. convex function, and one should then try to compute F^* and ∂F in terms of f^* and ∂f. We use the obvious notations:

$$f^*(\omega;\zeta) = f(\omega,\cdot)^*(\zeta) \tag{10}$$

$$\partial f(\omega;\xi) = \partial f(\omega,\cdot)(\xi)\ . \tag{11}$$

Henceforth we use the duality pairing $\left(L^\alpha, L^\beta\right)$, with $\alpha^{-1}+\beta^{-1}=1$.

Theorem 2. *Assume that Ω has finite measure, that f satisfies* (1), (3), (9), *and that there exists $\overline{x} \in L^\infty$ such that:*

$$\int_\Omega f\left(\omega, \overline{x}(\omega)\right) d\omega < +\infty\ . \tag{12}$$

Then:

$$F^*\left(x^*\right) = \int_\Omega f^*\left(\omega; x^*(\omega)\right) d\omega \quad \forall x^* \in L^\beta\ . \tag{13}$$

□

Proof. Introduce the sequence f_n defined by:

$$f_n(\omega,\xi) := f(\omega,\xi) + \frac{1}{n}\left\|\xi\right\|^\alpha\ . \tag{14}$$

We define accordingly

$$F_n(x) := \int_\Omega f_n\left(\omega, x(\omega)\right) d\omega\ . \tag{15}$$

We note that, for $x^* \in L^\beta$, the integrals $\int_\Omega f^*\left(\omega; x^*(\omega)\right) d\omega$ and $\int_\Omega f_n^*\left(\omega; x^*(\omega)\right)$ are well-defined, with values in $\mathbb{R}\cup\{+\infty\}$. Indeed, by assumption (12) we have:

$$f^*\left(\omega; x^*(\omega)\right) \geq f_n^*\left(\omega; x^*(\omega)\right) \geq \left(\overline{x}(\omega), x^*(\omega)\right) - f\left(\omega, \overline{x}(\omega)\right) - \frac{1}{n}\left\|\overline{x}(\omega)\right\|^\alpha \tag{16}$$

and the right-hand side is an integrable function.

The proof now goes in two steps. We first prove the result for the f_n, and then we take limits.

For each fixed $n \in \mathbb{N}$ and $x^* \in L^\beta$, we define $x_n(\omega)$ by:

$$f_n^*\left(\omega, x^*(\omega)\right) = \left(x_n(\omega), x^*(\omega)\right) - f_n\left(\omega, x_n(\omega)\right)\ . \tag{17}$$

In other words, $x_n(\omega)$ is the point where the function $\xi \to \left(\omega, x^*(\omega)\right) - f_n(\omega,\xi)$ attains its maximum. It is well-defined, since f_n is strictly convex and $f_n(\xi) \geq \frac{1}{n}\left\|\xi\right\|^\alpha$. From the latter inequality and the fact that $f\left(\omega, \overline{x}(\omega)\right) \leq c$ it also follows that $\left\|x_n(\omega)\right\|$ is bounded independently of ω. Hence

$$x_n \in L^\infty\left(\Omega; \mathbb{R}^N\right) \tag{18}$$

and $L^\infty \subset L^\alpha$ since Ω has finite measure.

Using the definitions, we have:

$$
\begin{aligned}
F_n^*(x^*) &= \operatorname*{Sup}_{x\in L^\alpha} \{\langle x, x^*\rangle - F_n(x)\} \\
&= \operatorname*{Sup}_{x\in L^\alpha} \int_\Omega \{(x(\omega), x^*(\omega)) - f_n(\omega, x(\omega))\}\, d\omega \\
&\geq \int_\Omega \{(x_n(\omega), x^*(\omega)) - f_n(\omega, x_n(\omega))\}\, d\omega \\
&= \int_\Omega f_n(\omega, x^*(\omega))\, d\omega = \int_\Omega \operatorname*{Sup}_{\xi\in\mathbb{R}^N} \{\langle \xi, x^*(\omega)\rangle - f_n(\omega,\xi)\}\, d\omega\ .
\end{aligned}
\tag{19}
$$

Now the last term in this string is clearly larger than the first ones. It follows that the inequality is in fact an equality, and formula (13) holds for F_n.

We now take limits. Clearly:

$$
\forall x \in L^\alpha\ , \quad F(x) = \operatorname*{Inf}_n F_n(x)\ . \tag{20}
$$

Dually, we have:

$$
\forall x^* \in L^\beta\ , \quad F^*(x^*) = \operatorname*{Sup}_n F_n^*(x^*)\ . \tag{21}
$$

Indeed, the inequality $F \leq F_n$ becomes $F^* \geq F_n^*$ by duality. The function $\Phi := \operatorname{Sup}_n F_n^*$ is convex and l.s.c. If $\Phi^* \neq F^*$, then there would be a convex function $\Phi := \Phi^{**}$ such that $\Phi \neq F$ and $F \leq \Phi \leq F_n$ for all n, thereby contradicting (20).

Replacing F_n^* by its value, we get:

$$
F^*(x^*) = \operatorname*{Sup}_n \int_\Omega f_n^*(\omega; x^*(\omega))\, d\omega\ . \tag{22}
$$

Now, for each fixed $\omega \in \Omega$, we have

$$
f(\omega,\cdot) = \operatorname*{Inf}_n f_n(\omega,\cdot) \quad \text{and} \quad f_{n+1} \leq f_n \tag{23}
$$

and hence, as we just saw:

$$
f^*(\omega,\cdot) = \operatorname{Sup} f_n^*(\omega,\cdot) \quad \text{and} \quad f_{n+1}^* \geq f_n^*\ . \tag{24}
$$

Since

$$
\int f_1^*(\omega; x^*(\omega))\, d\omega > -\infty\ , \tag{25}
$$

we may apply Lebesgue's monotone convergence theorem. We get:

$$
\int_\Omega f^*(\omega; x^*(\omega))\, d\omega = \operatorname*{Sup}_n \int_\Omega f_n^*(\omega; x^*(\omega))\, d\omega\ . \tag{26}
$$

Formula (13) now follows from (22) and (26). □

The non-negativity condition (1) may seem stringent. It can easily be weakened. Theorem 1 will still hold if it is replaced by the following:

$$\text{there exists } \overline{x}^* \in L^\beta \text{ such that } \int_\Omega |f^*(\omega; \overline{x}^*(\omega)) d\omega < \infty\,. \tag{27}$$

Indeed, if such is the case, we will have, by Fenchel's inequality:

$$f(\omega,\xi) + f^*(\omega; \overline{x}^*(\omega)) - (\xi, \overline{x}^*(\omega)) \geq 0\,. \tag{28}$$

Denoting the left-hand side by $\overline{f}(\omega,\xi)$, we find that $\overline{f}$ satisfies the non-negativity assumption. Applying Theorem 1 to $\overline{f}$, we then get:

$$\overline{F}(x) := F(x) - \langle x, x^* \rangle + \int_\Omega f^*(\omega; \overline{x}^*(\omega))\, d\omega \tag{29}$$

$$\overline{F}^*(x^*) = \int_\Omega \overline{f}^*(\omega; x^*(\omega))\, d\omega\,. \tag{30}$$

Replace $\overline{F}^*$ and $\overline{f}^*$ by their values in terms of F^* and f^*:

$$\begin{aligned} &F^*(x^* + \overline{x}^*) - \int_\Omega f^*(\omega; \overline{x}^*(\omega))\, d\omega \\ &= \int_\Omega \{f^*(\omega; x^*(\omega) + \overline{x}^*(\omega)) - f^*(\omega, \overline{x}^*(\omega))\}\, d\omega \end{aligned} \tag{31}$$

and formula (13) follows immediately. Note that, by (16) and (27), the integral $\int_\Omega f^*(\omega; \overline{x}^*(\omega))\, d\omega$ is in fact finite.

This leads us to the following:

Corollary 3. *Assume that Ω has finite measure, that f satisfies* (3), (9), *and that there exists $\overline{x} \in L^\infty$ and $\overline{x}^* \in L^\infty$ such that:*

$$\int_\Omega |f(\omega, \overline{x}(\omega))|\, d\omega < \infty \tag{32}$$

$$\int_\Omega |f^*(\omega; \overline{x}^*(\omega))|\, d\omega < \infty\,. \tag{33}$$

Then:

$$\partial F(x) = \{x^* \in L^\beta | x^*(\omega) \in \partial f(\omega; x(\omega)) \quad \text{a.e.}\}\,. \tag{34}$$ □

Proof. By Proposition 1.15, we know that $x^* \in \partial F(x)$ if and only if $x^* \in L^\beta$ and:

$$F(x) + F^*(x^*) - \langle x, x^* \rangle = 0\,. \tag{35}$$

Rewriting this with integrals, we get

$$\int_\Omega \{f(\omega, x(\omega)) + f^*(\omega; x^*(\omega)) - (x(\omega), x^*(\omega))\}\, d\omega = 0\,. \tag{36}$$

By Fenchel's inequality, the integrand is non-negative almost everywhere. It follows that it vanishes:

$$f(\omega, x(\omega)) + f^*(\omega; x^*(\omega)) = (x(\omega), x^*(\omega)) \quad \text{a.e.} \tag{37}$$

By proposition 1.15 again, this means that $x^*(\omega) \in \partial f(\omega; x(\omega)\}$ a.e. □

Note that Theorem 2 and Corollary 3 cover the cases $\alpha = 1$ and $\alpha = \infty$, provided $\sigma(L^\infty, L^1)$ is used as a weak topology on L^∞ instead of $\sigma(L^\infty, L^{\infty *})$. Then convex l.s.c. functions on L^∞ need not be weakly continuous. However, in the case of integral functionals, defined by formula (2), Theorem 2 enables us to state that if f satisfies (3), (9), and if there exists $\overline{x} \in L^\infty$ such that

$$\int_\Omega |f(\omega; \overline{x}(\omega))| \, d\omega < \infty \tag{38}$$

then F is l.s.c. for $\sigma(L^\infty, L^1)$. Indeed, we then have $F = G^*$ with

$$G(x) = \int_\Omega f^*(\omega; x(\omega)) \, d\omega \; . \tag{39}$$

To conclude this section, we turn to situations where F is C^1. We shall need the following continuity result, which is due to Krasnoselskii.

Theorem 4. *Assume Ω has finite measure, and $f : \Omega \times \mathbb{R}^N \to \mathbb{R}$ is borelian, with $f(\omega, \cdot)$ continuous for a.e. $\omega \in \Omega$. Suppose there are constants a, c and $\gamma > 0$ such that:*

$$|f(\omega, \xi)| \leq \text{Max}\,\{a, c\,\|\xi\|^\gamma\} \; . \tag{40}$$

Take α such that $\alpha/\gamma > 1$, and define a map $\Phi : L^\alpha \to L^{\alpha/\gamma}$ by:

$$[\Phi(x)](\omega) = f(\omega, x(\omega)) \; . \tag{41}$$

Then Φ is continuous in the norm topologies. □

Proof. Clearly $\Phi(x) \in L^{\alpha/\gamma}$ whenever $x \in L^\alpha$. Let $x_n \to x$ be a converging sequence in L^α. We claim that $\Phi(x_n) \to \Phi(x)$ in $L^{\alpha/\gamma}$.

We argue by contradiction. If this is not the case, we may extract a subsequence $x_{n'}$, such that $\Phi(x_{n'}) - \Phi(x)$ is bounded away from 0 in $L^{\alpha/\gamma}$, and a further subsequence $x_{n''}$ which converges a.e. on Ω. It follows that $f(\omega, x_{n''}(\omega))$ converges a.e. to $f(\omega, x(\omega))$.

In addition, we have

$$|f(\omega, x_{n''}(\omega)) - f(\omega, x(\omega))|^{\alpha/\gamma} \leq 2\text{Max}\,\{a, c\,\|x(\omega)\|^\alpha, c\,\|x_{n''}(\omega)\|^\alpha\} \; . \tag{42}$$

Since $x_{n''}$ converges to x in $L^\alpha(\Omega, \mathbb{R}^N)$, then $\|x_{n''}\|^\alpha$ converges to $\|x\|^\alpha$ in $L^1(\Omega, \mathbb{R})$. This proves that the right-hand side is equiintegrable. The left-hand side must therefore be relatively compact in L^1, and since it converges to 0 a.e., it must converge to 0 inL^1:

$$\int_\Omega |f(\omega, x_{n''}(\omega)) - f(\omega, x(\omega))|^{\alpha/\gamma} \to 0 \; . \tag{43}$$

But this just means that $\Phi(x_{n''}) - \Phi(x) \to 0$ in $L^{\alpha/\gamma}$, which is the desired contradiction. □

Corollary 5. *Assume that Ω has finite measure, that $f(\omega, \xi)$ is non-negative convex and C^1 with respect to ξ for a.e. ω, and that* (32) *and* (33) *are satisfied. Take $\alpha \in (1, \infty)$ and $\beta = \frac{\alpha}{\alpha - 1}$ the conjugate exponent. Assume that, for some constants a and c, we have:*

$$\|f'(\omega, x(\omega))\| \leq \operatorname{Max}\left\{a, c\|x\|^{\alpha-1}\right\} \; . \tag{44}$$

Then $F : L^\alpha \to L^\beta$ is C^1, and

$$[F'(x)](\omega) = f'(\omega, x(\omega)) \; . \tag{45}$$

□

Proof. First apply Corollary 3. We get

$$\partial F(x) = \left\{x^* \in L^\beta \,|x^*(\omega) = f'(\omega, x(\omega)) \text{ a.e.}\right\} \; . \tag{46}$$

By inequality (44), whenever $x \in L^\alpha$, then $f'(\cdot, x(\cdot)) \in L^\beta$. So the right-hand side of formula (46) is a singleton, and F is Gâteaux-differentiable everywhere by Proposition 2.9, with $F'(x)$ given by formula (45). It then follows from Theorem 4 that $F' : L^\alpha \to L^\beta$ is continuous. □

Notes and Comments. Krasnoselskiĭ's theorem (to be found in [Kra], together with its converse) does not hold for maps into L^∞. This has been the source of many mistakes. For instance, if a map $F : L^2 \to \mathbb{R}$ defined by the formula

$$F(x) := \int_\Omega f(\omega, x(\omega)\, d\omega \tag{47}$$

happens to be C^2, then it must be exactly quadratic, that is, we must have $f(\omega, x) = \langle A(\omega)x, x\rangle$ for every (ω, x) (see [Skr1] and [Skr2]).

4. The Clarke Duality Formula

Let X be a Banach space and $F : X \to \mathbb{R} \cup \{+\infty\}$ a convex l.s.c. function on X.

We are given a closed linear operator A from X to X^*. Recall that this means that A maps a linear subspace $\operatorname{dom} A \subset X$ into X^*, and that graph$A \subset X \times X^*$ is closed. We assume that A is self-adjoint:

$$\langle x, Ay\rangle = \langle Ax, y\rangle \quad \forall (x,y) \in (\operatorname{dom} A)^2 \ . \tag{1}$$

We define $\Phi : \operatorname{dom} A \to \mathbb{R}\{+\infty\}$ by:

$$\Phi(x) := \frac{1}{2}\langle x, Ax\rangle + F(x) \ . \tag{2}$$

Φ is the sum of two terms: a quadratic form and a convex l.s.c. function. The quadratic form is not required to be non-negative definite, so Φ need not be a convex function. Indeed, in the cases which shall be of interest to us, Φ will turn out to be quite complicated; for instance, it will have infinitely many critical points.

A *critical point* of Φ is defined to be a point $\overline{x} \in \operatorname{dom} A$ where:

$$A\overline{x} + \partial F(\overline{x}) \ni 0 \tag{3}$$

and $\Phi(\overline{x})$ is then called a *critical value.* If Φ is C^1, these are standard definitions. If Φ is not smooth, a little work may be needed, as in:

Proposition 1. *Assume that the restriction of Φ to any straight line running through $\overline{x}$ has a local minimum or a local maximum at $\overline{x}$. Then $\overline{x}$ is a critical point of Φ.*

Proof. Assume that $\overline{x}$ is not a critical point, so that $-A\overline{x}$ does not belong to $\partial F(\overline{x})$. This means that there exists some point $x_1 \in X$ and some $\epsilon > 0$ such that:

$$F(x_1) \le F(\overline{x}) - \langle A\overline{x}, x_1 - \overline{x}\rangle - \epsilon \ . \tag{4}$$

Set $x_t := tx_1 + (1-t)\overline{x}$. The function $t \to F(x_t)$ defined on the real line is convex, so that its slope $[F(x_t) - F(\overline{x})]/t$ is a decreasing function of t. Hence, for all $t \in (0,1)$:

$$\frac{1}{t}\left[F(x_t) - F(\overline{x})\right] \le F(x_t) - F(\overline{x}) \le -\langle A\overline{x}, x_1 - \overline{x}\rangle - \epsilon \ . \tag{5}$$

Write $a := \langle A(x_1 - \overline{x}), x_1 - \overline{x}\rangle$ so that

$$\frac{1}{2}\langle Ax_t, x_t\rangle = \frac{1}{2}\langle A\overline{x}, \overline{x}\rangle + t\langle A\overline{x}, x_1 - \overline{x}\rangle + a\frac{t^2}{2} \ . \tag{6}$$

Combining the last two relations, we get for $0 < t < 1$

$$F(x) + \frac{1}{2}\langle Ax_t, x_t\rangle \le F(\overline{x}) + \frac{1}{2}\langle A\overline{x}, \overline{x}\rangle - \epsilon t + a\frac{t^2}{2} \tag{7}$$

and for $t < 0$

$$F(x_t) + \frac{1}{2}\langle Ax_t, x_t\rangle \ge F(\overline{x}) + \frac{1}{2}\langle A\overline{x}, \overline{x}\rangle - \epsilon t + a\frac{t^2}{2} \ . \tag{8}$$

If we choose $|t| \le 2\epsilon|a|^{-1}$, then $-\epsilon t + a\frac{t^2}{2}$ has the same sign as $-\epsilon t$, negative in the first case and positive in the second. This proves that

(9) $$\Phi(x_t) < \Phi(\overline{x}) \quad \text{for } 0 < t < 2\epsilon|a|^{-1}$$

(10) $$\Phi(x_t) > \Phi(\overline{x}) \quad \text{for } -2\epsilon|a|^{-1} < t < 0 .$$ □

We are interested in finding the critical points of Φ. This can be quite a formidable problem. We therefore resort to a standard trick of convex optimization: perform a change of variables to bring the problem at hand into a more tractable form. The function Φ, however, is not convex; the idea of formulating a dual problem in this kind of context is due to Clarke.

Consider the (dual) functional $\Psi : \operatorname{dom} A \to \mathbb{R}$ defined by:

(11) $$\Psi(x) = \frac{1}{2}\langle Ax, x\rangle + F^*(-Ax) .$$

Recall that $F^* \circ (-A) : X \to \mathbb{R} \cup \{+\infty\}$ is defined by

(12) $$F^*(-Ax) = \begin{cases} F^*(-Ax) & \text{if } x \in \operatorname{dom} A \\ +\infty & \text{otherwise} . \end{cases}$$

With this convention, the critical points of Ψ are the solutions of:

(13) $$x \in \operatorname{dom} A , \quad Ax + \partial\left[F^* \circ (-A)\right](x) \ni 0 .$$

Theorem 2. *If x is a critical point of Φ, then all $x' \in \operatorname{Ker} A + x$ are critical points of Ψ. Conversely, assume that:*

(14) $$0 \in \operatorname{int}(\operatorname{dom} F^* - A(\operatorname{dom} A)) .$$

Then, if x' is a critical point of Ψ, there is some $x \in \operatorname{Ker} A + x'$ which is a critical point of Φ. We have:

(15) $$\Phi(x) + \Psi(x') = 0 .$$ □

Proof. Note first that Ψ is invariant by translations in $\operatorname{Ker} A$:

(16) $$\forall \xi \in \operatorname{Ker} A , \quad \Psi(x + \xi) = \Psi(x) \quad \forall x .$$

Let x be a critical point of Φ. We have:

(17) $$-Ax \in \partial F .$$

By the Legendre reciprocity formula, this becomes:

(18) $$x \in \partial F^*(-Ax) .$$

By formula (1.28)

(19) $$Ax \in A\partial F^*(-Ax) \subset -\partial\left[F^* \circ (-A)\right](x) .$$

This means that x is a critical point of Ψ, and so are all the points in $x + \operatorname{Ker} A$ since Ψ is invariant.

Conversely, let x' be a critical point of Ψ. By assumption (14) and Corollary 2.6, we have:

$$-Ax' \in \partial\left[F^* \circ (-Ax)\right](x') = -A\partial F^*(-Ax') \ . \tag{20}$$

Rewrite this as a system of two equations:

$$Ay = 0 \quad \text{and} \quad x - \partial F^*(-Ax') = -y \ . \tag{21}$$

Applying the Legendre reciprocity formula, we get:

$$-Ax' \in \partial F(x' + y) \tag{22}$$

and hence:

$$-A(x' + y) = -Ax' \in \partial F(x' + y) \ . \tag{23}$$

So $x = x' + y$ is a critical point of Φ. By the definition of the subgradient, we have:

$$F(x) + F^*(-Ax') = \langle x, -Ax' \rangle \ . \tag{24}$$

Since $Ax = Ax'$, we may rewrite this as:

$$\frac{1}{2}\langle Ax, x\rangle + F(x) = -\frac{1}{2}\langle Ax', x'\rangle - F^*(-Ax') \ . \tag{25}$$

This is formula (15). □

It should be noted that, by a trick due to Lasry, Theorem 2 also covers some situations when F is not convex:

Corollary 3. *Let $A = A^*$ be as above and $F : X \to \mathbb{R} \cup \{+\infty\}$ be a l.s.c. function. Assume that condition* (14) *is satisfied, and that there is another closed self-adjoint operator $B = B^*$ such that the function F_B defined by:*

$$F_B(x) := \frac{1}{2}\langle Bx, x\rangle + F(x) \tag{26}$$

is convex. Then the functions:

$$\Phi(x) = \frac{1}{2}\langle Ax, x\rangle + f(x) \tag{27}$$

$$\Psi(x) = \frac{1}{2}\langle (A - B)x, x\rangle + F_B^*(-Ax) \tag{28}$$

have the same critical points modulo Ker A. □

Proof. Write:

$$\Phi(x) = \frac{1}{2}\langle Ax, x\rangle - \frac{1}{2}\langle Bx, x\rangle + F_B(x) \tag{29}$$

and apply Theorem 2. □

We will apply this theory to the case of convex Hamiltonian systems with linear boundary conditions, for instance, periodicity conditions. This was the original contribution of Clarke, and will be treated in some detail. We will then give a few further examples, before concluding this section.

Endow $\mathbb{R}^{2n}$ with the standard symplectic structure, associated with the matrix:

$$J = \begin{bmatrix} 0 & I_n \\ -I_n & 0 \end{bmatrix} \tag{30}$$

and take a symplectic matrix M:

$$M^* J M = J \; . \tag{31}$$

Now choose $\alpha \in [1, \alpha)$, set $\beta = \frac{\alpha}{\alpha - 1}$, and define an operator $A : L^\alpha \to L^\beta$ by:

$$\operatorname{dom} A = \left\{ x \;\middle|\; \begin{matrix} x \quad \text{is absolutely continuous} \\ \dot{x} \in L^\beta \quad \text{and} \quad x(T) = Mx(0) \end{matrix} \right\} \tag{32}$$

$$Ax = J\dot{x} \qquad \text{for} \quad x \in \operatorname{dom} A \; . \tag{33}$$

Absolute continuity of x means that:

$$\forall t \; , \quad x(t) = x(0) + \int_o^t \dot{x}(s) ds \; . \tag{34}$$

$\operatorname{dom} A$ is just a subspace of the standard Sobolev space $W^{1,\beta}\left(0, T; \mathbb{R}^{2n}\right)$, with codimension $2n$:

$$\operatorname{dom} A = \left\{ x \in W^{1,\beta} \mid x(T) = Mx(0) \right\} \; . \tag{35}$$

Lemma 4. *A is a closed self-adjoint operator from L^α to L^β.* □

Proof. Take a sequence $(x_n, y_n) \to (x, y)$ in $L^\alpha \times L^\beta$, with $x_n \in \operatorname{dom} A$ and $y_n = Ax_n$. Define an absolutely continuous function z by:

$$z(t) := \int_o^t (-Jy) ds \; . \tag{36}$$

Since $y_n \to y$ in L^β, we have:

$$x_n(t) - x_n(0) = \int_o^t (-Jy_n)\, ds \to z(t) \quad \text{uniformly} \; . \tag{37}$$

But $x_n \to x$ in L^α. It follows that $x_n(0)$ must converge to some $\xi \in \mathbb{R}^{2n}$ such that:

$$x(t) - \xi = z(t) = \int_o^t (-Jy)dt \ . \tag{38}$$

Hence $y = J\dot{x}$, and $\xi = x(0)$. So $x_n(T)$ must converge to $x(T)$, and we get $x(T) = Mx(0)$ by taking limits in $Mx_n(0) = x_n(T)$. Finally $(x, y) \in \text{graph}A$, which is thus closed.

We now check that A is self-adjoint. Take $(x^*, y^*) \in \text{graph}A^*$, that is:

$$\langle x, x^* \rangle - \langle Ax, y^* \rangle = 0 \quad \forall x \in \text{dom}\, A \ . \tag{39}$$

This means:

$$\int_o^T [(x(t), x^*(t)) - (J\dot{x}(t), y^*(t))] = 0 \ . \tag{40}$$

Integrating by parts, we get:

$$\int_o^T \left(-\int_o^t x^*(s)ds + Jy^*(t), \dot{x}(t) \right) dt + \left(x(T), \int_o^T x^*(t)dt \right) = 0 \ . \tag{41}$$

Now $x \in \text{dom}\, A$ means that:

$$x(T) = x(0) + \int_o^T \dot{x}\, ds = Mx(0) \tag{42}$$

and hence:

$$(M - I)x(0) = \int_o^T \dot{x}\, ds \ . \tag{43}$$

Denote by L_o^β, the subspace of L^β consisting of all functions with mean value zero:

$$L_o^\beta = \left\{ y \in L^\beta | \int_o^T y\, ds = 0 \right\} \ . \tag{44}$$

It follows from the above that, whenever $y \in L_o^\beta$, the function $x(t) = \int_o^t y(s)ds$ belongs to $\text{dom}\, A$. Plugging this into relation (41), we get:

$$\int_o^T \left[-\int_o^t x^*(s)ds + Jy^*(t), y(t) \right] dt = 0 \quad \forall y \in L_o^\beta \ . \tag{45}$$

In other words, the function $t \to -\int_o^t x^*(s)ds + Jy^*(t)$ is orthogonal to L_o^β. But $\left(L_o^\beta\right)^\perp = \mathbb{R}^{2n}$, the subspace consisting of all constant functions. Hence:

$$\exists \xi \in \mathbb{R}^{2n} \ : \quad -\int_o^t x^*(s)ds + Jy^*(t) = \xi \quad \text{a.e.} \tag{46}$$

It follows immediately that y^* is absolutely continuous, with:

(47) $$\dot{y}^* = Jx^* \ .$$

This enables us to integrate formula (40) by parts the other way. Taking (47) into account, we get:

(48) $$(Jx(T), y^*(T)) - (Jx(0), y^*(0)) = 0 \quad \forall x \in \operatorname{dom} A \ .$$

By formula (32), this amounts to:

(49) $$(JMx(0), y^*(T)) - (Jx(0), y^*(0)) = 0 \quad \forall x(0) \in \mathbb{R}^{2n}$$

(50) $$(x(0), -M^* Jy^*(T) + Jy^*(0)) = 0 \quad \forall x(0) \in \mathbb{R}^{2n} \ .$$

Hence $y^*(T) = -J\,(M^*)^{-1}\,Jy^*(0)$. By formula (31), $-J\,(M^*)^{-1}\,J = M$. The boundary condition for A^* is $y^*(T) = My^*(0)$. Hence $A^* = A$. □

We now introduce the Hamiltonian $H : [0,T] \times \mathbb{R}^{2n} \to \mathbb{R} \cup \{+\infty\}$, with the following assumptions:

(51) $$H \quad \text{is Borelian on} \quad [0,T] \times \mathbb{R}^{2n}$$

(52) $$\forall t \in [0,T] \ , \quad H(t,\cdot) \quad \text{is convex l.s.c. on} \quad \mathbb{R}^{2n}$$

(53) $$\exists \overline{x} \in L^\infty\left([0,T]; \mathbb{R}^{2n}\right) \ : \quad \int_o^T H\left(t, \overline{x}(t)\right) dt < \infty$$

(54) $$\exists \epsilon > 0 \ : \quad \left(\xi \in \mathbb{R}^{2n}, \|\xi\| \leq \epsilon\right) \Rightarrow \int_o^T H^*\left(t, \xi\right) dt < \infty \ .$$

They will be satisfied if, for instance, the Hamiltonian H does not depend on t, and:

(55) $$H : \mathbb{R}^{2n} \to \mathbb{R} \cup \{+\infty\} \quad \text{is convex l.s.c. and proper}$$

(56) $$\exists \epsilon > 0 \ , \ \exists c \in \mathbb{R} \ : \quad H(x) \geq \epsilon \|x\| - c \quad \forall x \in \mathbb{R}^{2n} \ .$$

We now state the boundary-value problem:

(57) $$\begin{cases} \dot{x} \in JH(t,x) \\ x(T) = Mx(0) \end{cases}$$

where $T > 0$ and the symplectic matrix M are prescribed. We associate with it the operator A defined by formula (33) and the functional $\Phi : \operatorname{dom} A \to \mathbb{R} \cup \{+\infty\}$ given by:

(58) $$\begin{aligned} \Phi(x) &= \frac{1}{2}\langle Ax, x\rangle + \int_o^T H(t,x)dt \\ &= \int_o^T \left[\frac{1}{2}\left(J\dot{x}, x\right) + H(t,x)\right] dt \ . \end{aligned}$$

Proposition 5 (Least action principle). *The critical points of Φ are exactly the solutions of problem* (57). □

Proof. For $x \in L^\alpha$, set $\mathcal{H}(x) := \int_o^T H(t, x(t))\, dt$. It is a convex l.s.c. function. The assumption of Theorem 4.2 and Corollary 4.3 are satisfied, with (1) replaced by (27), so that

$$\mathcal{H}^*(x^*) = \int_o^T H^*(t; x^*(t))\, dt \quad \forall x^* \in L^\beta \tag{59}$$

$$\partial\mathcal{H}(x) = \left\{x^* \in L^\beta | x^* \in \partial H(t, x(t)) \text{ a.e.}\right\} . \tag{60}$$

By definition, x is a critical point of Φ if and only if $Ax \in \partial\mathcal{H}(x)$, that is:

$$\dot{x} \in L^\beta \quad \text{and} \quad J\dot{x} + \partial H(t, x(t)) \ni 0 \quad \text{a.e.} \tag{61}$$

Hence $\dot{x} \in J\partial H(t, x)$, as desired. □

We now wish to apply Theorem 2. Introduce a new functional Ψ on L^α:

$$\begin{aligned} \Psi(x) &= \frac{1}{2}\langle Ax, x\rangle + \mathcal{H}^*(Ax) \\ &= \int_o^T \left[\frac{1}{2}(Jx, x) + H^*(t, -Jx)\right] dt . \end{aligned} \tag{62}$$

Proposition 6 (Dual action principle). *x is a critical point of Ψ if and only if there is some constant $\xi \in \mathbb{R}^{2n}$ such that $x(t) + \xi$ solves problem* (57). □

Proof. Defining L_o^β by formula (44), we have:

$$L_o^\beta \subset A(\operatorname{dom} A) \subset L^\beta . \tag{63}$$

By condition (54), denoting by $B\left(\mathbb{R}^{2n}\right)$ the unit ball, we have:

$$\epsilon B\left(\mathbb{R}^{2n}\right) \subset \operatorname{dom}\mathcal{H}^* . \tag{64}$$

Hence:

$$A(\operatorname{dom} A) + \operatorname{dom}\mathcal{H}^* \supset L_o^\beta + \epsilon B\left(\mathbb{R}^{2n}\right) = L^\beta . \tag{65}$$

Condition (14) is satisfied, and we may apply Theorem 2. The result follows. □

If $M = I$, and H is T-periodic ($H(t+T, x) = H(t, x)$ for all t and x), or autonomous ($H(x)$ independent of t), problem (57) becomes

$$\begin{cases} \dot{x} \in JH(t, x) \\ x(0) = x(T) \end{cases} \tag{66}$$

and x is now a T-periodic solution. We shall dwell a length on this particular case in subsequent chapters. For the time being, we conclude this section with a few more examples, which we treat in lesser detail.

Example 1: The Bolza Problem

We are still dealing with the same equation $\dot{x} \in J\partial H(t,x)$, where the Hamiltonian H satisfies conditions (51) to (54), but this time we consider a fixed-endpoints problem. Split x into (p,q), with p and $q \in \mathbb{R}^n$. Take two points $q_o, q_1 \in \mathbb{R}^n$ and some time $T > 0$. We are interested in the boundary value problem:

$$\begin{cases} \dot{q} = H'_p(q,p) \\ \dot{p} = -H'_q(q,p) \\ q(0) = q_o \\ q(T) = q_1 \ . \end{cases} \tag{67}$$

Introduce the function $\overline{q}(t) := (q_1 - q_o)\, t/T$, and the operator $A : L^\alpha \to L^\beta$ defined by:

$$\begin{cases} \operatorname{dom} A = \{(p,q) | \dot{p} \in L^\beta, \dot{q} \in L^\beta, q(0) = 0 = q(T)\} \\ A(p,q) = (-\dot{q}, \dot{p}) \ . \end{cases} \tag{68}$$

Now define the action functional Φ on the linear space $L^\alpha \times \operatorname{dom} A$ by:

$$\Phi(q,p) := \int_o^T \left[-p\left(\dot{q} + \frac{d\overline{q}}{dt}\right) + H\left(p, q + \overline{q}\right) \right] dt \ . \tag{69}$$

If (p,q) is a critical point of Φ, then $(p, q+\overline{q})$ is a solution of problem (67), and conversely. Now apply Theorem 2. We write Φ as

$$\Phi(x) = \frac{1}{2}\langle Ax, x\rangle + \mathcal{H}(x) \tag{70}$$

where A is defined by formula (68) and $\mathcal{H}$ by:

$$\mathcal{H}(p,q) = \int_o^T \left[-p\frac{d\overline{q}}{dt} + H\left(p, q + \overline{q}\right) \right] dt \ . \tag{71}$$

The operator A is closed and self-adjoint. The conjugate $\mathcal{H}^*$ is given by:

$$\mathcal{H}^*(p,q) = \int_o^T \left[H^*\left(p + \frac{d\overline{q}}{dt}, q\right) - q\overline{q} \right] dt \ . \tag{72}$$

The dual functional Ψ given by Theorem 2 now is:

$$\begin{aligned} \Psi(p,q) &:= \int_o^T \left[-p\dot{q}\dot{p}\overline{q} + H\left(\dot{q} + \frac{d\overline{q}}{dt}, -\dot{p}\right)\right] dt \\ &= \int_o^T \left[-p\left(\dot{q} + \frac{d\overline{q}}{dt}\right) + H\left(\dot{q} + \frac{d\overline{q}}{dt}, -\dot{p}\right)\right] dt + q_1 p(T) - q_o p(0) \ . \end{aligned} \tag{73}$$

The kernel of A consists of the constant functions $(\zeta, 0)$, with $\zeta \in \mathbb{R}^n$. So (p, q) is a critical point of Ψ if and only if there is some constant $\zeta \in \mathbb{R}^n$ such that $(\zeta + p, \overline{q} + q)$ solves problem (67).

Of course, in the above formulas, $\frac{d\overline{q}}{dt}$ should be replaced by the constant $(q_1 - q_o)/T$.

Example 2: Lagrangian Formulation

In classical mechanics, the Hamiltonian splits in two parts, kinetic and potential energy. Typically, we have:

$$H(t, p, q) = \frac{1}{2}\left(\Lambda(t)p, q\right) + V(t, q) \ , \tag{74}$$

where $\Lambda(t) = \Lambda(t)^*$ is a positive definite matrix, depending continuously on t, and $V : [0, T] \times \mathbb{R}^n \to \mathbb{R} \cup \{+\infty\}$, the potential, is measurable in t and convex l.s.c. in q.

The Hamiltonian system $x = JH'(t, x)$ becomes a second-order system for q in $\mathbb{R}^n$:

$$\frac{d}{dt}\left(\Lambda(t)\dot{q}\right) + \partial V(t, q) = 0 \tag{75}$$

with $p = \Lambda(t)\dot{q}$.

To study Eq. (75) it is better to use the Lagrangian formalism. Instead of the Hamiltonian action, we shall use the functional:

$$\Phi(q) = \int_o^T \left[-\frac{1}{2}\left(\Lambda(t)\dot{q}, \dot{q}\right) + V(t, q)\right] dt \ . \tag{76}$$

Assumptions (51) to (54) on the Hamiltonian translate into assumptions on the potential V:

$$\forall t \in [0, T] \ , \quad V(t, \cdot) \ \text{ is convex l.s.c.} \tag{77}$$

$$\exists \overline{q} \in L^\infty \ , \quad \int_o^T V\left(t, \overline{q}(t)\right) dt < \infty \tag{78}$$

$$\exists \epsilon > 0 \ , \quad \left(\zeta \in \mathbb{R}^n, \|\zeta\| \leq \epsilon\right) \Rightarrow \int_o^T V^*(t, \zeta) dt < \infty \ . \tag{79}$$

Consider for instance the nonlinear Sturm-Liouville problem:

$$\begin{cases} \dfrac{d}{dt}\left(\Lambda(t)\dot{q}\right) + \partial V(t, q) = 0 \\ \begin{bmatrix} q(T) \\ \dot{q}(T) \end{bmatrix} = M \begin{bmatrix} q(0) \\ \dot{q}(0) \end{bmatrix} \ , \end{cases} \tag{80}$$

where $M \in \mathcal{L}\left(\mathbb{R}^{2n}\right)$ satisfies

$$M^* \begin{pmatrix} 0 & \Lambda(T) \\ -\Lambda(T) & 0 \end{pmatrix} M = \begin{pmatrix} 0 & \Lambda(0) \\ \Lambda(0) & 0 \end{pmatrix} . \tag{81}$$

For instance, if $\Lambda(t) \equiv I$, the matrix M is symplectic.

If condition (81) is satisfied, the operator

$$A := -\frac{d}{dt}\left(\Lambda(t)\frac{d}{dt}\right) \tag{82}$$

is closed and self-adjoint on the space

$$E := \left\{ q \in L^\alpha(0,T;\mathbb{R}^n) \,\middle|\, \frac{d}{dt}(\Lambda(t)\dot{q}) \in L^\beta ,\ \begin{pmatrix} \dot{q}(T) \\ q(0) \end{pmatrix} = M \begin{pmatrix} \dot{q}(0) \\ q(0) \end{pmatrix} \right\} . \tag{83}$$

The solutions of problem (80) are exactly the critical points of Φ and E. Applying Theorem 2, we find the dual functional:

$$\begin{aligned} \Psi(q) : &= \frac{1}{2}\langle -Aq, p\rangle + \int_o^T V^*(t, Aq)dt \\ &= \int_o^T \left[-\frac{1}{2}(\Lambda(t)\dot{q}, \dot{q}) + V^*\left(t, \frac{d}{dt}(\Lambda(t)\dot{q})\right)\right] dt . \end{aligned} \tag{84}$$

The kernel of A is finite-dimensional, with dimension at most $2n$. By Theorem 2, q will be a critical point of Ψ on E if and only if there is some $\overline{q} \in \operatorname{Ker} A$ such that $q + \overline{q}$ solves problem (80).

There are many possible extensions of this basic example. We can consider other operators on the real line:

$$A = -\frac{d^2}{dt^2} + C(t) , \quad \text{with } C(t) = C(t)^* \in \mathcal{L}\left(\mathbb{R}^{2n}\right) \tag{85}$$

$$A = -\frac{d^2}{dt^2} + K(t)\frac{d}{dt} , \quad \text{with } K(t) + K(t)^* = 0 \tag{86}$$

$$A = (-1)^p \frac{d^{2p}}{dt^{2p}} . \tag{87}$$

The first example enables us to treat some non-convex problems (see Corollary 3). The second example is the case of so-called gyroscopic forces: they depend on the velocity and do not work since $(K(t)\dot{x}, \dot{x}) = 0$.

We can also consider extensions to higher dimensions. Note particularly

$$A = -\Delta \quad \text{(Laplace operator)} \tag{88}$$

$$A = \frac{d^2}{dt^2} - \Delta \quad \text{(wave operator)} \tag{89}$$

$$A = J\frac{d}{dt} - \Delta \quad \text{(Schrödinger operator)} . \tag{90}$$

Notes and Comments. The dual action principle in the Hamiltonian framework was first stated by Clarke ([Cla1], [Cla2], [Cla3]) in the fixed-energy case, and developed in [ClaE1], [ClaE2] for the fixed period case (Proposition 6). It was adapted to the wave operator in [BreCN]. The general formulation (Theorem 2) was given in [EkeL2], [EkeL3].

Chapter III. Fixed-Period Problems: The Sublinear Case

Introduction

With this chapter, the preliminaries are over, and we begin the search for periodic solutions to Hamiltonian systems. All this will be done in the convex case; that is, we shall study the boundary-value problem

$$\left\{\begin{array}{l}\dot{x} = JH'(t,x) \\ x(0) = x(T)\end{array}\right. \tag{1}$$

with $H(t,\cdot)$ a convex function of x, going to $+\infty$ when $\|x\| \to \infty$.

One may ask for the physical meaning of convexity. This is best answered in the autonomous case, with $x = (p,q)$ and a classical Hamiltonian:

$$H(x) = \frac{1}{2}p^2 + V(q)\ . \tag{2}$$

It follows from the growth assumption on H that V attains its minimum at a point $\overline{q}$, which is an equilibrium of the motion. It then follows from the convexity assumption that

$$(V'(q), \overline{q} - q) \leq V(\overline{q}) - V(q) \leq 0\ . \tag{3}$$

Throwing in the equation of the motion, $\ddot{q} = -V'(q)$, this becomes

$$(\ddot{q}, q - \overline{q}) \leq 0 \tag{4}$$

which means that at every point the acceleration is directed toward the equilibrium point. In other words, the system is centripetal. It behaves like a spring, which is always pulled back to its equilibrium position. This is precisely the kind of system we expect to oscillate, that is, to have periodic solutions.

To push the analogy further, a linear spring obeys Hooke's law, $\ddot{q} = -kq$, with $k > 0$. Its potential $V(q) = \frac{k}{2}q^2$ is quadratic. Nonlinear springs can be modelled by homogeneous potentials, $V(q) = \frac{k}{\alpha}|q|^\alpha$, and they are classified into weak spring (for $1 < \alpha < 2$) and strong springs (for $\alpha > 2$). We shall classify nonlinear Hamiltonian systems in much the same way.

To put the matter in another light, for linear oscillators, i.e. quadratic Hamiltonians, there are only finitely many possible periods, corresponding to linear families of solutions, the so-called fundamental modes. On the other hand, for nonlinear oscillators, the period will depend on the amplitude, so that they can vibrate with any frequency. It is therefore reasonable to expect that convex Hamiltonians fall in two classes separated by the quadratic Hamiltonians: the sublinear class, corresponding to subquadratic Hamiltonians, and the superlinear class, corresponding to superquadratic Hamiltonians.

This is precisely what the mathematics confirm: the sublinear and superlinear case require different tools. We begin with the simplest one, the sublinear case, which can be treated by minimization.

1. Subquadratic Hamiltonians

A lienar Hamiltonian system:

$$\dot{x} = JA(t)x \ , \quad A(t) = A^*(t) \in \mathcal{L}\left(\mathbb{R}^{2n}\right) \tag{1}$$

corresponds to a quadratic Hamiltonian, namely $\frac{1}{2}(A(t)x, x)$.

In this chapter, we shall study sublinear systems, corresponding to subquadratic Hamiltonians. The precise assumptions are as follows. Recall that the matrix inequality $A \leq B$, with A and B symmetric, means that $B - A$ is positive semi-definite.

Definition 1. Let $A_\infty(t)$ and $B_\infty(t)$ be symmetric operators in $\mathbb{R}^{2n}$, depending continuously on $t \in [0, T]$, such that $A_\infty(t) \leq B_\infty(t)$ for all t.

A Borelian function $H : [0,T] \times \mathbb{R}^{2n} \to \mathbb{R}$ is called (A_∞, B_∞)-*subquadratic at infinity* if there exists a function $N(t,x)$ such that:

$$H(t,x) = \frac{1}{2}(A_\infty(t)x, x) + N(t,x) \tag{2}$$

$$\forall t \ , \quad N(t,x) \quad \text{is convex with respect to} \ \ x \tag{3}$$

$$N(t,x) \geq n(\|x\|) \quad \text{with} \ \ n(s)s^{-1} \to +\infty \ \ \text{as} \ \ s \to +\infty \tag{4}$$

$$\exists c \in \mathbb{R} \ : \quad H(t,x) \leq \frac{1}{2}(B_\infty(t)x, x) + c \quad \forall x \ . \tag{5}$$

□

If $A_\infty(t) = a_\infty I$ and $B_\infty(t) = b_\infty I$, with $a_\infty \leq b_\infty \in \mathbb{R}$, we shall say that H is (a_∞, b_∞)-subquadratic at infinity. As an example, the function $\|x\|^\alpha$, with $1 \leq \alpha < 2$, is $(0, \epsilon)$-subquadratic at infinity for every $\epsilon > 0$. Similarly, the Hamiltonian

$$H(t,x) = \frac{1}{2}k\,\|x\|^2 + \|x\|^\alpha \tag{6}$$

is $(k, k+\epsilon)$-subquadratic for every $\epsilon > 0$. Note that, if $k < 0$, it is not convex.

Suppose H is (A_∞, B_∞)-subquadratic at infinity. We are interested in solving the boundary-value problem:

$$\begin{cases} \dot{x} \in JA_\infty(t)x + J\partial N(t,x) \quad \text{a.e.} \\ x(0) = x(T)\ . \end{cases} \tag{7}$$

If H is C^1 and T-periodic with respect to t, for instance, if the system is autonomous, this amounts to finding T-periodic solutions of the Hamiltonian system:

$$\dot{x} = JH'(t,x)\ . \tag{8}$$

Arguing as in Sect. 4 of the preceding chapter, we find that the solutions of problem (8) are the critical points of the action functional:

$$\Phi(x) = \int_o^T \left[\frac{1}{2}(J\dot{x} + A_\infty(t)x, x) + N(t,x)\right] dt$$

on the space $W^{1,2}\left(\mathbb{R}/T\mathbb{Z}; \mathbb{R}^{2n}\right)$ of T-periodic absolutely continuous functions with square-integrable derivative.

Introduce the operator Λ on L^2 defined by:

$$\operatorname{dom}\Lambda = W^{1,2}\left(\mathbb{R}/T\mathbb{Z}; \mathbb{R}^{2n}\right) \tag{9}$$

$$\Lambda = J\frac{d}{dt} + A_\infty\ . \tag{10}$$

Its main properties are summarized in the following:

Proposition 2. *The operator Λ is closed and self-adjoint. Its range $R(\Lambda) := \Lambda(\operatorname{dom}\Lambda)$ is closed in L^2, its kernel* Ker Λ *is $2n$-dimensional at most, and L^2 splits orthogonally:*

$$L^2 = \operatorname{Ker}\Lambda \oplus R(\Lambda)\ . \tag{11}$$

Endow dom Λ *with its standard (graph) norm and $R(\Lambda)$ with the L^2-norm, and consider the double restriction Λ_o of Λ:*

$$\Lambda_o : \operatorname{dom}\Lambda \cap R(\Lambda) \to R(\Lambda)\ . \tag{12}$$

Then Λ_o is a Hilbert space isomorphism, and its inverse Λ_o^{-1} is compact. □

Before proceeding with the proof, note that if $\operatorname{Ker}\Lambda = \{0\}$, then $\Lambda = \Lambda_o$, and $\Lambda : W^{1,2} \to L^2$ is invertible with compact inverse.

Proof. All these are well-known properties of the differential operator $\Lambda = J\frac{d}{dt} + A_\infty(t)$ over T-periodic functions. To invert Λ, we have to solve the equation

$$J\dot{x} + A_\infty(t)x = u \in L^2 \tag{13}$$

which has the explicit solution (variation of constants formula)

$$x(t) = R(t)x(0) - \int_o^t R(s)^{-1}Ju(s)ds \ . \tag{14}$$

Here $R(t)$ is the matrizant

$$\begin{cases} \frac{d}{dt}R(t) = JA_\infty(t)R(t) \\ R(0) = I \ . \end{cases} \tag{15}$$

Throwing in the boundary condition $x(T) = x(0)$, we get a relation between $x(0) \in \mathbb{R}^{2n}$ and $u \in L^2$

$$(R(T) - I)\, x(0) = \int_o^T R(t)^{-1}Ju(t)dt \ . \tag{16}$$

This determines $x(0)$ in termes of u, provided $R(T)$ does not have the eigenvalue 1: the operator Λ then is invertible.

If $R(T)$ has the eigenvalue 1, the right-hand side u of Eq. (16) must satisfy the relation

$$\int_o^T R(t)^{-1}Ju(t)dt \in (R(T) - I)\left(\mathbb{R}^{2n}\right) \tag{17}$$

which characterizes the range of Λ. Its codimension in L^2 must therefore be less than or equal to $2n$.

We now check that the range of Λ is the orthogonal of its kernel. Take any $v \in \operatorname{Ker} \Lambda$. We have:

$$v(t) = R(t)\xi \ , \qquad \text{with } \ \xi \in \operatorname{Ker}\,(R(T) - I) \tag{18}$$

and:

$$(u, v,)_{L^2} = \int_o^T (u(t), R(t)\xi)\, dt = \int_o^T (R^*(t)u(t), \xi)\, dt \ . \tag{19}$$

Remember that $R(t)$ is symplectic:

$$(u, v)_{L^2} = \int_o^T (-JR(t)^{-1}Ju(t), \xi)\, dt = \left(\int_o^T R(t)^{-1}Ju\, dt, J\xi\right) \ . \tag{20}$$

So $u \in L^2$ is orthogonal to $\operatorname{Ker} \Lambda$ if and only if:

$$\int_o^T R(t)^{-1} Ju \, dt \in [J \operatorname{Ker}(R(T) - I)]^{\perp} \ . \tag{21}$$

Since $R(T)$ is symplectic, we have the identity

$$(R(T)\zeta - \zeta, J\xi) = (\zeta, R(T)^* J\xi - J\xi) = \left(J\zeta, \xi - R(T)^{-1}\xi\right) \tag{22}$$

from which it follows that the right-hand sides of Eqs. (17) and (21) coincide. So $u \in (\operatorname{Ker} \Lambda)^{\perp}$ if and only if $u \in R(\Lambda)$, as announced.

From the orthogonal splitting

$$L^2 = \operatorname{Ker} \Lambda \oplus R(\Lambda) \tag{23}$$

it follows that Λ_o is one-to-one and continuous. By the closed graph theorem, it must be an isomorphism.

Denote by $x_u \in W^{1,2}$ the preimage $\Lambda_o^{-1} u$. Since Λ_o^{-1} is continuous, and $W^{1,2}$ is contained in C^o, the map $u \to x_u(0)$ from $R(\Lambda)$ into $\mathbb{R}^{2n}$ is continuous. It then follows from the integral representation (14) that the map $u \to x_u = \Lambda_o^{-1} u$ is compact. □

Now define a convex l.s.c. map $\mathcal{N}$ on L^2 by:

$$\mathcal{N}(x) = \int_o^T N(t, x(t)) \, dt \ . \tag{24}$$

Set:

$$C_\infty(t) := B_\infty(t) - A_\infty(t) \ . \tag{26}$$

By assumption (4), $n^*(\rho)$ is finite everywhere. We have:

$$\frac{1}{2}\left(C_\infty(t)^{-1} y, y\right) - c \leq N^*(t, y) \leq n^*(\|y\|) \tag{27}$$

from which it follows that:

$$\mathcal{N}^*(y) = \int_o^T N^*(t; y(t)) \, dt \tag{28}$$

$$\partial \mathcal{N}^*(y) = \left\{ z \in L^2 \mid z(t) \in \partial N^*(t, y(t)) \ \text{a.e.} \right\} \tag{29}$$

(Theorem II.3.2 and Corollary II.3.3). We find that the domain of $\mathcal{N}^*$ contains L^∞. Since the kernel of Λ consists of C^1 functions, it is contained in L^∞, and condition (14) of Theorem II.4.2 is satisfied.

It then follows from Theorem II.4.2 that solutions of problem (7) are related to critical points of the dual functional

$$\Psi(x) := \frac{1}{2}(\Lambda x, x)_{L^2} + N^*(-\Lambda x)$$
(30)
$$= \int_o^T \left[\frac{1}{2}(J\dot{x} + A_\infty(t)x, x) + N^*(t, -J\dot{x} - A_\infty(t)x)\right] dt\ .$$

Here $x \in \operatorname{dom}\Lambda = W^{1,2}(\mathbb{R}/T\mathbb{Z}; \mathbb{R}^{2n})$. In accordance with Proposition 2, it will be convenient to make the change of variables $\Lambda x = u$, thereby introducing the reduced functional

(31) $$\psi : R(\Lambda) \to \mathbb{R} \cup \{+\infty\}$$

defined by:

$$\psi(u) := \frac{1}{2}(\Lambda_o^{-1}u, u)_{L^2} + \mathcal{N}^*(-u)$$
(32)
$$= \int_o^T \left[\frac{1}{2}(\Lambda_o^{-1}u, u) + N^*(t, -u)\right] dt\ .$$

The results of Theorem II.4.2 are summarized in the following:

Theorem 3. *If $u \in R(\Lambda)$ is a critical point of ψ, there will exist some $x_o \in \operatorname{Ker}\Lambda$ such that*

(33) $$x := \Lambda_o^{-1}u + z_o$$

solves problem (7). Conversely, if $x \in W^{1,2}$ solves problem (7), then $u := \Lambda x$ is a critical point of ψ on $R(\Lambda)$. □

The question now is to find critical points of ψ on $\operatorname{dom}\Lambda$. Clearly, the spectrum of Λ_o will play an important role. By construction:

(34) $$\operatorname{Spec}\Lambda_o = (\operatorname{Spec}\Lambda) \setminus \{0\}\ .$$

Proposition 4. *The spectrum of Λ_o consists of a doubly infinite sequence λ_n, $n \in \mathbb{Z}$ of reals:*

(35) $$\lambda_n \le \lambda_{n+1} \quad \forall n$$

(36) $$\lambda_n \to +\infty \quad \text{as } n \to +\infty$$

(37) $$\lambda_n \to -\infty \quad \text{as } n \to -\infty\ .$$

Each eigenvalue λ_n has multiplicity at most 2n. □

Proof. Since Λ_o^{-1} is compact and self-adjoint, its spectrum consists of a sequence μ_k, $k \in \mathbb{N}$, of reals, with $|\mu_k| \to 0$. The spectrum of λ_o consists of the sequence $\lambda_k := \mu_k^{-1}$ so that $|\lambda_k| \to \infty$.

We now check that the spectrum of Λ_o contains arbitrarily large positive and negative terms. If this were not the case, it would be bounded from above or from below: assume for instance it is bounded from below:

$$\exists \overline{\lambda} : \quad \lambda_k \geq \overline{\lambda} \quad \forall k \in \mathbb{N} . \tag{38}$$

Then $\Lambda_o - \overline{\lambda}$ would be positive semi-definite:

$$(\Lambda_o x, x)_{L^2} - \overline{\lambda} \|x\|_{L^2} \geq 0 \quad \forall x \in \operatorname{dom} \Lambda_o . \tag{39}$$

If $x \in \operatorname{dom} \Lambda$, we can write $x = x_o + y$, with $x_o \in \operatorname{dom} \Lambda_o$ and $y \in \operatorname{Ker} \Lambda_o$. So the above relation immediately extends to

$$(\Lambda x, x)_{L^2} - \overline{\lambda} \|x\|_{L^2} \geq 0 \quad \forall x \in \operatorname{dom} \Lambda . \tag{40}$$

Now define a test function x_p as follows:

$$x_p(t) := e^{2p\pi Jt/T} \xi \tag{41}$$

with $p \in \mathbb{Z}$ and $\xi \in \mathbb{R}^{2n}$ with $\|\xi\| = 1$. We have, since the operator $e^{p2\pi Jt/T}$ is orthogonal:

$$\|x_p\|_{L^2} = T \tag{42}$$

$$\begin{aligned} (\Lambda x_p, x_p)_{L^2} &= -2\pi p + \int_o^T (A_\infty(t) x_p(t), x_p(t))\, dt \\ &\geq -2\pi p + a_\infty^+ T , \end{aligned} \tag{43}$$

where a_∞^+ is a constant, chosen in such a way that

$$(A_\infty(t)\zeta, \zeta) \leq a_\infty^+ \|\zeta\|^2 \tag{44}$$

for $0 \leq t \leq T$ and $\|\zeta\| = 1$. This is possible by compactness.

If we choose $p > \left(a_\infty^+ - \overline{\lambda}\right)/2\pi$, inequality (43) contradicts (39). So condition (38) cannot hold.

We have proved that the spectrum of Λ_o is unbounded on both sides. Proposition 4 now is a matter of ordering the λ_k. □

For instance, if $A_\infty = 0$, we have:

$$\Lambda = J \frac{d}{dt} \tag{45}$$

$$R(\Lambda) = \left\{ u \in L^2\left(0, T; \mathbb{R}^{2n}\right) \mid \int_o^T u\, dt = 0 \right\} \tag{46}$$

$$\Lambda_o^{-1} u = x \Leftrightarrow \left[u = J\dot{x} \;\text{ and }\; \int_o^T x\, dt = 0 \right] . \tag{47}$$

In other words, $\Lambda_o^{-1} u$ is just the primitive of u with mean value zero. In this particular case, the spectrum can be computed explicitly:

$$\text{Spec}\left(J\frac{d}{dt}\right) = \frac{2\pi}{T}\mathbb{Z} \tag{48}$$

$$\text{Spec}\, \Lambda_o = \frac{2\pi}{T}\mathbb{Z} \setminus \{0\} . \tag{49}$$

Each eigenvalue has multiplicity $2n$. If $\lambda = \frac{2\pi}{T}k$ is an eigenvalue, the corresponding eigenspace is

$$E_T^\lambda = \left\{ e^{-2k\pi Jt/T}\xi \,\middle|\, \xi \in \mathbb{R}^{2n} \right\} . \tag{50}$$

2. An Existence Result

In this section, we consider a Hamiltonian

$$H(t,x) = \frac{1}{2}\left(A_\infty(t)x, x\right) + N(t,x) \tag{1}$$

which is (A_∞, B_∞)-subquadratic at infinity, and we want to solve the boundary-value problem:

$$\begin{cases} \dot{x} \in JA_\infty(t)x + J\partial N(t,x) \\ x(0) = x(T) . \end{cases} \tag{2}$$

We recall that the time-dependent matrix:

$$C_\infty(t) := B_\infty(t) - A_\infty(t) \tag{3}$$

is assumed to be positive semi-definite. Denote by $0 \leq \gamma_1(t) \leq \ldots \leq \gamma_{2n}(t)$ its eigenvalues, and set:

$$\gamma := \text{Min}\, \{\gamma_1(t) | 0 \leq t \leq T\} . \tag{4}$$

Recall also that the eigenvalues of Λ_o form an increasing sequence λ_n, $n \in \mathbb{Z}$, with $\lambda_n \to \pm\infty$ as $n \to \pm\infty$, and $\lambda_n \neq 0$ for all n. Denote by λ the largest negative eigenvalue:

$$\lambda := \text{Max}\, \{\lambda_n < 0\} . \tag{5}$$

Theorem 1. *If $|\lambda| > \gamma$, then problem* (2) *has one solution at least.* □

Before proving Theorem 1, let us give some consequences in various settings: first, the case when $A_\infty = 0$; then the case when $C_\infty = 0$; finally a case when the assumptions are given in terms of the Hessian $H''(t,x)$.

Proposition 2. *Let $H(t,x)$ be a Hamiltonian, Borelian with respect to all variables, convex with respect to x for fixed t, and such that:*

$$\rho^{-1}\text{Min}\, \{H(t,x) | 0 \leq t \leq T, \|x\| = \rho\} \to +\infty \quad \text{as } \rho \to \infty \tag{6}$$

(7) $$H(t,x) \leq \frac{1}{2} b_\infty \|x\|^2 + c \quad \forall (t,x)$$

for some $b_\infty > 0$ *and* $c \in \mathbb{R}$. *Then, provided:*

(8) $$T b_\infty < 2\pi$$

the problem:

(9) $$\dot{x} \in J \partial H(t,x)$$

(10) $$x(0) = x(T)$$

has a solution. □

Proof. Here $A_\infty = 0$, so $\Lambda = J\frac{d}{dt}$ and the spectrum of Λ_o is $\frac{2\pi}{T}\mathbb{Z} \setminus \{0\}$, as we pointed out at the end of the preceding section. So

(11) $$\lambda = -\frac{2\pi}{T} .$$

On the other hand, $B_\infty = C_\infty = b_\infty I$, so $\gamma = b_\infty$. The result now follows from Theorem 1. □

Proposition 3. *Assume the Hamiltonian* $H(t,x)$ *is given as:*

(12) $$H(t,x) = \frac{1}{2} (A_\infty(t) x, x) + N(t,x)$$

with $A_\infty(t) = A_\infty^*(t)$, *continuous with respect to* t, *and* $N(t,x)$, *Borelian in* (t,x) *such that*

(13) $$N(t,\cdot) \quad \text{is convex in } x \quad \text{for fixed } t$$

(14) $$N(t,x) \geq n(\|x\|) \quad \text{with } n(s)s^{-1} \to +\infty \quad \text{as } s \to \infty$$

(15) $$N(t,x) \leq m(\|x\|) \quad \text{with } m(s)s^{-2} \to 0 \quad \text{as } s \to \infty .$$

Then the problem:

(16) $$\begin{cases} \dot{x} \in J A_\infty(t) x + \partial N(t,x) \\ x(0) = x(T) \end{cases}$$

has a solution for every $T > 0$. □

Proof. Take any $T > 0$. Condition (15) means that, for every $\epsilon > 0$, we will be able to find some constant $c \in \mathbb{R}$ such that:

(17) $$N(t,x) \leq \frac{\epsilon}{2} \|x\|^2 + c .$$

From this it follows that $H(t,x)$ is $(A_\infty, A_\infty + \epsilon I)$-subquadratic at infinity for every $\epsilon > 0$. In other words, we may take $C_\infty = \epsilon I$, and hence $\gamma = \epsilon$. Taking $\epsilon < \lambda$, we are in a position to apply Theorem 1. □

Proposition 4. *Let the Hamiltonian $H(t,x)$ be C^2. Assume that two real numbers a and b can be found with*

$$\frac{2\pi}{T}k_o \le a \le b < \frac{2\pi}{T}(k_o+1) \qquad \text{for some} \quad k_o \in \mathbb{Z} \tag{18}$$

$$aI \le H''(t,x) \le bI \quad \forall (t,x) \,. \tag{19}$$

Then the boundary-value problem

$$\begin{cases} \dot{x} = JH'(t,x) \\ x(0) = x(T) \end{cases} \tag{20}$$

has one solution at least. □

Proof. It follows from the mean value theorem that $H(t,x)$ is (a,b')-subquadratic at infinity for every $b' > b$. We may then take $\gamma = b' - a$ in Theorem 1.

On the other hand, $\Lambda = J\frac{d}{dt} + a$, so

$$\operatorname{Spec} \Lambda = a + \frac{2\pi}{T}\mathbb{Z} \,. \tag{21}$$

The largest negative eigenvalue λ is $\left(a + \frac{2\pi}{T}k_1\right)$ where $k_1 \in \mathbb{Z}$ is defined by:

$$-\frac{2\pi}{T}(k_1+1) \le a < -\frac{2\pi}{T}k_1 \,. \tag{22}$$

The condition $|\lambda| > \gamma$ of Theorem 1 now becomes

$$-\frac{2\pi}{T}k_1 - a > b' - a \tag{23}$$

which amounts to:

$$b' < -\frac{2\pi}{T}k_1 \,. \tag{24}$$

Setting $k_o := -(k_1+1)$, we find that formulas (22) and (24) reduce to condition (18). Hence the result. □

We now proceed to the proof of Theorem 1. We shall show that the dual action functional:

$$\psi(u) = \int_o^T \left[\frac{1}{2}\left(\Lambda_o^{-1}u, u\right) + N^*(t,-u)\right] dt \tag{25}$$

attains its minimum on $R(\Lambda)$, and the result will then follow from Theorem 1.3.

Recall first that Λ_o^{-1} is a compact self-adjoint operator from $R(\Lambda)$ into itself, where $R(\Lambda) \subset L^2$ is a closed subspace with the induced topology. Its eigenvalues are the λ_n^{-1}. Defining $\lambda > 0$ by (5), we get

(26) $$\left(\Lambda_o^{-1}u, u\right) \geq \lambda^{-1} \|u\|^2 \quad \forall u \in R(\Lambda) .$$

On the other hand, we have, since $H(t,x)$ is (A_∞, B_∞)-subquadratic at infinity:

(27) $$N(t,x) \leq \frac{1}{2}\left(C_\infty(t)x, x\right) + c \quad \forall (t,x) .$$

Here $C_\infty(t) = B_\infty(t) - A_\infty(t)$ is positive semi-definite. Remembering that $N(t,x)$ is superlinear with respect to x:

(28) $$n\left(\|x\|\right) \leq N(t,x) \leq \frac{1}{2}\left(C_\infty(t)x, x\right) + c ,$$

with $n(s)s^{-1} \to +\infty$ as $s \to \infty$, we see that $C_\infty(t)$ is in fact positive definite, and hence invertible. Taking conjugates in inequality (28) then yields:

(29) $$N^*(t;x) \geq \frac{1}{2}\left(C_\infty(t)^{-1}x, x\right) - c \quad \forall (t,x) .$$

So the number γ defined by (4) is in fact positive, and we have:

(30) $$N^*(t,x) \geq \frac{1}{2}\gamma^{-1} \|x\|^2 - c .$$

Writing inequalities (26) and (30) into the functional ψ, we get:

(31) $$\psi(u) \geq \frac{1}{2}\left(\lambda^{-1} + \gamma^{-1}\right) \|u\|^2 - c .$$

Since $|\gamma| > \lambda$, we have $\gamma^{-1} + \lambda^{-1} > 0$, so that $\varphi(u) \to +\infty$ when $\|u\| \to \infty$. Now take a minimizing sequence u_n for ψ:

(32) $$\psi(u_n) \to \mathrm{Inf}\psi .$$

We have, for n large enough

(33) $$\psi(u_n) \leq 1 + \mathrm{Inf}\psi .$$

Writing in inequality (31), we get:

(34) $$\frac{1}{2}\left(\gamma^{-1} + \lambda^{-1}\right) \|u_n\|^2 \leq c + 1 + \mathrm{Inf}\psi$$

from which is follows that the sequence u_n is bounded:

(35) $$\|u_n\| \leq 2\left(\gamma^{-1} + \lambda^{-1}\right)\left(c + 1 + \mathrm{Inf}\psi\right) .$$

We may extract a weakly convergent subsequence; which we denote by u_{n_k}:

(36) $$\exists u \in R(\Lambda) : u_{n_k} \to \overline{u} \in R(\Lambda)\ .$$

Since Λ_o^{-1} is a compact operator, it changes sequences which are bounded and weakly convergent into convergent sequences:

(37) $$\Lambda_o^{-1} u_{n_k} \to \Lambda_o^{-1} \overline{u}\ .$$

Therefore:

(38) $$\left(\Lambda_o^{-1} u_{n_k}, u_{n_k}\right)_{L^2} \to \left(\Lambda_o^{-1} \overline{u}, \overline{u}\right)_{L^2}\ .$$

On the other hand, the function $\mathcal{N}^* : L^2 \to \mathbb{R} \cup \{+\infty\}$ defined by

(39) $$\mathcal{N}^*(u) = \int_o^T N^*(t; u(t))dt$$

is known to be convex and l.s.c. (see Sect. II.3). It is therefore weakly l.s.c. (Theorem II.2.1), and so is its restriction to the closed subspace $R(\Lambda)$.

Writing all this into formula (32), we get:

(40) $$\begin{aligned} \mathrm{Inf}\psi &= \lim_k \psi\left(u_{n_k}\right) = \left(\Lambda_o^{-1}\overline{u}, \overline{u}\right)_{L^2} + \lim_k \mathcal{N}^*\left(u_{n_k}\right) \\ &\geq \left(\Lambda_o^{-1}\overline{u}, \overline{u}\right)_{L^2} + \mathcal{N}^*(\overline{u}) = \psi(\overline{u})\ . \end{aligned}$$

So $\psi(\overline{u}) \leq \mathrm{Inf}\psi$. Since $\overline{u} \in R(\Lambda)$, the converse inequality is trivial, and $\overline{u}$ is a minimizer. It follows that there is some $z_o \in \mathrm{Ker}\,\Lambda$ such that $\overline{x} := \Lambda_o^{-1}\overline{u} + z_o$ solves problem (2). □

Notes and Comments. Loosely speaking, Proposition 4 means that the Hamiltonian does not interfere with the spectrum of the operator $J\frac{d}{dt}$. If the inequality $\frac{2}{T}k_o \leq a$ in formula (18) is strict, the result can actually be proved by using some global version of the inverse function theorem. There is a vast amount of literature on the subject.

The problems becomes much more interesting when no bound is assumed on H''. The first results were obtained by Benci and Rabinowitz in [BenR], who put a growth condition on H'. Their method relies on the direct action principle, and their results therefore hold in non-convex situations. Clarke and Ekeland, in [ClaE3], [ClaE4] then used the dual action principle to obtain results in the convex case, with growth conditions on H only.

3. Autonomous Systems

In this section, we will consider the case when the Hamiltonian $H(x)$ is autonomous. For the sake of simplicity, we shall also assume that it is C^1.

We shall first consider the question of nontriviality, within the general framework of (A_∞, B_∞)-subquadratic Hamiltonians. In the second subsection, we shall look into the special case when H is $(0, b_\infty)$-subquadratic, and we shall try to derive additional information.

A. The General Case: Nontriviality

We assume that H is (A_∞, B_∞)-subquadratic at infinity, for some constant symmetric matrices A_∞ and B_∞, with $B_\infty - A_\infty$ positive definite. Set:

$$\gamma := \text{smallest eigenvalue of } B_\infty - A_\infty \tag{1}$$

$$\lambda := \text{largest negative eigenvalue of } J\frac{d}{dt} + A_\infty . \tag{2}$$

Theorem 2.1 tells us that if $\lambda + \gamma < 0$, the boundary-value problem:

$$\begin{cases} \dot{x} = JH'(x) \\ x(0) = x(T) \end{cases} \tag{3}$$

has at least one solution $\overline{x}$, which is found by minimizing the dual action functional:

$$\psi(u) = \int_o^T \left[\frac{1}{2}\left(\Lambda_o^{-1} u, u\right) + N^*(-u)\right] dt \tag{4}$$

on the range of Λ, which is a subspace $R(\Lambda) \subset L^2$ with finite codimension. Here

$$N(x) := H(x) - \frac{1}{2}\left(A_\infty x, x\right) \tag{5}$$

is a convex function, and

$$N(x) \le \frac{1}{2}\left(\left(B_\infty - A_\infty\right) x, x\right) + c \quad \forall x . \tag{6}$$

It should be noted that problem (3) may have – and, in general, does have – constant solutions (equilibria). If $\xi \in \mathbb{R}^{2n}$ is a critical point of H:

$$H'(\xi) = 0 \tag{7}$$

then $x(t) := \xi$ solves the differential equation $\dot{x} = JH'(x)$, and is T-periodic for any T. Such solutions are easy to find directly, and should therefore be eliminated from consideration. In other words, our existence theorem is not very interesting unless we can state $\overline{x}$ is nontrivial, that is to say, non-constant.

If ξ is the equilibrium we are interested in, we can always perform the change of variables $y = x - \xi$, thereby bringing ξ to the origin. Adding a constant to H does not affect the equations. So $\xi = 0$ and $H(\xi) = 0$ is the general case. Note that we then have $N'(0) = 0$; since N is convex, 0 must be a local minimum for N.

Proposition 1. *Assume $H'(0) = 0$ and $H(0) = 0$. Set:*

$$\delta := \liminf_{x \to 0} 2N(x) \left\|x\right\|^{-2} . \tag{8}$$

If $\gamma < -\lambda < \delta$, the solution $\overline{x}$ is non-zero:

$$\overline{x}(t) \neq 0 \quad \forall t \,. \tag{9}$$ □

Proof. Condition (8) means that, for every $\delta' > \delta$, there is some $\epsilon > 0$ such that

$$\|x\| \leq \epsilon \Rightarrow N(x) \leq \frac{\delta'}{2} \|x\|^2 \,. \tag{10}$$

It is an exercise in convex analysis, into which we shall not go, to show that this implies that there is an $\eta > 0$ such that

$$\|y\| \leq \eta \Rightarrow N^*(y) \leq \frac{1}{2\delta'} \|y\|^2 \,. \tag{11}$$

We will use this inequality to show that 0 cannot be a local minimizer for ψ. Indeed, let $u_1 \in R(\Lambda)$ be an eigenvector of Λ_o associated with the eigenvalue λ:

$$\Lambda_o u_1 = \lambda u_1 \,. \tag{12}$$

Let $h \in \mathbb{R}$ be a parameter, and consider $\psi\,(hu_1)$:

$$\psi(hu_1) = \frac{h^2}{2} \int_o^T \left(\Lambda_o^{-1} u_1, u_1\right) dt + \int_o^T N^*(hu_1) dt \,. \tag{13}$$

Since u_1 is a smooth function, we will have $\|hu_1\|_\infty \leq \eta$ for h small enough, and inequality (11) will hold, yielding thereby:

$$\psi(hu_1) \leq \frac{h^2}{2} \frac{1}{\lambda} \|u_1\|_2^2 + \frac{h^2}{2} \frac{1}{\delta'} \|u_1\|^2 \,. \tag{14}$$

If we choose δ' close enough to δ, the quantity $\left(\frac{1}{\lambda} + \frac{1}{\delta'}\right)$ will be negative, and we end up with

$$\psi(hu_1) < 0 \qquad \text{for } \; h \neq 0 \; \text{ small} \,. \tag{15}$$

On the other hand, we check directly that $\psi(0) = 0$. This shows that 0 cannot be a minimizer of ψ, not even a local one. So $\overline{u} \neq 0$ and $\overline{u} \neq \Lambda_o^{-1}(0) = 0$. □

If we assume H to be C^2, we can introduce the linearized equations around an equilibrium $\xi \in \mathbb{R}^{2n}$. They are:

$$\dot{y} = JH''(\xi) y = J\left(A_\infty + N''(\xi)\right) y \,. \tag{16}$$

If in addition we have $A_\infty = a_\infty I$ and $B_\infty = b_\infty I$ we can make the condition $\gamma < -\lambda < \delta$ in Proposition 1 quite explicit.

Corollary 2. *Assume H is C^2 and (a_∞, b_∞)-subquadratic at infinity. Let $\xi_1, \ldots, \xi_N$ be the equilibria, that is, the solutions of $H'(\xi) = 0$. Denote by ω_k the smallest eigenvalue of $H''(\xi_k)$, and set:*

(17) $$\omega := \text{Min}\,\{\omega_1, \ldots, \omega_k\} \ .$$

If:

(18) $$\frac{T}{2\pi} b_\infty < -E\left[-\frac{T}{2\pi} a_\infty\right] < \frac{T}{2\pi}\omega$$

then minimization of ψ yields a non-constant T-periodic solution $\overline{x}$. □

We recall once more that by the integer part $E[\alpha]$ of $\alpha \in \mathbb{R}$, we mean the $a \in \mathbb{Z}$ such that $a < \alpha \leq a + 1$. For instance, if we take $a_\infty = 0$, Corollary 2 tells us that $\overline{x}$ exists and is non-constant provided that:

(19) $$\frac{T}{2\pi} b_\infty < 1 < \frac{T}{2\pi}$$

or

(20) $$T \in \left(\frac{2\pi}{\omega}, \frac{2\pi}{b_\infty}\right) \ .$$

Proof. The spectrum of Λ is $\frac{2\pi}{T}\mathbb{Z} + a_\infty$. The largest negative eigenvalue λ is given by $\frac{2\pi}{T}k_o + a_\infty$, where

(21) $$\frac{2\pi}{T}k_o + a_\infty < 0 \leq \frac{2\pi}{T}(k_o + 1) + a_\infty \ .$$

Hence:

(22) $$k_o = E\left[-\frac{T}{2\pi} a_\infty\right] \ .$$

The condition $\gamma < -\lambda < \delta$ now becomes:

(23) $$b_\infty - a_\infty < -\frac{2\pi}{T}k_o - a_\infty < \omega - a_\infty$$

which is precisely condition (18). □

Corollary 3. *Assume $H(x) = \frac{1}{2}a_\infty \|x\|^2 + N(x)$, with $N : \mathbb{R}^{2n} \to \mathbb{R}$ convex and C^2, such that:*

(24) $$N(x)\,\|x\|^{-2} \to 0 \quad \text{as} \quad \|x\| \to \infty$$

(25) $$H'(\xi) = 0 \Rightarrow N''(\xi) \quad \text{positive definite} \ .$$

Then there is some $T_o > 0$ such that, for every $T > T_o$, minimization of ψ yields a non-constant T-periodic solution $\overline{x}$. □

Proof. The assumption implies that H is (a_∞, b)-subquadratic at infinity for every $b > a_\infty$. It is easily checked that $\frac{T}{2\pi}a_\infty < -E\left[-\frac{T}{2\pi}a_\infty\right]$. As soon as:

(26) $$a_\infty < b < -\frac{2\pi}{T} E\left[-\frac{T}{2\pi} a_\infty\right]$$

the left inequality in (18) will be satisfied.

On the other hand, the right inequality in the same formula can be rewritten as:

(27) $$-\frac{2\pi}{T} E\left[-\frac{T}{2\pi a_\infty}\right] < \omega \, .$$

As $T \to \infty$, the left-hand side goes to a_∞, while the right-hand side ω is the lowest eigenvalue of $a_\infty + N''(\xi)$, when ξ is an equilibrium. It then follows from assumption (25) that $a_\infty < \omega$, so that inequality (18) is satisfied for large T. □

For instance, if $a_\infty = 0$, that is if $H(x)$ is convex C^2 and such that

(28) $$H(x) \|x\|^{-2} \to 0 \qquad \text{as} \quad \|x\| \to \infty$$

(29) $$H'(x) = 0 \Rightarrow H''(x) \geq \omega I \qquad \text{with} \quad \omega > 0$$

then we find a non-constant T-periodic solution whenever $\frac{T}{2\pi}\omega > 1$, that is $T > \frac{2\pi}{\omega}$. Note that $\frac{2\pi}{\omega}$ is an upper bound for the periods of the solutions to the linearized system at equilibrium, $\dot{y} = JH''(\xi)y$.

In fact, in this example, $\overline{x}$ has minimal period T (which implies non-constancy), as we now show.

B. Minimal Period

Minimizing the dual action functional $\Psi(x)$ over $\operatorname{dom} \Lambda$ (or equivalently, minimizing $\Psi(u)$ over $R(\Lambda)$) yields a T-periodic solution $\overline{x}$ of $\dot{x} = JH'(x)$:

(30) $$\Psi(\overline{x}) \leq \Psi(x) \quad \forall x \in \operatorname{dom} \Lambda \, .$$

We would like to know what the actual period of $\overline{x}$ is. It is T-periodic, to be sure, but does it have a lower period: $T/2$, perhaps $T/10^6$? In other words, it is just the k-th iterate of some T/k-periodic solution ? If such is not the case, that is if $\overline{x}$ is never T/k-periodic for any integer $k \geq 2$, we shall say that $\overline{x}$ has minimal period T.

To simplify matters, we shall assume that H is $(0, b_\infty)$-subquadratic, that is:

(31) $$\begin{cases} H : \mathbb{R}^{2n} \to \mathbb{R} & \text{is convex ,} \\ H(x) \|x\|^{-1} \to +\infty & \text{when} \quad \|x\| \to \infty \\ H(x) \leq \frac{1}{2} b_\infty \|x\|^2 + c & \end{cases}$$

and that it attains its minimum at the origin

$$\text{(32)} \qquad \begin{cases} H(x) > H(0) = 0 & \forall x \neq 0 \\ \liminf\limits_{\|x\| \to 0} 2H(x)\, \|x\|^{-2} := \delta > 0\,. \end{cases}$$

Note that H^* also attains its minimum at the origin, and that this minimum is zero:

$$\text{(33)} \qquad H^*(0) = \operatorname*{Max}_x \{(0, x) - H(x)\} = 0$$

$$\text{(34)} \qquad 0 \in \partial H(0) \Leftrightarrow 0 \in \partial H^*(0)$$

by the Legendre reciprocity formula.

The dual action functional then is:

$$\text{(35)} \qquad \Psi(x) := \int_o^T \left[\frac{1}{2}\left(J\frac{dx}{dt}, x\right) + H^*\left(-J\frac{dx}{dt}\right)\right] dt$$

to be minimized on the space:

$$\text{(36)} \qquad W^{1,2}\left(\mathbb{R}/T\mathbb{Z}; \mathbb{R}^{2n}\right) = \left\{x | \dot{x} \in L^2\left(0, T; \mathbb{R}^{2n}\right), x(0) = x(T)\right\}\,.$$

The operator $\Lambda = J\frac{d}{dt}$ has a kernel, namely the constants $\xi \in \mathbb{R}^{2n}$. Any minimizer of Ψ is of the form $\overline{x}(t) + \xi$, with $\xi \in \mathbb{R}^{2n}$ and $\overline{x}$ a T-periodic solution of

$$\text{(37)} \qquad \dot{x} = J\partial H(x)\,.$$

Proposition 4. *Assume* (31) *and* (32), *together with*

$$\text{(38)} \qquad 2\pi\delta^{-1} < T < 2\pi b_\infty^{-1}\,.$$

Then $\overline{x}$ has minimal period T. □

Proof. The only equilibrium is the origin 0. By Corollary 2, we have:

$$\text{(39)} \qquad \Psi(\overline{x}) < \Psi(0)$$

so that $\overline{x} \notin \mathbb{R}^{2n}$: the solution $\overline{x}$ is non-constant. We have to show that it has minimal period T.

Assume otherwise, so that the minimal period is T/k for some $k \geq 2$. Define a new function x_k by

$$\text{(40)} \qquad x_k(t) := k\overline{x}(t/k)\,.$$

Then x_k is still T-periodic. We may therefore substitute into Ψ, yielding thereby:

$$\begin{aligned}\Psi(x_k) &= \int_o^T \left[\frac{1}{2}(J\dot{x}_k, x_k) + H^*(-J\dot{x}_k)\right] dt \\ &= \int_o^T \left[\frac{k}{2}\left(J\dot{x}\left(\frac{t}{k}\right), x\left(\frac{t}{k}\right)\right) + H^*\left(-J\dot{x}\left(\frac{t}{k}\right)\right)\right] dt \\ &= k\int_o^{T/k} \left[\frac{k}{2}\left(J\frac{d\overline{x}}{dt}, \overline{x}\right) + H^*\left(-J\frac{d\overline{x}}{dt}\right)\right] ds\,.\end{aligned} \tag{41}$$

Since x is T/k periodic, the right-hand side satisfies:

$$\begin{aligned}\Psi(x_k) &= \int_o^T \left[\frac{k}{2}\left(J\frac{d\overline{x}}{dt}, \overline{x}\right) + H^*\left(-J\frac{d\overline{x}}{dt}\right)\right] dt \\ &= (k-1)\int_o^T \frac{1}{2}\left(J\frac{d\overline{x}}{dt}, \overline{x}\right) dt + \Psi(\overline{x})\,.\end{aligned} \tag{42}$$

But we know that:

$$0 > \Psi(\overline{x}) = \int_o^T \left[\frac{1}{2}\left(J\frac{d\overline{x}}{dt}, \overline{x}\right) + H^*\left(-J\frac{d\overline{x}}{dt}\right)\right] dt \tag{43}$$

and since $H^* \geq 0$, we must have:

$$\int_o^T \frac{1}{2}\left(J\frac{d\overline{x}}{dt}, \overline{x}\right) dt < 0\,. \tag{44}$$

Writing this into formula (42) we get

$$\Psi(x_k) < \Psi(\overline{x}) \tag{45}$$

thereby contradicting the fact that $\overline{x}$ is a minimizer. □

This argument uses the fact that $\overline{x}$ is a global minimizer. One can do better by appealing to index theory. In fact:

Lemma 5. *Assume that H is C^2 on $\mathbb{R}^{2n}\backslash\{0\}$ and that $H''(x)$ is non-degenerate for any $x \neq 0$. Then any local minimizer $\widetilde{x}$ of Ψ has minimal period T.*

Proof. We know that $\widetilde{x}$, or $\widetilde{x} + \xi$ for some constant $\xi \in \mathbb{R}^{2n}$, is a T-periodic solution of the Hamiltonian system:

$$\dot{x} = JH'(x)\,. \tag{46}$$

There is no loss of generality in taking $\xi = 0$. So $\Psi(x) \geq \Psi(\widetilde{x})$ for all $\widetilde{x}$ in some neighbourhood of x in $W^{1,2}\left(\mathbb{R}/T\mathbb{Z};\mathbb{R}^{2n}\right)$.

Clearly $C^1 \subset W^{1,2}$ with a stronger topology. So the restriction of Ψ to $C^1\left(\mathbb{R}/T\mathbb{Z};\mathbb{R}^{2n}\right)$, which we denote by $\widetilde{\Psi}$, has $\widetilde{x}$ as a local minimizer.

On the other hand, $\widetilde{\Psi}$ is clearly C^2 on a neighbourhood of $\widetilde{x}$, and

$$\left(\widetilde{\Psi}''(\widetilde{x})y, y\right) = \int_o^T \left[(J\dot{y}, y) + \left((H^*)''\left(-J\frac{d\widetilde{x}}{dt}\right)J\dot{y}, J\dot{y}\right)\right] dt \tag{47}$$

for $y \in C^1$. Since $\widetilde{x}$ is a local minimizer, this quadratic form is semi-positive definite.

We have, by Proposition II.2.10, remembering that $-J\frac{d\widetilde{x}}{dt} = H'(\widetilde{x})$,

$$(H^*)''\left(-J\frac{d\widetilde{x}}{dt}(t)\right) = H''\left(\widetilde{x}(t)\right)^{-1} . \tag{48}$$

Both sides depend continuously on t. Therefore, C^1 being dense in $W^{1,2}$, the quadratic form of formula (47) extends to a quadratic form on $W^{1,2}$ given by:

$$\int_o^T \left[(J\dot{y}, y) + \left(H''\left(\widetilde{x}(t)\right)^{-1} J\dot{y}, J\dot{y}\right)\right] dt \tag{49}$$

which is also semi-positive definite. In other words, its index is zero.

But this index is precisely the index $i_T(\widetilde{x})$ of the T-periodic solution $\widetilde{x}$ over the interval $(0, T)$, as defined in Sect. I.6. So

$$i_T(\widetilde{x}) = 0 . \tag{50}$$

Now if $\widetilde{x}$ has a lower period, T/k say, we would have, by Corollary I.6.5:

$$i_T(\widetilde{x}) = i_{kT/k}(\widetilde{x}) \geq k i_{T/k}(\widetilde{x}) + k - 1 \geq k - 1 \geq 1 . \tag{51}$$

This would contradict (50), and thus cannot happen. □

For future reference, we spell out on easy consequence of Proposition 4.

Corollary 6. *Let $H : \mathbb{R}^{2n} \to \mathbb{R}$ be a convex function such that*

$$H(x) \geq H(0) = 0 \tag{52}$$

$$H(x) \|x\|^{-2} \to \infty \quad \text{when} \quad \|x\| \to 0 \tag{53}$$

$$H(x) \|x\|^{-2} \to 0 \quad \text{when} \quad \|x\| \to \infty . \tag{54}$$

Then, for every $T > 0$, the system

$$\dot{x} \in JH(x) \tag{55}$$

$$x(0) = x(T) \tag{56}$$

has a solution with minimal period T. □

Notes and Comments. The results in this section are a refined version of [ClaE3]; the minimality result of Proposition 4 was the first of its kind.

To understand the nontriviality conditions, such as the one in formula (20), one may think of a one-parameter family x_T, $T \in \left(2\pi\omega^{-1}, 2\pi b_\infty^{-1}\right)$ of periodic solutions, $x_T(0) = x_T(T)$, with x_T going to 0 when $T \to 2\pi\omega^{-1}$, which is the smallest period of the linearized system at 0.

Unfortunately, although this picture is correct in the homogeneous case, $H(x) = \|x\|^\alpha$ with $1 < \alpha < 2$, I have been unable to support it by a priori estimates in the general case. As far as we know, the x_T could be bounded away from 0 and infinity when T ranges over $\left(2\pi\omega^{-1}, 2\pi b_\infty^{-1}\right)$. It would be interesting to settle this question.

It would also be interesting to know what happens for $T > 2\pi b_\infty^{-1}$. The minimization method fails, but one expects that, for most if not all systems, there will be periodic solutions for arbitrarily large minimal periods.

4. Nonautonomous Systems

Consider a nonautonomous Hamiltonian $H(t,x)$, T-periodic in t:

$$H(t+T,x) = H(t,x) \quad \forall (t,x) \in \mathbb{R} \times \mathbb{R}^{2n} \tag{1}$$

and the associated system:

$$\dot{x} = JH'(t,x)\ . \tag{2}$$

Such systems have a natural period, namely T. Except in special cases (an example of which will be seen presently), there are no constant solutions, so we may apply the existence results of Sect. 2 without worrying about showing that the periodic solutions we obtain are non-trivial. The question of minimality is also much simplified: it will generally be the case that, if T is the minimal period of the Hamiltonian H, it will also be the minimal period of any T-periodic solution.

There remains, however, the possibility of subharmonics. Take an integer k, and let us seek a kT-periodic solution x_k of the Hamiltonian system:

$$\begin{cases} \dot{x}_k = JH'\left(t, x_k(t)\right) \\ x_k(t+kT) = x_k(t)\ . \end{cases} \tag{3}$$

Clearly, if x_1 exists, it is T-periodic and hence kT-periodic for all integer k, so that $x_k = x_1$ solves problem (3). Any $x_k \neq x_1$ for $k \neq 1$ is called a *subharmonic.*

The question then is whether we can find for each $k \geq 2$ a subharmonic x_k, such that the x_k are pairwise distinct. No such result is known in the subquadratic case.

Failing that, one may ask whether there exists an infinite sequence x_k, $k \in A \subset \mathbb{N}$, of pairwise distinct subharmonics. An easy answer is provided by taking $A \subset P$, where P denotes the set of all primes. To prove that the x_k, $k \in A$, are pairwise distinct, it will then be enough to prove that $x_k \neq x_1$ for all $k \in A$.

We give a stronger result, which will imply that non-trivial subharmonics x_k exist also for non-prime k.

Proposition 1. *Assume $H(t,x)$ is $(0,\epsilon)$-subquadratic at infinity for all $\epsilon > 0$, and T-periodic in t*

$$H(t,\cdot) \quad \text{is convex } \forall t \tag{4}$$

$$H(\cdot,x) \quad \text{is } T\text{−periodic } \forall x \tag{5}$$

$$H(t,x) \geq n(\|x\|) \quad \text{with } n(s)s^{-1} \to \infty \text{ as } s \to \infty \tag{6}$$

$$\forall \epsilon > 0\,, \quad \exists c\,:\, H(t,x) \leq \frac{\epsilon}{2}\|x\|^2 + c\,. \tag{7}$$

Assume also that H is C^2, and $H''(t,x)$ is positive definite everywhere. Then there is a sequence x_k, $k \in \mathbb{N}$, of kT-periodic solutions of the system

$$\dot{x} = JH'(t,x) \tag{8}$$

such that, for every $k \in \mathbb{N}$, there is some $p_o \in \mathbb{N}$ with:

$$p \geq p_o \Rightarrow x_{pk} \neq x_k\,. \tag{9}$$

□

Proof. For each integer k, we consider the functional Ψ_k on $W^{1,2}\left(\mathbb{R}/T\mathbb{Z};\mathbb{R}^{2n}\right)$ defined by:

$$\Psi_k(x) := \int_o^{kT}\left[\frac{1}{2}(J\dot{x},x) + H^*(t,-J\dot{x})\right]dt\,. \tag{10}$$

By the results of Sect. 2, the minimization of Ψ_k is possible for every k, and yields a kT-periodic solution x_k. Arguing as in Proposition 3.4, we see that x_k has index 0 on the interval $(0,kT)$.

If $x_{pk} = x_k$, then x_k still has index 0 on the interval $(0,pkT)$, since it minimizes Ψ_{pk}:

$$i_{pkT}(x_k) = 0\,. \tag{11}$$

On the other hand, by Proposition I.5.20, we have:

$$\lim_{p\to\infty} p^{-1} i_{pkT}(x_k) > 0\,. \tag{12}$$

This shows that equality (11) ceases to hold when p is large enough. Hence the result. □

Let us illustrate all this in two simple situations.

Example 1 (External forcing). Consider the system:

$$\dot{x} = JH'(x) + f(t) \tag{13}$$

where the Hamiltonian H is $(0,b_\infty)$-subquadratic, and the forcing term is a distribution on the circle:

(14) $$f = \frac{d}{dt}F + f_o \quad \text{with } F \in L^2\left(\mathbb{R}/T\mathbb{Z}; \mathbb{R}^{2n}\right) ,$$

where $f_o := T^{-1}\int_o^T f(t)dt$. For instance,

(15) $$f(t) = \sum_{k\in\mathbb{Z}} \delta_k \xi ,$$

where δ_k is the Dirac mass at $t = k$ and $\xi \in \mathbb{R}^{2n}$ is a constant, fits the prescription. This means that the system $\dot{x} = JH'(x)$ is being excited by a series of identical shocks at interval T.

Formally, the Hamiltonian corresponding to Eq. (13) is:

(16) $$H(t,x) := H(x) - (Jf(t,x), x) .$$

Hence the dual action functional:

(17) $$\Psi(x) = \int_o^T \left[\frac{1}{2}(J\dot{x}, x) + H^*\left(-J\dot{x} + Jf(t)\right)\right] dt .$$

Change variables by setting $x = F(t) + y$. We get $\Psi(x) = \widetilde{\Psi}(y)$ with:

(18) $$\widetilde{\Psi}(y) = \int_o^T \left[\frac{1}{2}(J\dot{y} + J(f - f_o), y + F) + H^*\left(-J\dot{y} + Jf_o\right)\right] dt .$$

Throwing away the term $\int (J(f - f_o), F)$, which is constant and therefore plays no role in the minimization, and integrating by parts the term $\int (J(f - f_o), y)$, we are led to the following lemma, which we leave to the reader:

Lemma 2. *Consider the functional* $\overline{\Psi}$ *on* $W^{1,2}\left(\mathbb{R}/T\mathbb{Z}; \mathbb{R}^{2n}\right)$ *defined by:*

(19) $$\overline{\Psi}(y) = \int_o^T \left[\frac{1}{2}(J\dot{y}, y + 2F) + H^*\left(-J\dot{y} + Jf_o\right)\right] dt .$$

If y *is a critical point of* $\overline{\Psi}$, *then there is a constant* $\xi \in \mathbb{R}^{2n}$ *such that* $x = y + F(t) + \xi$ *is a* T*-periodic solution of system* (13). □

As a corollary, we see that if H is $(0, b_\infty)$-subquadratic at infinity, with

(20) $$Tb_\infty < 2\pi$$

then system (13) has a T-periodic solution (see Proposition 2.2).

Example 2 (Parametric excitation). Consider the system

(21) $$\dot{x} = Ja(t)H(x)$$

where the Hamiltonian H is $(0, \epsilon)$-subquadratic at infinity for all $\epsilon > 0$, and where $a : \mathbb{R} \to \mathbb{R}$ is a continuous T-periodic function:

$$0 < \alpha \leq a(t) \leq \beta \tag{22}$$

$$a(t+T) = a(t) \ . \tag{23}$$

Assume moreover that H is C^2 and has its unique equilibrium at the origin:

$$H(x) > H(0) = 0 \quad \text{for } x \neq 0 \tag{24}$$

$$H''(x) \quad \text{positive definite } \forall x \tag{25}$$

and denote by ω the lowest eigenvalue of $H''(0)$:

$$H''(0) \geq \omega I \ . \tag{26}$$

So $x = 0$ is a solution of the autonomous problem $\dot{x} = JH'(x)$. If $T < \frac{2\pi}{\omega}$, it is to be expected that it is the only one.

Clearly, the nonautonomous system (21) also has the trivial solution $x = 0$. We now look for subharmonics:

$$\begin{cases} \dot{x}_k = Ja(t)H'(x_k) \\ x_k(t+kT) = x_k(t) \ . \end{cases} \tag{27}$$

We find x_k by minimizing, for each $k \in \mathbb{N}$, the dual action functional Ψ_{kT}. Proposition 1 then applies and tells us, for every $k \in \mathbb{N}$, there is some p_o with $x_{pk} \neq x_k$ for all $p \geq p_o$. In particular, there will be some k_o such that $x_k \neq 0$ for all $k \geq k_o$.

Again, if H is subquadratic, the analysis in the preceding sections can be carried over, with suitable changes in the numerical constants. Here is a typical result.

Proposition 2. *Assume that:*

$$H''(x) \quad \text{is positive definite for all } x \in \mathbb{R}^{2n} \ . \tag{28}$$

$$H(x) \geq n(\|x\|) \quad \text{with } n(s)s^{-1} \to \infty \text{ as } s \to \infty \ . \tag{29}$$

$$H(x) \leq m(\|x\|) \quad \text{with } m(s)s^{-2} \to 0 \text{ as } s \to \infty \ . \tag{30}$$

Then problem (21), with $a(t)$ T-periodic and bounded away from zero and $+\infty$, has an infinite sequence of pairwise distinct T-periodic solutions, the minimal period of which goes to infinity. □

Notes and Comments. The first results on subharmonics were obtained by Rabinowitz in [Rab3], who showed the existence of infinitely many subharmonics both in the subquadratic and superquadratic case, with suitable growth conditions on H'. Again the duality approach enabled Clarke and Ekeland in [ClaE3] to treat the same problem in the convex-subquadratic case, with growth conditions on H only.

Recently, Michalek and Tarantello (see [MicT] and [Tar]) have obtained lower bound on the number of subharmonics of period kT, based on symmetry considerations and on pinching estimates, as in Sect. 5.2 of this book.

5. Other Problems

As we mentioned earlier, the duality method will apply to other situations. We mentioned two specific examples in Sect. II.4, and we pick them up again.

A. Second-Order Systems

We consider the boundary-value problem:

$$\begin{cases} \ddot{q} + V'(t,q) = 0 \\ q(0) = q(T)\ , \quad \dot{q}(0) = \dot{q}(T)\ . \end{cases} \tag{1}$$

According to Example 2 of Sect. II.4, the direct functional in the Lagrangian formalism is:

$$\Phi(q) = \int_0^T \left[-\frac{1}{2}\dot{q}^2 + V(t,q) \right] dt \tag{2}$$

while the dual is:

$$\Phi(q) = \int_o^T \left[-\frac{1}{2}\dot{q}^2 + V^*(t,\ddot{q}) \right] dt\ . \tag{3}$$

All the analysis in the preceding sections can be carried over to this case, with different values for the numerical constants, leading to different bounds for T. To be precise, we have, by Wirtinger' inequality:

$$\int_o^T \dot{q}^2 dt \leq \frac{T^2}{4\pi^2} \int_o^T \ddot{q}^2 dt \tag{4}$$

which will replace the inequality:

$$\int_o^T (J\dot{x}, x)\, dt \leq \frac{T}{2\pi} \int_o^T \dot{x}^2 dt\ . \tag{5}$$

Here is a typical result.

Proposition 1. *Consider the autonomous system:*

$$\ddot{x} + \partial V(x) \ni 0 \tag{6}$$

where $V : \mathbb{R}^n \to \mathbb{R}$ is convex and satisfies:

$$V(x) > V(0) = 0 \quad \forall x \neq 0 \tag{7}$$

(8) $$V(x) \geq v(\|x\|) \quad \text{with } v(s)s^{-1} \to \infty \text{ as } s \to \infty$$

(9) $$\exists k > 0\ ,\ \exists c\ :\ V(x) \geq \frac{k}{2}\|x\|^2 + c \quad \forall x\ .$$

Assume that:

(10) $$K := \liminf_{x \to 0} V(x)\,\|x\|^{-2} > k\ .$$

Then, for all T such that

(11) $$2\pi K^{-1/2} < T < 2\pi k^{-1/2}$$

problem (7) has a non-trivial T-periodic solution. □

B. Bolza Problem

We consider the boundary-value problem of Example II.4.1:

(12) $$\begin{cases} \dot{q} = H'_p(q,p) \\ \dot{p} = -H'_q(q,p) \\ q(0) = q_o\ , \quad q(T) = q_1\ . \end{cases}$$

After a suitable change of variables, this problem can be rewritten as:

$$\begin{cases} \dot{q} = H'_p\left(q + \frac{t}{T}(q_1 - q_o), p\right) \\ \dot{p} = -H'_q\left(q + \frac{t}{T}(q_1 - q_0), p\right) \\ q(0) = 0\ , \quad q(T) = 0\ . \end{cases}$$

We gave the direct functional as:

(13) $$\Phi(p,q) := \int_o^T \left[-p\left(\dot{q} + \frac{q_1 - q_o}{T}\right) + H\left(p, q + \frac{t}{T}(q_1 - q_o)\right)\right] dt$$

and the dual as:

(14) $$\Psi(p,q) := \int_o^T \left[-p\dot{q} + \dot{p}\overline{q} + H^*\left(\dot{q} + \frac{d\overline{q}}{dt}, -\dot{p}\right)\right] dt\ .$$

It should be noted that

(15) $$\Psi(p + \eta, q) = \Psi(p,q) \quad \forall \eta \in \mathbb{R}^{2n}$$

so that we can replace p by $\overline{p}$:

(16) $$\overline{p}(t) := p(t) - \frac{1}{T}\int_o^T p\,dt$$

thereby enjoying Wirtinger's inequality

$$\|\overline{p}\|_{L^2} \leq \frac{T}{2\pi}\left\|\frac{d\overline{p}}{dt}\right\|_{L^2} . \tag{17}$$

If H is subquadratic, the analysis in the preceding sections can be carried on. Here is a typical result:

Proposition 2. *Assume H is $(0, b_\infty)$-subquadratic at infinity, and*

$$T < 2\pi b_\infty^{-1} \tag{18}$$

Then problem (12) *has a solution.* □

Notes and Comments. These results were proved in [ClE$_3$] and [ClE$_4$].

Chapter IV. Fixed-Period Problems: The Superlinear Case

Introduction

We now consider the case when the Hamiltonian H is superquadratic, that is, behaves like $\|x\|^\alpha$ for some $\alpha > 2$. The dual action ψ can no longer be minimized (it is easy to check that $\inf \psi = -\infty$), so a subtler procedure is required. It will be described in the opening section.

1. Mountain-Pass Points

The aim of this section is to find critical points of functionals in some simple situations where minimization procedures are not available. The proof will make repeated use of the following variational principle of Ekeland:

Theorem 1. *Let (X, d) be a complete metric space and $\Phi : X \to \mathbb{R} \cup \{+\infty\}$ a lower semi-continuous function which is bounded from below:*

(1) $$\inf \Phi > -\infty \ .$$

Let $z \in X$ and $\epsilon \geq 0$ be such that:

(2) $$\Phi(z) \leq \inf \Phi + \epsilon \ .$$

Then there is some point $y \in X$ such that:

(3) $$\Phi(y) + \epsilon d(y, z) \leq \Phi(z)$$

(4) $$\forall x \in X \ , \quad \Phi(x) \leq \Phi(y) - \epsilon d(x, y) \ .$$ □

As an immediate consequence of inequality (3), we have:

(5) $$d(y, z) \leq 1$$

$$\Phi(y) \leq \inf \Phi + \epsilon \ . \tag{6}$$

Note also that we can define a new distance on X by taking $\delta(x_1, x_2) = \lambda^{-1} d(x_1, x_2)$ for some fixed $\lambda > 0$. Then (X, δ) is still a complete metric space, with the same property as (X, d). Applying Theorem 1 to (X, δ) gives us a point $y_\lambda \in X$ such that:

$$d(y_\lambda, z) \leq \lambda \tag{7}$$

$$\forall x \in X \ , \quad \Phi(x) \geq \Phi(y_\lambda) - \frac{\epsilon}{\lambda} d(x, y_\lambda) \ . \tag{8}$$

For instance, if $\epsilon > 0$, we may take $\lambda = \sqrt{\epsilon}$. Let us rephrase the result in this case:

Corollary 2. *Let (X, d) be a complete metric space, and $\Phi : X \to \mathbb{R} \cup \{+\infty\}$ a l.s.c. function on X with* $\inf F > -\infty$. *Let $\epsilon > 0$ be given, and choose $z \in X$ such that:*

$$\Phi(z) \leq \inf \Phi + \epsilon \ . \tag{9}$$

Then there is a point $y \in X$ such that:

$$\Phi(y) \leq \Phi(z) \tag{10}$$

$$d(z, y) \leq \sqrt{\epsilon} \tag{11}$$

$$\forall x \in X \ , \quad \Phi(x) \geq \Phi(y) - \sqrt{\epsilon}\, d(x, y) \ . \tag{12}$$

□

We will be interested in the particular case when X is a Banach space and $\Phi : X \to \mathbb{R}$ is Gâteaux-differentiable. We immediately derive from Theorem 1 the existence of minimizing sequences of a particular type.

Corollary 3. *Assume X is a Banach space and $\Phi : X \to \mathbb{R}$ is Gâteaux-differentiable, l.s.c. and bounded from below. Then there exists a sequence x_n such that:*

$$\Phi(x_n) \to \inf \Phi \tag{13}$$

$$\|\Phi'(x_n)\|_* \to 0 \ . \tag{14}$$

□

Proof. Apply Corollary 2 with $\epsilon = 1/n^2$. We get a point $y = x_n$ such that

$$\Phi(x_n) \leq \inf \Phi + \frac{1}{n^2} \tag{15}$$

$$\forall x \in X \ , \quad \Phi(x) \geq \Phi(x_n) - \frac{1}{n} \|x - x_n\| \ . \tag{16}$$

Condition (13) follows from inequality (15).

For condition (14) take any $u \in X$, and substitute $x = x_n + tu$, $t > 0$, into inequality (16). We get:

$$\frac{1}{t}\left[\Phi\left(x_n + tu\right) - \Phi(x_n)\right] \geq -\frac{1}{n}\|u\| \ . \tag{17}$$

Now let $t \to 0$ in the left-hand side. Since Φ is Gâteaux-differentiable, it becomes:

$$\langle \Phi'(x_n), u\rangle \geq -\frac{1}{n}\|u\| \ . \tag{18}$$

Changing u to $-u$ gives the reverse inequality:

$$|\langle \Phi'(x_n), u\rangle| \leq \frac{1}{n}\|u\| \ . \tag{19}$$

Since $u \in X$ is arbitrary, this means that:

$$\|\Phi'(x_n)\|_* \leq \frac{1}{n} \ . \tag{20}$$

□

We can do slightly better, by changing the metric on X. We shall take the geodesic distance associated with the (formal) Riemannian structure $\|x\|^{-2} dx^2$. To be precise, define the *geodesic length* $\ell(c)$ of a curve $c \in C^1([0,1]; X)$ to be:

$$\ell(c) := \int_o^1 \frac{\|\dot{c}(t)\|}{1 + \|c(t)\|}\, dt \ . \tag{21}$$

Then define the geodesic distance δ between two points x_1 and x_2 in X to be

$$\delta(x_1, x_2) := \inf\left\{\ell(c) | c \in C^1, c(0) = x_1, c(1) = x_2\right\} \ . \tag{22}$$

Clearly $\delta(x_1, x_2) \leq \|x_1 - x_2\|$. Conversely, for every norm-bounded set B in X, there is some constant $\gamma > 0$ such that $\delta(x_1, x_2) \geq \gamma \|x_1 - x_2\|$ whenever x_1 and x_2 belong to B.

By radial symmetry, we find that, when $x_1 = 0$, the infimum on the right-hand side of formula (22) must be achieved by the line segment from 0 to $x_2 = x$. Hence:

$$\delta(0, x) = \int_o^1 \frac{\|x\|}{1 + t\|x\|}\, dt = \operatorname{Log}(1 + \|x\|) \ . \tag{23}$$

It follows that δ-bounded and norm-bounded subsets of X are the same. Any Cauchy sequence for δ is δ-bounded, and hence norm-bounded. By the preceding considerations it is also a Cauchy sequence for the norm, and hence norm- and δ-convergent. So (X, δ) is a complete metric space. Applying Theorem 1, we get:

Corollary 4. *Assume X is a Banach space and $\Phi : X \to \mathbb{R}$ is Gâteaux-differentiable, l.s.c. and bounded from below. Then there exists a sequence x_n such that:*

$$\Phi(x_n) \to \inf \Phi \tag{24}$$

$$(1 + \|x_n\|) \|\Phi'(x_n)\|_* \to 0 \; . \tag{25}$$ □

Proof. Apply Corollary 2 to (X, δ) with $\epsilon = \frac{1}{n^2}$. We get a sequence x_n such that

$$\Phi(x_n) \leq \inf \Phi + \frac{1}{n^2} \tag{26}$$

$$\forall x \; , \quad \Phi(x) \geq \Phi(x_n) - \frac{1}{n}\delta(x, x_n) \; . \tag{27}$$

Write $x = x_n + tu$, for $t > 0$ and $u \in X$:

$$\Phi(x_n + tu) - \Phi(x_n) \geq -\frac{1}{n}\delta(x_n, x_n + tu) \; . \tag{28}$$

Devide both sides by t, and recall the definition of the geodesic distance δ:

$$\frac{1}{t}\left(\Phi(x_n + tu) - \Phi(x_n)\right) \geq -\frac{1}{n}\|u\| \frac{1}{t}\int_o^t \frac{ds}{(1 + \|x_n + su\|)} \tag{29}$$

Letting $t \to 0$, we get:

$$\langle \Phi'(x_n), u \rangle \geq -\frac{1}{n}\left(1 + \|x_n\|\right)^{-1} \|u\| \; . \tag{30}$$

Changing u to $-u$, we get the reverse inequality, as in Corollary 3, and hence:

$$\|\Phi'(x_n)\|_* \leq \frac{1}{n}\left(1 + \|x_n\|\right)^{-1} \tag{31}$$

from which the result follows immediately. □

Of course, the existence of sequences x_n satisfying (14) or (25) relies heavily on the fact that Φ is bounded from below. We now turn to another situation where such sequence exist:

Definition 5. A closed subset F *separates* two points z_o and z_1 in X if z_o and z_1 belong to disjoint connected components in $X \setminus F$. □

Since $X \setminus F$ is locally connected, any component will be an open and closed subset of $X \setminus F$, and hence open in X. Denoting by Ω_o the component containing z_o, and by Ω_1 the union of all components not containing z_o, we get a partition of $X \setminus F$ into two disjoint open sets, with $z_i \in \Omega_i$, $i = 0, 1$.

Theorem 6 (Ghoussoub-Preiss). *Let X be a Banach space and $\Phi : X \to \mathbb{R}$ a continuous, Gâteaux-differentiable function, such that $\Phi' : X \to X$ is continuous from the norm topology of X to the weak* topology of X^*. Take two points (z_o, z_1) in X and consider the set Γ of all continuous paths from z_o to z_1:*

$$\Gamma := \{c \in C^o([0,1]; X) \,|c(0) = z_o, c(1) = z_1\} \ . \tag{32}$$

Define a number γ by:

$$\gamma := \inf_{c \in \Gamma} \max_{0 \leq t \leq 1} \Phi(c(t)) \ . \tag{33}$$

Assume there is a closed subset F of X such that:

$$F \cap \Phi_\gamma \qquad \text{separates} \quad z_o \quad \text{and} \quad z_1 \tag{34}$$

with $\Phi_\gamma := \{x \in X \,|\, \Phi(x) \geq \gamma\}$. Then there is a sequence x_n in X such that:

$$\delta(x_n, F) \to 0 \tag{35}$$

$$\Phi(x_n) \to \gamma \tag{36}$$

$$(1 + \|x_n\|) \, \|F'(x_n)\|_* \to 0 \ . \tag{37}$$

□

Proof. Write $F_\gamma := \Phi_\gamma \cap F$. Since F_γ is a closed subset separating z_o and z_1, we may partition $X \setminus F_\gamma$ into two open subsets Ω_o and Ω_1, with $z_i \in \Omega_i$ for $i = 0, 1$.

Choose ϵ such that:

$$0 < \epsilon < \frac{1}{2}\text{Min } (1, \delta(z_o, F_\gamma), \delta(z_1, F_\gamma)\} \tag{38}$$

and find a path $c \in \Gamma$ such that:

$$\underset{0 \leq t \leq 1}{\text{Max}} \ \Phi(c(t)) < \gamma + \frac{\epsilon^2}{4} \ . \tag{39}$$

Such a path c always exists because of the definition of γ. We then define two numbers t_o and t_1, with $0 \leq t_o < t_1 < 1$, by:

$$t_o = \sup \{t \in [0,1] | c(t) \in \Omega_o, \delta\,(c(t), F_\gamma) \geq \epsilon\} \tag{40}$$

$$t_1 = \inf\,(t \in [t_o, 1] | c(t) \in \Omega_1, \delta\,(c(t), F_\gamma) \geq \epsilon\} \ . \tag{41}$$

Note that $\delta\,(c(t), F_\gamma) \leq \epsilon$ for $t_o \leq t \leq t_1$.

Consider the space:

$$\Gamma(t_o, t_1) := \{f \in C^o([t_o, t_1]; X) \,| f(t_o) = c(t_o), f(t_1) = c(t_1)\} \tag{42}$$

and endow it with the uniform distance

(43) $$d(f_1, f_2) = \operatorname*{Max}_{t_o \le t \le t_1} \delta\left(f_1(t), f_2(t)\right) .$$

Set:

(44) $$\Psi(x) := \operatorname{Max}\left\{0, \epsilon^2 - \epsilon\delta(x, F_\gamma)\right\}$$

and define a function $\varphi : \Gamma(t_o, t_1) \to \mathbb{R}$ by:

(45) $$\varphi(f) := \operatorname*{Max}_{t_o \le t \le t_1} \left(\Phi(f(t)) + \Psi(f(t))\right)$$

Since $f(t_o) = c(t_o) \in \Omega_o$ and $f(t_1) = c(t_1) \in \Omega_1$, there must be some point $t_f \in (t_o, t_1)$ where $f(t_f) \in \partial\Omega_o \subset F_\gamma$, and hence:

(46) $$\varphi(f) \ge \Phi\left(f(t_f)\right) + \Psi\left(f(t_f)\right) \ge \gamma + \epsilon^2$$

for every $f \in \Gamma(t_o, t_1)$. On the other hand, consider the restriction $\widehat{c}$ of the path c to $[t_0, t_1]$. We have:

(47) $$\varphi(\widehat{c}) \le \operatorname*{Max}_{0 \le t \le 1} \left\{\Phi\left(c(t)\right) + \Psi\left(x(t)\right)\right\} \le \left(\gamma + \frac{\epsilon^2}{4}\right) + \epsilon^2 .$$

$\Gamma(t_o, t_1)$ is a complete metric space, and the function φ is l.s.c. and bounded from below. We have found a point $\widehat{c}$ where $\varphi(\widehat{c}) \le \operatorname{Inf}\varphi + \epsilon^2/4$. We may therefore apply Corollary 2 to find a path $\widehat{f} \in \Gamma(t_o, t_1)$ such that:

(48) $$\varphi\left(\widehat{f}\right) \le \varphi(\widehat{c}) , \quad d\left(\widehat{f}, \widehat{c}\right) \le \epsilon/2$$

(49) $$\forall f \in \Gamma(t_o, t_1) , \quad \varphi(f) \ge \varphi\left(\widehat{f}\right) - \frac{\epsilon}{2} d\left(f, \widehat{f}\right) .$$

Now introduce the set M consisting of all points $t \in [t_o, t_1]$ where $(\Phi + \Psi) \circ \widehat{f}$ attains its maximum:

(50) $$M := \left\{t \in [t_o, t_1] \,|\, \Phi\left(f(t)\right) + \Psi\left(f(t)\right) = \varphi(f)\right\} .$$

It is compact, non-empty, and it contains neither t_o nor t_1. Indeed, from the definition of t_o and t_1 it follows that $\delta\left(c(t_i), \widehat{F}_\gamma\right) = \epsilon$ for $i = 1, 2$, so that $\psi\left(c(t_i)\right) = 0$ and hence:

(51) $$\Phi\left(\widehat{f}(t_i)\right) + \Psi\left(\widehat{f}(t_i)\right) \le \Phi\left(c(t_i)\right) + \Psi\left(c(t_i)\right) \le \gamma + \frac{\epsilon^2}{4}$$

which is strictly less than $\varphi\left(\widehat{f}\right)$ by inequality (46).

I claim that M contains a point t where:

(52) $$\left\|\Phi'\left(\widehat{f}(t)\right)\right\|_* \left(1 + \left\|\widehat{f}(t)\right\|\right) \le \frac{3\epsilon}{2} .$$

Assume otherwise, that is:

$$\forall t \in M\,, \quad \left\|\Phi'\left(\widehat{f}(t)\right)\right\|_* \left(1+\left\|\widehat{f}(t)\right\|\right) > \frac{3\epsilon}{2}\,. \tag{53}$$

It follows that for every $t \in M$ there will be a vector $u(t) \in X$ such that

$$\|u(t)\| = \left(1+\left\|\widehat{f}(t)\right\|\right)^{-1} \tag{54}$$

$$\langle \Phi'\left(\widehat{f}(t)\right), u(t)\rangle < -\frac{3\epsilon}{2}\,. \tag{55}$$

By the continuity assumption on F', there will be an open neighbourhood $\mathcal{N}(t)$ of t in $[t_o, t_1]$ such that, for every $s \in \mathcal{N}(t)$, we have:

$$\langle \Phi'\left(\widehat{f}(s)\right), u(t)\rangle < -\frac{3\epsilon}{2}\,. \tag{56}$$

We now take a covering of M by finitely many of the $\mathcal{N}(t)$, say $\mathcal{N}(t_k)$ for $0 \leq k \leq K$, and we associate with this covering a continuous partition of unity ρ_k, $0 \leq k \leq K$. Setting:

$$v(t) := \sum_{k=0}^{K} \rho_k(t)u(t_k) \tag{57}$$

we get a continuous map $v : M \to X$ such that

$$\forall t \in M\,, \quad \langle \Phi'\left(\widehat{f}(t)\right), v(t)\rangle < -\frac{3\epsilon}{2} \tag{58}$$

$$\forall t \in M\,, \quad \|v(t)\| \leq \left(1+\left\|\widehat{f}(t)\right\|\right)^{-1}\,. \tag{59}$$

Since $M \subset [t_o, t_1]$ does not contain the endpoints t_o and t_1, we may extend v to a continuous map on the whole of $[t_o, t_1]$, still denoted by v, and such that:

$$v\left(t_o\right) = 0 = v\left(t_1\right) \tag{60}$$

$$\|v(t)\| \leq \left(1+\left\|\widehat{f}(t)\right\|\right)^{-1} \quad \forall t \in [0,1]\,. \tag{61}$$

Because of the first condition, the path $\widehat{f} + hv$ belongs to Γ for every number $h > 0$, and we may substitute $\widehat{f} + hv$ for f in formula (49). We get:

$$\forall h > 0\,, \quad \varphi\left(\widehat{f} + hv\right) \geq \varphi\left(\widehat{f}\right) - \frac{\epsilon}{2} d\left(f, \widehat{f}\right)\,. \tag{62}$$

Choose $t_h \in [t_o, t_1]$ such that:

$$\varphi\left(\widehat{f} + hv\right) = (\Phi + \Psi)\left(\widehat{f}(t_h) + hv(t_h)\right)\,. \tag{63}$$

We have of course $\varphi\left(\widehat{f}\right) \geq (\Phi+\Psi)\left(\widehat{f}(t_h)\right)$, so that inequality (62) implies the following:

$$(64)\ \forall h>0\,,\quad (\Phi+\Psi)\left(\widehat{f}(t_h)+hv(t_h)\right)\geq(\Phi+\Psi)\left(\widehat{f}(t_h)\right)-\frac{\epsilon}{2}d\left(\widehat{f}+hv,\widehat{f}\right)$$

which we rewrite as:

$$(65)\quad \begin{aligned}\forall h>0\,,\quad &\Phi\left(\widehat{f}(t_h)+hv(t_h)\right)-\Phi\left(\widehat{f}(t_h)\right)\\ &\geq-\Psi\left(\widehat{f}(t_h)+hv(t_h)\right)+\Psi\left(\widehat{f}(t_h)\right)-\frac{\epsilon}{2}d\left(\widehat{f}+hv,\widehat{f}\right)\,.\end{aligned}$$

The function Ψ, defined in formula (44) is clearly ϵ-Lipschitizian: $|\Psi(x)-\Psi(y)|\leq\epsilon\|x-y\|$. So inequality (65) implies the following:

$$(66)\qquad \forall h>0\,,\quad \Phi\left(\widehat{f}(t_h)+hv(t_h)\right)-\Phi\left(\widehat{f}(t_h)\right)\geq-\frac{3\epsilon}{2}d\left(\widehat{f}+hv,\widehat{f}\right)\,.$$

Devide both sides by h, and use the mean-value theorem on the left-hand side:

$$(67)\ \exists\theta_h\in(0,1)\ :\quad \left\langle\Phi'\left(\widehat{f}(t_h)+\theta_h hv(t_h)\right),v(t_h)\right\rangle\geq-\frac{3\epsilon}{2}\frac{1}{h}d\left(\widehat{f}+hv,\widehat{f}\right)\,.$$

Now let $h\to 0$. By extracting a subsequence if need be, we may assume that

$$(68)\qquad t_h\to\tau\in M\,.$$

Using the continuity of F' and of v, and going back to the definition of d and δ, we pass to the limit in inequality (67):

$$(69)\qquad \left\langle\Phi'\left(\widehat{f}(\tau)\right),v(\tau)\right\rangle\geq-\frac{3\epsilon}{2}\operatorname*{Max}_t\left(\frac{\|v(t)\|}{1+\left\|\widehat{f}(t)\right\|}\right)=-\frac{3\epsilon}{2}\,.$$

Here we have used inequality (59) on the right-hand side. Now use inequality (58) on the left-hand side:

$$(70)\qquad \left\langle\Phi'\left(\widehat{f}(\tau)\right),v(\tau)\right\rangle<-\frac{3\epsilon}{2}$$

which, together with (69), yields a contradiction. So assumption (53) cannot be true, and we have proved that M contains a point t where

$$(71)\qquad \left\|\Phi'\left(\widehat{f}(t)\right)\right\|_*\left(1+\left\|\widehat{f}(t)\right\|\right)\leq\frac{3\epsilon}{2}\,.$$

Remember that $d\left(\widehat{f},\widehat{c}\right)\leq\epsilon/2$. It follows that:

$$(72)\qquad \delta\left(\widehat{f}(t),F_\gamma\right)\leq\frac{\epsilon}{2}+\delta\left(c(t),\widehat{F}_\gamma\right)\,,$$

and the last term on the right is no greater than ϵ, because of the fact that $t_o<t<t_1$ and the definition of t_o and t_1. Finally:

$$(73)\qquad \delta\left(\widehat{f}(t),\widehat{F}_\gamma\right)\leq\frac{3\epsilon}{2}\,.$$

Remember also that $\varphi(\widehat{f}) \le \varphi(\widehat{c})$, which implies that:

$$\gamma + \epsilon^2 \le \Phi\left(\widehat{f}(t)\right) + \Psi\left(\widehat{f}(t)\right) \le \gamma + \frac{5}{4}\epsilon^2 . \tag{74}$$

Now set $x := \widehat{f}(t)$. The point x we have found in this way has the following properties:

$$\|\Phi'(x)\|_* (1 + \|x\|) \le \frac{3\epsilon}{2} \tag{75}$$

$$\delta\left(x, \widehat{F}_\gamma\right) \le \frac{3\epsilon}{2} \tag{76}$$

$$\gamma \le \Phi(x) \le \gamma + \frac{5}{4}\epsilon^2 . \tag{77}$$

Setting $\epsilon = 1/n$, and letting $n \to \infty$, we get a sequence x_n with the desired properties, (35), (36) and (37). □

If we now add an assumption that will make the sequence x_n converge, we will be able to prove the existence of critical points for Φ.

Definition 7. *We shall say that a Gâteaux-differentiable map* $\Phi : X \to \mathbb{R}$ satisfies condition (C) *at the level* γ *for the closed subset* $F \subset X$ *if every sequence* x_n *such that*

$$\delta(x_n, F) \to 0 \tag{78}$$

$$\Phi(x_n) \to \gamma \tag{79}$$

$$(1 + \|x_n\|) \|\Phi'(x_n)\|_* \to 0 \tag{80}$$

has a subsequence x_n, $n \in A$, *which converges to a critical point of* Φ *on the level* γ:

$$x_n \to \overline{x} \in F \quad \text{when } n \to \infty , \quad n \in A \tag{81}$$

$$\Phi(\overline{x}) = \gamma \quad \text{and } \Phi'(\overline{x}) = 0 . \tag{82}$$

□

We shall say that Φ satisfies condition (C) if it satisfies it at every level γ and for $F = X$. Historically, the first ones to introduce a condition of this type were Palais and Smale:

Definition 8. *We shall say that a* C^1 *function* $\Phi : X \to \mathbb{R}$ *satisfies condition (PS) if every sequence* x_n *such that* $|\Phi(x_n)|$ *is uniformly bounded and* $\|\Phi'(x_n)\|_* \to 0$ *has a convergent subsequence.* □

Clearly, condition (PS) implies condition (C). Using these notions, we get some easy corollaries of the Ghoussoub-Preiss theorem. First of all, a celebrated result which goes back to Ambrosetti and Rabinowitz:

Corollary 9. *Let X be a Banach space and $\Phi : X \to \mathbb{R}$ a continuous Gâteaux-differentiable function with $\Phi' : X \to X^*$ norm-to-weak* continuous. Take two points (z_o, z_1) in X, and define*

$$\Gamma := \{c \in C^o([0,1]; X) \,|\, c(0) = z_o, c(1) = z_1\} \tag{83}$$

$$\gamma := \operatorname*{Inf}_{c\in\Gamma} \operatorname*{Max}_{0\le t\le 1} \Phi(c(t)) \ . \tag{84}$$

Assume that Φ satisfies condition (C) on the level γ, and that:

$$\gamma > \operatorname{Max}\{\Phi(z_o), \Phi(z_1)\} \ . \tag{85}$$

Then there is a critical point of Φ on the level γ

$$\exists \overline{x} : \quad \Phi'(\overline{x}) = 0 \quad \text{and} \quad \Phi(\overline{x}) = \gamma \ . \tag{86}$$ □

Proof. Take $F = X$ in Theorem 6. Assumption (85) implies that Φ_γ separates z_o and z_1, and the result follows. □

Then a situation when equality may hold in (85). Defining X, Φ, Γ and γ as in Corollary 9, we have

Corollary 10. *Let $F \subset X$ be a closed subset separating z_o and z_1. Assume that:*

$$\gamma = \operatorname{Max}\{\Phi(z_o), \Phi(z_1)\} \tag{87}$$

$$\forall x \in F\,, \quad \Phi(x) \ge \gamma \tag{88}$$

$$\Phi \text{ satisfies condition (C) on the level } \gamma \text{ for the set } F\ . \tag{89}$$

Then F contains a critical point of Φ on the level γ:

$$\exists \overline{x} \in F : \quad \Phi'(\overline{x})\,, \quad \Phi(\overline{x}) = \gamma \ . \tag{90}$$ □

The existence of a critical point $\overline{x}$ will not be enough for our purposes. The feeling we get from our construction is that $\overline{x}$ should be some type of saddle-point, or mountain-pass. It is this kind of information we are after, and it turns out that Theorem 6 provides it.

First some notations and definitions. We define X, Φ, Γ, γ, F as in Theorem 6, and we assume that Φ satisfies condition (C) for F on the level γ. Set:

$$K := \{x \in K \mid \Phi(x) = \gamma, \Phi'(x) = 0\} \tag{91}$$

$$M := \{x \in K \mid x \text{ is a local minimum for } \Phi\} \ . \tag{92}$$

It follows from condition (C) that $K \cap F$ is compact and non-empty.

Definition 11 (Hofer). Let $\overline{x} \in K$, so that $\gamma = \Phi(\overline{x})$. We say that $\overline{x}$ is a *mountain-pass point* if, for every open neighbourhood $\mathcal{U}$ of $\overline{x}$ in X, the set

$$\Phi_{\mathcal{U}}^{\gamma} := \{x \in \mathcal{U} \,|\, \Phi(x) < \gamma\} \tag{93}$$

is neither empty nor connected. □

Saying that $\Phi_{\mathcal{U}}^{\gamma}$ is never empty means that $\overline{x}$ is not a local minimum. Note also that since $\Phi_{\mathcal{U}}^{\gamma}$ is open in X, it is connected if and only if it is path connected. For $\mathcal{U} = X$, we shall write Φ^{γ} instead of $\Phi_{\mathcal{U}}^{\gamma}$:

$$\Phi^{\gamma} := \{x \,|\, \Phi(x) < \gamma\} \ . \tag{94}$$

For the sake of brevity, we shall henceforth write mp instead of mountain-pass. Introduce another subset of K:

$$P := \{x \in K \,|\, x \ \text{ is a mp-point for } \Phi\} \tag{95}$$

The following theorem tells us that K contains either a mp-point or a local minimum.

Theorem 12. *Assume that $F \cap \Phi_{\gamma}$ separates z_o and z_1 and Φ satisfies condition (C) at the level γ. Then:*

$$F \cap \overline{M} \neq \emptyset \quad \text{or} \quad F \cap P \neq \emptyset \ . \tag{96}$$

Proof. As before, noting that $F_{\gamma} := F \cap \Phi_{\gamma}$ separates z_o and z_1, we find two open sets Ω_o and Ω_1 with $\Omega_o \cup \Omega_1 = X \setminus F$, $\Omega_o \cap \Omega_1 = \emptyset$, and $z_i \in \Omega_i$.

Suppose $F \cap P = \emptyset$, that is, F contains no mp-point. We claim that there is an $\epsilon_1 > 0$ such that the set

$$\{x | \delta\left(x, F_{\gamma} \cap K\right) < \epsilon_1\} \tag{97}$$

meets only finitely many components of Φ_{γ} say $\mathcal{U}_1, \ldots, \mathcal{U}_N$. Indeed, otherwise we could find a sequence x_n in $F_{\gamma} \cap K$ and a sequence $\mathcal{U}_n$ of pairwise disjoint components in Φ_{γ} such that $\delta\left(x_n, \mathcal{U}_n\right) \to 0$. But then, any cluster point $\overline{x}$ of the sequence x_n would be a mp-point belonging to $F_{\gamma} \subset F$, which contradicts our initial assumption.

Consider now the sets Q_i, $1 \leq i \leq N$, defined by:

$$Q_i := F_{\gamma} \cap K \cap \overline{\mathcal{U}}_i \ . \tag{98}$$

Again, any point in $Q_i \cap \left(\cup_{j \neq i} \overline{\mathcal{U}}_j\right)$ would be a mp-point. It follows that there is an $\epsilon_2 \leq \epsilon_1$ such that:

$$\{x \,|\, \delta\left(x, Q_i\right) < \epsilon_2\} \bigcap \left(\bigcup_{j \neq i} \overline{\mathcal{U}}_j\right) = \emptyset \ . \tag{99}$$

We may of course assume that $\epsilon_2 < \delta(z_i, F_\gamma)$ for $i = 0, 1$. With every $\epsilon \in (0, \epsilon_2)$ we associate the set

$$R(\epsilon) := \left[\Omega_o \bigcup \bigcup_{i \in A} \{x \mid \delta(x, Q_i) < \epsilon\}\right] \bigcup_{j \in B} \{x \mid \delta(x, Q_j) \leq \epsilon\} \ , \tag{100}$$

where the set of indices $i = 1, \ldots, N$ is partitioned into:

$$A := \{i \mid \mathcal{U}_i \subset \Omega_o\} \tag{101}$$

$$B := \{i \mid \mathcal{U}_i \cap \Omega_o = \emptyset\} \ . \tag{102}$$

$R(\epsilon)$ is an open subset of X containing z_o, while $z_1 \notin \overline{R(\epsilon)}$. Its boundary $\partial R(\epsilon)$ therefore separates z_o and z_1. It follows from Definition (100) that:

$$\bigcup_{i \in A} \mathcal{U}_i \subset R(\epsilon) \tag{103}$$

$$\bigcup_{i \in B} \mathcal{U}_i \subset X \setminus R(\epsilon) \tag{104}$$

and hence:

$$\partial R(\epsilon) \cap \Phi^\gamma = \partial R(\epsilon) \cap \left(\bigcup_{i=1}^{N} \mathcal{U}_i\right) = \emptyset \ . \tag{105}$$

Consequently, $\Phi \geq \gamma$ on $R(\epsilon)$. Therefore, we may use Theorem 6 to find a critical point x_ϵ for Φ in $\partial R(\epsilon)$:

$$x_\epsilon \in K \cap \partial R(\epsilon) \ . \tag{106}$$

We claim that, for each ϵ, the point x_ϵ is a local minimum for Φ. Indeed, if such is not the case, then $x_\epsilon \in \overline{\mathcal{U}}_i$ for some $i \in \{1, \ldots, N\}$. If $i \in A$, then $x_\epsilon \in Q_i \subset R(\epsilon)$, and since $R(\epsilon)$ is open, $x_\epsilon \notin \partial R(\epsilon)$, yielding a contradiction. If $i \in B$, then $x_\epsilon \in Q_i \subset X \setminus \overline{R(\epsilon)}$, yielding $x_\epsilon \notin \partial R(\epsilon)$, still a contradiction.

Finally observe that $\delta(x_\epsilon, F_\gamma) \to 0$ when $\epsilon \to 0$ and since K is compact, this shows that $F_\gamma \cap \overline{M}_c \neq \emptyset$, as announced in formula (95). □

Taking $F = X$, we get an earlier result of Hofer:

Corollary 13. *Let X, Φ, Γ, γ be as in Corollary 9. Then:*

$$K \cap (\overline{M} \setminus M) \neq \emptyset \quad \text{or} \quad K \cap P \neq \emptyset \ . \tag{107}$$

□

In other words, on the level γ there is either a mp-point, or a point which is not a local minimum, but which is the limit of a sequence of local minima on the level γ.

Proof. Consider the set:

$$\Phi_\gamma := \{x \mid \Phi(x) \geq \gamma\} \ . \tag{108}$$

Let F be its boundary, and apply Theorem 12 to F. We get $F \cap \overline{M} \neq \emptyset$ or $F \cap P \neq \emptyset$. On the other hand, any local minimum of Φ on the level γ must be interior to Φ_γ, so $F \cap M = \emptyset$. The result follows. □

Notes and Comments. The chronology is as follows. Palais and Smale introduced their celebrated condition (PS) to extend Morse and Linsternik-Schnirelman theory to the infinite-dimensional setting (see [PalS], [Pal1], [Pal2]). Using the deformation lemma (see Sect. V.2 for an equivariant version) they were then able to give results on the existence of critical points, provided the functional was regular (slightly better than C^1). That condition (C) is enough was first remarked by Cerami in [Cer]; see [Ben2] for other weights.

Then came the celebrated theorem of Ambrosetti and Rabinowitz [AmbR] (here Corollary 9), which gave a simple procedure to find a critical point which is not a local minimum. They had, of course, more stringent regularity conditions than we assume here.

Theorem 1 is due to Ekeland ([Eke1], [Eke2]) and throw a new light on the whole thing. One could avoid the deformation lemma, and weaken considerably the regularity assumptions.

Theorem 12 is due to Hofer ([Hof2] and [EkeH1]), relying on the deformation lemma. Ghoussoub and Preiss, in [GhoP], found a way to localize these results of Ekeland's variational principle. Theorem 6 is due to them, and has among other consequences several earlier results of Pucci and Serrin ([PucS1], [PucS2]) on the topological structure of the set of critical points.

2. A Preliminary Existence Result

This section serves as an introduction to superquadratic Hamiltonians, and to set up the moutain-pass situation which will be exploited further in the following sections. We will prove an existence result for periodic solutions which will be considerably improved later on.

Theorem 1. *Let* $H : [0,T] \times \mathbb{R}^{2n} \to \mathbb{R}$ *be such that:*

$$\forall t \in [0,T] \ , \quad H(t,\cdot) \ \text{ is strictly convex and } \ C^1 \ \text{ on } \ \mathbb{R}^{2n} \tag{1}$$

$$H \ \text{ and } \ H' \ \text{ are continuous on } \ [0,T] \times \mathbb{R}^{2n} \ . \tag{2}$$

Assume that:

$$\forall (t,x) \ , \quad H(t,x) \geq H(t,0) = 0 \tag{3}$$

and that there are constants $\alpha > 2$ *and* $\omega \in \left(0, 2\pi T^{-1}\right)$ *satisfying:*

$$\exists r > 0 \ : \quad H(t,\lambda x) \geq \lambda^\alpha H(t,x) \qquad \text{for } \ \|x\| \geq r \ \text{ and } \ \lambda \geq 1 \tag{4}$$

(5)
$$\limsup_{0\le t\le T, \|x\|\to\infty} H(t,x)\,\|x\|^{-\alpha} < \infty$$

(6)
$$\exists \epsilon > 0 : \quad H(t,x) \le \frac{\omega}{2}\,\|x\|^2 \qquad \text{for} \quad \|x\| \le \epsilon \,.$$

Then, the boundary-value problem

(7)
$$\begin{cases} \overline{x} = JH'(t,x) \\ x(0) = x(T) \end{cases}$$

has a non-trivial solution. □

By assumption (3), the trivial solution $x = 0$ (equilibrium at the origin) is always present. The theorem asserts the existence of another one provided $T\omega < 2\pi$.

Condition (6) gives us the behaviour of H near the origin. For instance, if

(8)
$$H(t,x)\,\|x\|^{-2} \to 0 \qquad \text{when} \quad \|x\| \to 0$$

then it is satisfied for every $\omega > 0$; and in particular for $\omega < 2\pi T^{-1}$. In the autonomous case, this gives us an existence result for every $T > 0$:

Corollary 2. *Let $H \in C^1\left(\mathbb{R}^{2n}, \mathbb{R}\right)$ be a strictly convex function such that:*

(9)
$$\forall x \,, \quad H(x) \ge H(0) = 0$$

(10)
$$\exists r > 0 : \quad H(\lambda x) \ge \lambda^{\alpha} H(x) \qquad \text{for} \quad \|x\| \ge r \quad \text{and} \quad \lambda \ge 1$$

(11)
$$\limsup_{\|x\|\to\infty} H(x)\,\|x\|^{-\alpha} < \infty$$

for some $\alpha > 2$, and

(12)
$$\lim_{\|x\|\to 0} H(x)\,\|x\|^{-2} = 0 \,.$$

Then, for every $T > 0$, the system

(13)
$$\dot{x} = JH'(x)$$

has a non-trivial T-periodic solution. □

Of course, conditions (3) and (6) have dual versions in terms of the Fenchel conjugate $H^*(t;x)$ of H with respect to x, for fixed t as in formula (II.3.10).

Lemma 3. *We have:*

(14)
$$\forall (t,y) \,, \quad H^*(t,y) \ge H^*(t,0) = 0$$

$$\exists \eta > 0 \ : \quad H^*(t,y) \geq \frac{1}{2\omega} \|y\|^2 \qquad \text{for} \quad \|y\| \leq \eta \ . \tag{15}$$

Proof. Straightforward. We have:

$$\eta := \text{Min} \ \{\|H'(t,x)\| \mid \|x\| \geq \epsilon\} \ . \tag{16}$$

□

Condition (4) tells us that $H(t,x)$ grows faster than $\|x\|^\alpha$ at infinity. Setting

$$\frac{m^\alpha}{\alpha} := \text{Min} \ \{H(t,x) \mid 0 \leq t \leq T, \|x\| = r\} > 0 \tag{17}$$

we have:

$$H(t,x) \geq \frac{m^\alpha}{\alpha} \frac{\|x\|^\alpha}{r^\alpha} \qquad \text{for} \quad \|x\| \geq r \ . \tag{18}$$

It is equivalent to a condition on the derivative, which will be extremely useful in the sequel:

Lemma 4. *Condition* (4) *is equivalent to the following:*

$$(x, H'(t,x)) \geq \alpha H(t,x) \qquad \text{for} \quad \|x\| \geq r \ . \tag{19}$$

Proof. Assume (4) holds. Fix (t,x) with $\|x\| \geq r$ and consider the functions $\varphi(\lambda) := \lambda^\alpha H(t,x)$ and $\psi(\lambda) := H(t, \lambda x)$. We have $\varphi(1) = \psi(1)$ and $\varphi(\lambda) \leq \psi(\lambda)$ for $\lambda \geq 1$. Since φ and ψ are convex $\varphi'(1) \leq \psi'(1)$, which is exactly formula (19).

Conversely, assume (19) holds. Fix x with $\|x\| \geq r$, and consider the function $\psi(\lambda) := H(t, \lambda x)$. For $\lambda \geq 1$, we have $\|\lambda x\| \geq r$, and hence:

$$\frac{\psi'(\lambda)}{\psi(\lambda)} \geq \frac{\alpha}{\lambda} \ . \tag{20}$$

Formula (4) follows immediately. □

Lemma 5. *For every $t \in [0,T]$, the function $x \to H^*(t;x)$ is strictly convex and C^1. The functions $(t,x) \to H^*(t,x)$ and $(t,x) \to \nabla H^*(t,x)$ are continuous. We have:*

$$H^*(t; \lambda y) \leq \lambda^\beta H^*(y) \qquad \text{for} \quad \|y\| \geq R \quad \text{and} \quad \lambda \geq 1 \tag{21}$$

$$(y, \nabla H^*(t;y)) \leq \beta H^*(t;y) \qquad \text{for} \quad \|y\| \geq R \tag{22}$$

where β is the conjugate exponent of α:

$$\alpha^{-1} + \beta^{-1} = 1 \tag{23}$$

and

$$R := \operatorname{Max}\{\|H'(t,x)\| \mid \|x\| \leq r\} \ . \tag{24}$$

Proof. Fix $t \in [0,T]$. It follows from the estimate (5) on $H(t,\cdot)$ that $H^*(t,\cdot)$ is finite everywhere; its subdifferential at y is:

$$\partial H^*(t;y) = \{x \mid (x,y) - H(t,x) = H^*(t;y)\} \ . \tag{25}$$

Since $H(t,\cdot)$ is strictly convex, the function $x \to (x,y) - H(t,x)$ attains its maximum at a single point. So $\partial H^*(t;y)$ is a singleton, and hence (Proposition II.2.9) $H^*(t,\cdot)$ is Gâteaux-differentiable on $\mathbb{R}^{2n}$. By the Legendre reciprocity formula, its derivative $\nabla H^*(t,\cdot)$ is the inverse map of $H'(t,\cdot)$; it follows that it has closed graph. A direct argument based on convexity, or the estimate (33) which will be proved presently, show that $\nabla H^*(t,\cdot)$ maps bounded sets into bounded sets. Fitting the pieces together, we find that $\nabla H^*(t,\cdot)$ maps compact sets onto compact sets, and hence is continuous. So $H^*(t;\cdot)$ is C^1.

The same argument shows that $\nabla H^*(t,y)$, and hence $H^*(t,y)$, depend continuously on (t,y).

Now rewrite formula (19), setting $y = H'(t,x)$ and using the Legendre reciprocity formula $x = \nabla H^*(t,y)$.

We get, for $\|y\| \geq R$:

$$(\nabla H^*(t;y), y) \geq \alpha\, [H^*(t;y) - (y, \nabla H^*(t;y))] \tag{26}$$

and formula (22) follows immediately. Arguing as in Lemma 4, we get formula (21). □

Finally, we have condition (5). It tells us that $H(t,x)$ grows exactly like $\|x\|^\alpha$ at infinity. As the following result shows, this implies that $H'(t,x)$ grows like $\|x\|^{\alpha-1}$ at infinity.

Lemma 6. *There is a constant $c > 0$ such that:*

$$\frac{m^\alpha}{\alpha r^\alpha}\|x\|^\alpha \leq H(t,x) \leq \frac{c^\alpha}{\alpha}\|x\|^\alpha \qquad \text{for} \quad \|x\| \geq r \tag{27}$$

$$\|H'(t,x)\| \leq \frac{2}{\alpha}\left[\left(\frac{3c}{2}\right)^\alpha - \left(\frac{m}{r}\right)^\alpha\right]\|x\|^{\alpha-1} \ . \tag{28}$$

Proof. The left side of inequality (27) has already been given in formula (18). The right side follows immediately from condition (5).

To prove inequality (28), fix x with $\|x\| \geq 2r$ and take any y with $\|y\| = \|x\|/2$. Using the convexity of $H(t,\cdot)$, we have:

$$(H'(t,x), y) \leq H(t,x+y) - H(t,x) \ . \tag{29}$$

Note that $r \le \|x+y\| \le \frac{3}{2}\|x\|$. Applying formula (27), we get:

$$(H'(t,x),y) \le \frac{c^\alpha}{\alpha}\left(\frac{3}{2}\right)^\alpha \|x\|^\alpha - \frac{m^\alpha}{\alpha r^\alpha}\|x\|^\alpha \ . \tag{30}$$

Taking the supremum of the left-hand side over all y with $\|y\| = \|x\|/2$, we get formula (28). □

Dually, H^* will satisfy a similar condition:

$$\liminf_{\|y\|\to\infty} H^*(t;y)\,\|y\|^{-\beta} > 0 \tag{31}$$

from wich similar estimates will follow:

$$\frac{M^\beta}{\beta R^\beta}\|y\|^\beta \ge H^*(t;y) \ge \frac{1}{\beta k^\beta}\|y\|^\beta \qquad \text{for } \ \|y\| \ge R \tag{32}$$

$$\|\nabla H^*(t,y)\| \le \frac{2}{\beta}\left[\left(\frac{3M}{2R}\right)^\beta - \frac{1}{k^\beta}\right]\|y\|^{\beta-1} \ , \tag{33}$$

where R has been defined in Lemma 5. The existence of $k > 0$ (which can be taken arbitrarily large) follows from condition (11), and

$$M := \operatorname{Max}\left\{H^*(t;y) \mid \|y\| = R \ \text{ and } \ 0 \le t \le T\right\} \ . \tag{34}$$

Taking condition (15) into account, we find that we can choose k so large that:

$$\begin{cases} H^*(t;y) \ge \frac{1}{2\omega}\|y\|^2 & \text{for } \ \|y\| \le \eta \\ H^*(t;y) \ge \frac{1}{\beta k^\beta}\|y\|^\beta & \text{for } \ \|y\| \ge \eta \ . \end{cases} \tag{35}$$

We now introduce the dual action functional. The Hamiltonian H clearly satisfies conditions (51) to (54) of Sect. II.4. By Proposition II.4.6, problem (7) then is equivalent to finding critical points of the functional

$$\Psi(x) := \int_o^T \left[\frac{1}{2}(J\dot{x},x) + H^*(t,-J\dot{x})\right] dt \tag{36}$$

on the space:

$$W^{1,\beta} := \left\{x \in C^o\left([0,T];\mathbb{R}^{2n}\right) \mid \dot{x} \in L^\beta, x(0) = x(T)\right\} \ . \tag{37}$$

As in the preceding chapter, we take advantage of the fact that Ψ is invariant by space translations:

$$\Psi(x+\xi) = \Psi(x) \quad \forall \xi \in \mathbb{R}^{2n} \tag{38}$$

to perform the change of variables $\dot{x} = u$. We get a reduced functional:

$$\psi(u) := \int_o^T \left[\frac{1}{2}(Ju,\Pi u) + H^*(t,-Ju)\right] dt \tag{39}$$

on the space:

$$L^\beta_o\left(0,T;\mathbb{R}^{2n}\right) := \left\{u \in L^\beta \middle| \int_o^T u\,dt = 0\right\} . \tag{40}$$

Here Πu is defined to be the primitive of u with mean value 0:

$$\frac{d}{dt}(\Pi u) = u \quad \text{and} \quad \int_o^T (\Pi u)dt = 0 . \tag{41}$$

Lemma 7. $\Pi : L^\beta_o \to L^\alpha_o$ *is a compact operator.*

Proof. This was proved in Proposition III.1.2 when $\alpha = 2 = \beta$. For the general case, write Πu explicitly:

$$(\Pi u)(t) = \int_o^t u(s)ds - \frac{1}{T}\int_o^T dt \int_o^t u(s)ds . \tag{42}$$

Now let u_n converge to u in L^β. Then $\Pi u_n(t)$ is clearly uniformly bounded for $0 \le t \le T$ and $n \in \mathbb{N}$. By Cauchy-Schwarz, we have

$$|\Pi u_n(t) - \Pi u_k(s)| \le \|u_n - u_k\|_\beta \left(|t-s|^{1/\alpha} + \frac{\alpha}{\alpha+1}T^{1/\alpha}\right) \tag{43}$$

so that Πu_n, $n \in \mathbb{N}$, is an equicontinuous family of functions. By Ascoli's theorem, it must be relatively compact in C^o, and hence in L^α. □

We shall also need the differentiability properties of ψ. They are dealt with as follows.

Lemma 8. ψ *is a* C^1 *functional on* L^β_o, *and we have:*

$$\langle \psi'(u), v\rangle = \langle -J\Pi u + J\nabla H^*(t, -Ju), v\rangle \quad \forall v \in L^\beta_o \tag{44}$$

where $\langle\cdot,\cdot\rangle$ *denotes the duality pairing* $\left(L^\alpha, L^\beta\right)$. □

Proof. ψ is the sum of two terms. The first one is a continuous quadratic form, hence C^∞. The second one:

$$\mathcal{H}^*(u) := \int_o^T H^*\left(t; u(t)\right) dt \tag{45}$$

is strictly convex. Note that the integrand $H^*(t,\cdot)$ is C^1 and its derivative satisfies estimate (33). It then follows from Proposition II.3.5 that $\mathcal{H}^*$ is C^1, with

$$\langle \nabla\mathcal{H}^*(u), w\rangle = \langle \nabla H^*(\cdot; u(\cdot)), w\rangle . \tag{46}$$

Hence the result. □

We are now in a position to apply the Ambrosetti-Rabinowitz mountain-pass theorem to ψ. We first have to check the assumptions of Corollary 1.9, where we take $z_o := 0$ and z_1 will be defined presently. Note that, because of condition (14)

$$\psi(0) = 0 \,. \tag{47}$$

Lemma 8. *ψ has a local minimum at the origin. In fact*

$$\exists \rho > 0\,, \quad \exists a > 0 \;:\quad \|u\|_\beta = \rho \Rightarrow \psi(u) \geq a\,. \tag{48}$$ □

Proof. For any $u \in L^\beta_o$, set $A(u) := \{t|\;\|u(t)\| \leq \eta\}$, and write $u = v + w$ with:

$$v := u\mathbf{1}_{A(u)} \quad \text{and} \quad w := u\mathbf{1}_{[0,T]\setminus A(u)}\,. \tag{49}$$

Note that $v \in L^\infty$. By Wirtinger's inequality, we have, for a suitable constant $c_1 > 0$:

$$\|\Pi u\|_2 \leq \|\Pi v\|_2 + \|\Pi w\|_2 \leq \frac{T}{2\pi}\|v\|_2 + c_2\|w\|_\beta\,. \tag{50}$$

Write the decomposition $u = v + w$ into ψ, and take into account estimate (35):

$$\begin{aligned} \psi(u) &\geq -\frac{1}{2}\left|\|v\|_2\|\Pi u\|_2 + \|w\|_\beta\|\Pi u\|_\alpha\right| + \frac{1}{2\omega}\|v\|_2^2 + \frac{1}{\beta k^\beta}\|w\|_\beta^\beta \\ &\leq -\frac{1}{2}\left[\frac{T}{2\pi}\|v\|_2^2 + c_2\|w\|_\beta\|v\|_2 + c_3\|w_\beta\|^2\right] \\ &\quad + \frac{1}{2\omega}\|v\|_2^2 + \frac{1}{\beta k^\beta}\|w\|_\beta^\beta\,, \end{aligned} \tag{51}$$

where c_2 and c_3 are now positive constants. We have a rectangular term $\|w\|_\beta\|v\|_2$, which we transform by the inequality $ab \leq \frac{1}{2}\left(\frac{a^2}{\epsilon^2} + \epsilon^2 b^2\right)$, valid for any ϵ. We finally get:

$$\psi(u) \geq \varphi_\epsilon\left(\|v\|_2, \|w\|_\beta\right)\,, \tag{52}$$

where $\varphi_\epsilon : \mathbb{R}_+ \times \mathbb{R}_+ \to \mathbb{R}$ is given by

$$\varphi_\epsilon(s,t) = \frac{1}{2}\left(\frac{1}{\omega} - \frac{T}{2\pi} - c_2\frac{\epsilon^2}{2}\right)s^2 - \frac{1}{2}\left(c_2\frac{1}{2\epsilon^2} + c_3\right)t^2 + \frac{1}{\beta k^\beta}t^\beta\,. \tag{53}$$

Since $T < \frac{2\pi}{\omega}$, we may choose ϵ^2 so small that the coefficient of s^2 is positive. The coefficient of t^2 is negative, but since $\beta \in (1,2)$, behaviour near 0 is determined by the term in t^β, which is positive. So φ_ϵ has a strict local minimum at the origin, and there is a closed neighbourhood $\mathcal{N}$ of $(0,0)$ in $\mathbb{R}^2_+$ such that:

$$(0,0) \neq (s,t) \in \mathcal{N} \Rightarrow \varphi(s,t) > \varphi(0,0) = 0\,. \tag{54}$$

Finally, we have:

$$\eta^{\beta-2}\|v\|_2 \leq \|v\|_\beta \leq T^{\frac{2\beta}{2-\beta}}\|v\|_2 , \tag{55}$$

where the left inequality comes from $\|v\|_\infty \leq \eta$ and the right one from Hölder's inequality. That is, the L^2-norm and the L^β-norm for v are equivalent. It follows that, if $\|u\|_\beta = \rho$ is small enough, the pair $\left(\|v\|_2, \|w\|_\beta\right)$ will belong to $\mathcal{N}$, and formula (48) follows with

$$0 < a := \text{Min}\left\{\varphi(s,t) \mid (s,t) \in \mathcal{N},\ \eta^{\beta(\beta-2)}s^\beta + t^\beta \geq \varphi^\beta\right\} \tag{56}$$ □

Lemma 9. *Choose $u_1 \in L_o^\beta$ such that*

$$\int_o^T (Ju_1, \Pi u_1)\,dt < 0 . \tag{57}$$

Then $\psi(hu_1) \to -\infty$ when $h \to +\infty$. □

Proof. We have

$$\psi(hu_1) = \int_o^T \left[\frac{h^2}{2}(Ju_1, \Pi u_1) + H^*(t, -Jhu_1)\right] dt . \tag{58}$$

By formula (32), we have

$$H^*(t,y) \leq \frac{M^\beta}{\beta R^\beta}\|y\|^\beta + C , \tag{59}$$

where $C := \text{Max}\,\{H^*(t,y) | 0 \leq t \leq T \text{ and } \|y\| \leq R\}$. Writing this in the preceding integral, we get:

$$\psi(hu_1) \leq \frac{h^2}{2}\int_o^T (Ju_1, \Pi u_1)\,dt + \frac{h^\beta}{\beta}\frac{M}{R^\beta}\|u_1\|_\beta^\beta + CT \tag{60}$$

and the right-hand side goes to $-\infty$ when $h \to \infty$, since $1 < \beta < 2$. □

As we saw earlier, there are functions u_1 satisfying (57), the eigenfunctions of $-J\Pi$ associated with negative eigenvalues for instance. Pick one of them and choose $\overline{h}$ so large that

$$\left\|\overline{h}u_1\right\|_\beta > \rho \quad \text{and} \quad \psi(\overline{h}u_1) < 0 , \tag{61}$$

where ρ is given by the preceding lemma. Then the pair $\left(0, \overline{h}_1\right)$ satisfies the assumptions of Corollary 1.9. All that remains is to check condition (C).

Lemma 10. *The functional ψ satisfies condition (PS).*

Proof. Consider a sequence $u_n \in L_o^\beta$ such that

$$\psi(u_n) \leq b \tag{62}$$

$$\|\psi'(u_n)\|_* \to 0 . \tag{63}$$

We want to show that it has a convergent subsequence in L_o^β. To do this, we rephrase the last condition, using the explicit formula for $\varphi'(u)$:

$$J\Pi u_n - J\nabla H^*(t, -Ju_n) = \xi_n + \epsilon_n , \tag{64}$$

where equality holds in L^α, ξ_n is a constant vector in $\mathbb{R}^{2n}$, and $\epsilon_n \to 0$ in L^α. Hence:

$$\Pi u_n = \nabla H^*(t, -Ju_n) - J\xi_n - J\epsilon_n . \tag{65}$$

Writing this into $\psi(u_n)$, we get:

$$\begin{aligned} b \geq \psi(u_n) &= \int_o^T \left[\frac{1}{2}(Ju_n, \Pi u_n) + H^*(t, -Ju_n)\right] dt \\ &= \int_o^T \left[\frac{1}{2}(Ju_n, \nabla H^*(t, -Ju_n)) - \frac{1}{2}(u_n, \epsilon_n) + H^*(t, -Ju_n)\right] dt . \end{aligned} \tag{66}$$

Now write $u_n = v_n + w_n$ as in the proof of Lemma 8. Define $A_n := \{t | \, \|u_n(t)\| \leq \eta\}$ and $B_n := [0, T] \setminus A_n$, then set

$$v_n := u_n \mathbf{1}_{A_n} \quad \text{and} \quad w_n := u_n \mathbf{1}_{B_n} . \tag{67}$$

Split the integral in formula (66) in two parts. The first one is bounded uniformly since $\|v_n\|_\infty \leq \eta$:

$$\int_{A_n} \left[\frac{1}{2}(Jv_n, \nabla H^*(t, -Jv_n)) + H^*(t, -Jv_n)\right] dt + (v_n, \epsilon_n) \geq c_1 . \tag{68}$$

The second one we estimate by conditions (22) and (35). Note first that condition (22) implies that there exists a constant $c_2 \geq 0$ such that

$$(y, \nabla H^*(t, y)) \leq \beta H^*(t, y) + c_2 \quad \forall y . \tag{69}$$

Hence:

$$\begin{aligned} &\int_{B_n} \left[\frac{1}{2}(Ju_n, \nabla H^*(t, -Ju_n)) + H^*(t, -Ju_n)\right] dt + (w_n, \epsilon n) \\ &\geq \int_{B_n} \left(1 - \frac{\beta}{2}\right) H^*(t, -Ju_n) dt - Tc_2 - \|\epsilon_n\|_\alpha \|w_n\|_\beta \\ &\geq \left(1 - \frac{\beta}{2}\right) \frac{1}{\beta k^\beta} \|w_n\|_\beta^\beta - Tc_2 - \|\epsilon_n\|_\alpha \|w_n\|_\beta . \end{aligned} \tag{70}$$

Finally, setting $c := c_1 - Tc_2$, we get an estimate on w_n:

$$b \geq c + \left(1 - \frac{\beta}{2}\right) \frac{1}{\beta k^\beta} \|w_n\|_\beta^\beta - \|\epsilon_n\|_\alpha \|w_n\|_\beta \tag{71}$$

with $\|\epsilon_n\|_\alpha \to 0$ as $n \to \infty$. Since $\left(1 - \frac{\beta}{2}\right)$ is positive, w_n must be uniformly bounded in L^β. After extracting a subsequence, we may assume that it converges weakly in L_o^β.

On the other hand, since $\|v_n\|_\infty \leq \eta$, the sequence v_n also has a weakly convergent subsequence. Adding up, we set that $u_n = v_n + w_n$ is bounded and converges weakly to some u in L_o^β.

As we noticed earlier, the estimate (33) then implies that $\nabla H^*(u_n)$ is bounded in L^α. Writing this into formula (65), we find that ξ_n is bounded in $\mathbb{R}^{2n}$. A subsequence, still denoted by ξ_n, then converges to some $\xi \in \mathbb{R}^{2n}$. On the other hand, since Π is compact, the sequence Πu_n converges strongly in L^α. Finally:

$$\Pi u_n + J\xi_n + J\epsilon_n \to \Pi u + J\xi \quad \text{in } L^\alpha . \tag{72}$$

Now we apply to Eq. (65) the Fenchel reciprocity formula, thereby getting:

$$-Ju_n = H'\left(\Pi u_n + J\xi_n + \epsilon_n\right) . \tag{73}$$

By Theorem II.3.4, the estimate (33) implies that the nonlinear operator $v \to H'(v)$ is continuous from L^α to L^β. It then follows from (72) that

$$u_n \to JH'\left(\Pi u + J\xi\right) \quad \text{in } L^\alpha . \tag{74}$$

Since u_n is already known to converge weakly to u, we must have $u = JH'\left(\Pi u + J\xi\right)$. So u is a critical point of ψ, and the result is proved. □

By Corollary 1.9, there will be a critical point $\overline{u}$ of ψ such that

$$\psi(\overline{u}) \geq a > 0 \tag{75}$$

so $\overline{u} \neq 0$, and the corresponding solution $\overline{x}$ of problem (7), defined by $\overline{u} = \frac{d\overline{x}}{dt}$, is non-constant. This concludes the proof of Theorem 1.

Notes and Comment. As stated in the text, Theorem 1 will be improved in Sect. 4, that is, its assumptions will be considerably weakened. As it is, it is a weak version of a result of Rabinowitz in [Rab2]. Ekeland in [Eke3] gave a different proof for the convex case; it is the one we use here. This method was then adapted by Brezis, Coron and Nirenberg to the wave equation in [BreCN], where they reproved yet another result of Rabinowitz in [Rab1].

3. The Index at Mountain-Pass Points

In the preceding section, we have found a non-trivial T-periodic solution $\overline{x}$ of the superlinear Hamiltonian system

$$\dot{x} = JH'(t, x) \tag{1}$$

by applying the Ambrosetti-Rabinowitz theorem (Corollary 1.9) to the reduced functional $\psi(u)$. Hofer's theorem 1.13 tells us that the critical point $\overline{u} = \frac{d\overline{x}}{dt}$ we find either is a mp-point, or is a limit of local minima. We now ask what it means for the T-periodic solution $\overline{x}$ itself. In this (rather technical) section we shall answer this question by showing that $\overline{x}$ must have index 0 or 1.

This is continuation of Sect. 2. All the assumptions and notations of Sect. 2 remain valid. In addition, we assume that $H(t,\cdot)$ is C^2 in x for every t, the Hessian H'' being continuous in (t,x), and that there is a continuous function $\lambda : [0,T] \times \mathbb{R}^{2n} \to \mathbb{R}$ such that

$$H''(t,x) \geq \lambda(t,x)I \quad \forall (t,x) \tag{2}$$

$$x \neq 0 \Rightarrow \lambda(t,x) > 0 \ . \tag{3}$$

Then every non-trivial T-periodic solution of Eq. (1) is admissible in the sense of Definition I.6.1. We recall that the index $i_T(\overline{x})$ is defined to be the dimension of the negative subspace associated with the quadratic form q_T on L^2_o given by

$$\begin{aligned} q_T(u,u) :&= \int_o^T \frac{1}{2}\left[(Ju, \Pi u) + \left(H''(t,\overline{x}(t))^{-1} Ju, Ju\right)\right] dt \\ &= \int_o^T \frac{1}{2}\left[(Ju, \Pi u) + \left(|(H^*)|''\,(t, -J\overline{u}(t)) Ju, Ju\right)\right] dt \end{aligned} \tag{4}$$

and that the nullity $\nu_T(\overline{x})$ is the dimension of the kernel of q_T (see Sect. II.6).

The problem is now to relate q_T and ψ. Unfortunately, q_T is not the Hessian of ψ at $\overline{u}$; they are not defined on the same spaces (L^2_o and L^β_o respectively, with $1 < \beta < 2$), and anyway ψ is not C^2. The regularity of H^* is not in question; it is simply the fact that, since $\beta > 2$, if we differentiate twice we fall outside the range of Krasnoselskii's Theorem II.3.4. Put in another way, the integral (4) does not define a continuous quadratic form on L^β, for $\beta < 2$.

On the other hand, the restriction of ψ to $L^\infty_o \subset L^\beta_o$, denoted by ψ_∞, clearly is C^2. If $\overline{u}$ is a critical point of ψ on L^β_o, then $\overline{u} = \frac{d\overline{x}}{dt}$ and $\overline{x}$ solves Eq. (1), so $\overline{u} \in C^o \subset L^\infty$. The Hessian of ψ_∞ at $\overline{u}$ is the restriction of q_T to L^∞_o:

$$(\psi''_\infty(\overline{u})v, v) = q_T(v,v) \quad \forall v \in L^\infty_o \ . \tag{5}$$

So we wish to bring the whole mountain-pass situation down from L^β_o to L^β_o. It is the purpose of the following lemma.

Lemma 1. *Take $\overline{u} \in K(\psi,\gamma)$, that is, $\psi'(\overline{u}) = 0$ and $\psi(\overline{u}) = \gamma$, and set*

$$\psi^\gamma := \{u \in L^\beta_o \mid \psi(u) < \gamma\} \tag{6}$$

$$B_\beta(\overline{u},\delta) = \left\{u \mid \|u - \overline{u}\|_\beta \leq \delta\right\} \tag{7}$$

$$B_\infty(\overline{u},\epsilon) = \{u \mid \|u - \overline{u}\|_\infty \leq \epsilon\} \ . \tag{8}$$

Then, for any $\epsilon > 0$, there exists a $\delta > 0$ such that, whenever $\mathcal{P}$ is a path component of ψ^γ with $\mathcal{P} \cap B_\beta(\overline{u}, \delta) \neq \emptyset$, then:

$$\exists u_{\epsilon,\mathcal{P}} \in \mathcal{P} \bigcap B_\beta(\overline{u}, \delta) \bigcap B_\infty(\overline{u}, \epsilon) \ . \tag{9}$$

□

In other words, if a path component of ψ^γ comes arbitrarily close to $\overline{u}$ in L^β_o, then it comes arbitrarily close to $\overline{u}$ in L^∞_o.

Proof. Let $\mathcal{P}$ be a path component of ψ^γ with $\mathcal{P} \cap B_\beta(\overline{u}, \delta) \neq \emptyset$, and set:

$$\gamma_{\delta,\mathcal{P}} = \text{Inf} \left\{ \psi(u) | u \in \mathcal{P} \bigcap B_\beta(\overline{u}, \delta) \right\} \ . \tag{10}$$

Since $\mathcal{P} \subset \psi^\gamma$, it follows immediately that:

$$\gamma_{\delta,\mathcal{P}} < \gamma \ . \tag{11}$$

The proof now proceeds in several steps.

Step 1: The infimum is achieved at a point $\widetilde{u}_{\delta,\mathcal{P}}$

Take a minimizing sequence $u_n \in \mathcal{P} \cap B_\beta(\overline{u}, \delta)$. We have $\psi(u_n) \to \gamma_{\delta,\mathcal{P}}$. Since L^β_o is reflexive, we may choose the sequence to be weakly convergent:

$$u_n \to \widetilde{u} \quad \text{for } \sigma\left(L^\beta_o, L^\alpha\right) \ . \tag{12}$$

We wish to show that $\widetilde{u} \in \mathcal{P} \cap B_\beta(\overline{u}, \delta)$. Since the ball is convex and closed, it is clear that $\widetilde{u} \in B_\beta(\overline{u}, \delta)$. Since ψ is weakly l.s.c. (remember that it is the sum of a compact term and a convex term), we have

$$\psi(\widetilde{u}) \leq \liminf_{n\to\infty} \psi(u_n) = \gamma_{\delta,\mathcal{P}} \ . \tag{13}$$

We claim that there is some n_o such that the line segment connecting u to u_{n_o}:

$$\{h\widetilde{u} + (1-h)u_{n_o} | 0 \leq h \leq 1\} \tag{14}$$

lies entirely inside ψ^γ. This will prove that $\widetilde{u}$ lies in the same path component as u_{n_o}, that is $\widetilde{u} \in \mathcal{P}$.

Argue by contradiction, assuming that there are sequence h_k in $[0,1]$ and $u_{n(k)}$ in $\mathcal{P} \cap B_\beta(\overline{u}, \delta)$ such that:

$$\psi(v_k) \geq \gamma \quad \text{with } v_k := h_k\widetilde{u} + (1-h_k)u_{n(k)} \ . \tag{15}$$

We write this explicity:

$$\gamma \leq \int_o^T \frac{1}{2}(Jv_k, \Pi v_k)\, dt + \int_o^T H^*(t, -Jv_k) dt \ . \tag{16}$$

Remember that $H^*(t,\cdot)$ is convex, so that:

$$
\begin{aligned}
\gamma \leq & \int_o^T \frac{1}{2}\left(Jv_k, \Pi v_k\right) dt + h_k \int_o^T H^*\left(t, -J\widetilde{u}\right) dt \\
& + (1-h_k)\int_o^T H^*(t, -Ju_{n(k)})dt \\
= & (1-h_k)\psi(u_{n(k)}) - (1-h_k)\int_o^T \frac{1}{2}\left(Ju_{n(k)}, \Pi u_{n(k)}\right) dt \\
& + \int_o^T \frac{1}{2}\left(Jv_k, \Pi v_k\right) dt + h_k \int_o^T H^*(t, -J\overline{u})dt \ .
\end{aligned} \tag{17}
$$

Because of (12), we have $v_k \to \widetilde{u}$ for $\sigma\left(L_o^\beta, L^\alpha\right)$, and hence $\Pi v_k \to \Pi\widetilde{u}$ strongly, since Π is compact. We may assume that $h_k \to h_\infty$ in $[0,1]$, and we know that $\psi\left(u_{n(k)}\right) \to \gamma_{\delta,\mathcal{P}}$. Passing to the limit in inequality (17), we get:

$$
\gamma \leq (1-h_\infty)\,\gamma_{\delta,\mathcal{P}} + h_\infty \psi(\widetilde{u}) \ . \tag{18}
$$

Using (13), we get $\gamma \leq \gamma_{\delta,\mathcal{P}}$, which contradicts (11). This finally proves that $\widetilde{u} \in \mathcal{P}$ and concludes Step 1.

Step 2: The necessary conditions for optimality at $\widetilde{u}_{\delta/\mathcal{P}}$

$\mathcal{P}$ is an open subset of L_o^β, so $\widetilde{u}$ is in fact a local minimum of ψ on the ball $B_\beta(\widetilde{u},\delta)$. Since ψ is C^1, and the ball is defined by the constraint $\|u-\overline{u}\|^\beta \leq \delta$, the left-hand side of which is C^1, there is a Lagrange multiplier $\lambda = \lambda_{\delta,\mathcal{P}} \in \mathbb{R}$ such that:

$$
\langle \lambda\left(\widetilde{u}-\overline{u}\right)^{\beta-1} + \psi'(\widetilde{u}), w\rangle = 0 \quad \forall w \in L_o^\beta \ . \tag{19}
$$

Here $v^{\beta-1}$ denotes the function $v\,\|v\|^{\beta-2}$.

The Lagrane multiplier λ cannot be negative. Indeed, since $\mathcal{P}$ is open and $B_\beta(\overline{u},\delta)$ is convex, the point $(1-h)\widetilde{u}+h\overline{u}$ must belong to $\mathcal{P}\cap B_\beta(\overline{u},\delta)$ for small h. Hence:

$$
\frac{d}{dh}\psi\left((1-h)\widetilde{u}+h\overline{u}\right) = \langle \psi'(\widetilde{u}), \overline{u}, -\widetilde{u}\rangle \geq 0 \ . \tag{20}
$$

Taking $w = \overline{u} - u$ in Eq. (19), we get

$$
\lambda\,\|u-\overline{u}\|^\beta + \langle \psi'(\widetilde{u}), \widetilde{u}-\overline{u}\rangle = 0 \ . \tag{21}
$$

Comparing with (20), we derive:

$$
\lambda \geq 0 \ . \tag{22}
$$

Now replace $\psi'(\widetilde{u})$ by its value (Lemma 2.8) in formula (19). It becomes:

$$
\left(\lambda(\widetilde{u}-\overline{u})^{-1} - J\Pi\widetilde{u} + J\nabla H^*\left(t,-J\widetilde{u}\right), w\right) = 0 \quad \forall w \in L_o^\beta \ . \tag{23}
$$

This means that there is a constant vector $\xi = \widetilde{\xi}_{\delta,\mathcal{P}} \in \mathbb{R}^{2n}$ such that:

$$\lambda\left(\widetilde{u}-\overline{u}\right)^{\beta-1} - J\Pi\widetilde{u} + J\nabla H^*\left(t,-J\widetilde{u}\right) = \widetilde{\xi} \quad \text{a.e.} \tag{24}$$

On the other hand, we know that $\psi'(\overline{u}) = 0$, so that there exists $\overline{\xi} \in \mathbb{R}^{2n}$ with:

$$-J\Pi\overline{u} + J\nabla H^*(t,-J\overline{u}) = \overline{\xi} \quad \text{a.e.} \tag{25}$$

Substracting both equations, we get:

$$\lambda\left(\widetilde{u}-\overline{u}\right)^{\beta-1} + J\nabla H^*\left(t,-J\widetilde{u}\right) - J\nabla H^*\left(t,-J\overline{u}\right) - J\Pi\left(\widetilde{u}-u\right) = \widetilde{\xi} - \overline{\xi}\,. \tag{26}$$

Step 3: An L^∞ estimate for $\widetilde{u}_{\delta,\mathcal{P}}$ when $\delta \to 0$

We first derive a uniform estimate for $\widetilde{\xi}_{\delta,\mathcal{P}}$. Integrate both sides of Eq. (26) against $\widetilde{u}-\overline{u}$, and then divide by $\|\widetilde{u}-\overline{u}\|_\beta$. We get (not writing the variable t):

$$\begin{aligned}\lambda\left\|u-\overline{u}\right\|_\beta^{\beta-1} &+ \langle J\nabla H^*(-J\widetilde{u}) - J\nabla H^*(-J\overline{u}), \widetilde{u}-\overline{u}\rangle \left\|\widetilde{u}-\overline{u}\right\|_\beta^{-1} \\ &- \langle J\Pi(\widetilde{u}-\overline{u}), \widetilde{u}-\overline{u}\rangle \left\|\widetilde{u}-\overline{u}\right\|_\beta^{-1} = 0\,.\end{aligned} \tag{27}$$

By Krasnoselskii's theorem II.3.4, since ∇H^* satisfies the estimate (2.33), the map $u \to \nabla H^*(-Ju)$ is continuous from L^β to L^α. The second term on the left therefore goes to zero when $\delta \to 0$. The third term on the left also vanishes when $\delta \to 0$. So the remaining term has to vanish too. More precisely, if $\delta_n \to 0$ and $\mathcal{P}_n$, $n \in \mathbb{N}$, is a sequence of path components in ψ^γ such that $B_\beta\left(\widetilde{u},\delta_n\right) \cap \mathcal{P}_n \neq \emptyset$, then:

$$\lambda_{\delta_n,\mathcal{P}_n}\left\|\widetilde{u}_{\delta_n,\mathcal{P}_n} - \overline{u}\right\|_\beta^{\beta-1} \to 0 \qquad \text{when } n \to \infty\,. \tag{28}$$

Writing this back into Eq. (26), and using Krasnoselskii's theorem again, we find that

$$\widetilde{\xi}_{\delta_n,\mathcal{P}_n} \to \overline{\xi} \qquad \text{when } n \to \infty\,. \tag{29}$$

This will give us an L^∞ estimate for $\widetilde{u}_{\delta_n,\mathcal{P}_n}$. Indeed, simplify the notations:

$$\widetilde{u}_n := \widetilde{u}_{\delta_n,\mathcal{P}_n}\,, \quad \widetilde{\xi}_n := \widetilde{\xi}_{\delta_n,\mathcal{P}_n}\,, \quad \lambda_n := \lambda_{\delta_n,\mathcal{P}_n} \tag{30}$$

and rewrite Eq. (26) as follows:

$$J\nabla H^*\left(t,-J\widetilde{u}_n\right) + \lambda_n\left(\widetilde{u}_n-\overline{u}\right)^{\beta-1} = f_n \tag{31}$$

$$f_n(t) := J\nabla H^*\left(t,-J\overline{u}(t)\right) + J\Pi\left(\widetilde{u}_n-\overline{u}\right) + \widetilde{\xi}_n - \overline{\xi}\,. \tag{32}$$

The f_n are continuous and converge uniformly to $J\nabla H^*\left(t,-J\overline{u}(t)\right)$ as $n \to \infty$.

Note that the left-hand side of Eq. (31) is the derivative at $y = \widetilde{u}(t)$ of the convex function:

$$y \to H^*(t, -Jy) + \frac{\lambda_n}{\beta} \|y - \widetilde{u}(t)\|^\beta \quad . \tag{33}$$

Hence, writing that the function lies above its tangent hyperplane at $\widetilde{u}(t)$:

$$\begin{aligned} H^*(t; -J\overline{u}(t)) \geq H^*(t; -J\widetilde{u}_n(t)) + \frac{\lambda_n}{\beta} \|\widetilde{u}_n(t) - \overline{u}(t)\|^\beta \\ + (f_n(t), \overline{u}(t) - \widetilde{u}_n(t)) \quad . \end{aligned} \tag{34}$$

Using estimate (2.32), we get for every $t \in [0, T]$ and $n \in \mathbb{N}$ the alternative

$$\text{either} \quad \|\widetilde{u}_n(t)\| \leq R \tag{35}$$

$$\text{or} \quad H^*(t, -J\overline{u}(t)) \geq \frac{1}{\beta k^\beta} \|\widetilde{u}_n(t)\|^\beta + (f_n(t), \overline{u}(t) - \widetilde{u}_n(t)) \tag{36}$$

from which it follows that the $\widetilde{u}_n$ are essentially bounded:

$$\exists A : \quad \|\widetilde{u}_n(t)\| \leq A \quad \forall t \in [0, T] \quad \forall n \in \mathbb{N} \; . \tag{37}$$

Step 4: $\widetilde{u}_{\delta,\mathbb{P}}$ converges to $\overline{u}$ in L^∞ when $\delta \to 0$

We claim that:

$$\|\widetilde{u}_n - \overline{u}\|_\infty \to 0 \qquad \text{when } n \to \infty \; . \tag{38}$$

Assume otherwise. Then we can find a subsequence $\widetilde{u}_{n'}$, and a sequence $t_{n'} \in [0, 1]$, such that

$$\widetilde{u}_{n'}(t_{n'}) - \overline{u}(t_{n'}) \to \overline{y} \neq 0 \qquad \text{when } n' \to \infty \; . \tag{39}$$

Since $\lambda_n \geq 0$, inequality (34) will imply in the limit:

$$H^*(t; -J\overline{u}(t)) \geq H^*(t; -J\overline{y}) + (J\nabla H^*(t; -J\overline{u}(t)), \overline{u}(t) - \overline{y}) \tag{40}$$

which we rewrite as:

$$H^*(t; -J\overline{u}(t)) - H^*(t; -J\overline{y}) \geq (\nabla H^*(t; -J\overline{u}(t)), -J\overline{u}(t) + J\overline{y}) \quad . \tag{41}$$

On the other hand, since H^* is strictly convex with respect to y, we have

$$H^*(t; -J\overline{y}) - H^*(t; -J\overline{u}(t)) > (\nabla H^*(t; -J\overline{u}(t)), -J\overline{y} + J\overline{u}(t)) \tag{42}$$

whenever $\overline{y} \neq 0$. Adding up the two last inequalities, we get a contradiction.

Step 5: Conclusion

Assume Lemma 1 is false. Then there is an $\epsilon > 0$, a sequence $\delta_n \to 0$ and a sequence $\mathcal{P}_n$ of path components of ψ^γ such that:

(43) $$\mathcal{P}_n \cap B_\beta(\overline{u}, \delta_n) \neq \emptyset$$

(44) $$\mathcal{P}_n \cap B_\beta(\overline{u}, \delta_n) \cap B_\infty(\overline{u}, \epsilon) = \emptyset \ .$$

Because of condition (43), we find in $\mathcal{P}_n \cap B_\beta(\overline{u}, \delta_n)$ a minimizer $\widetilde{u}_n$ for ψ. By the preceding analysis $\|\widetilde{u}_n - \overline{u}\|_\infty \to 0$ when $n \to \infty$, which contradicts condition (44). □

Lemma 1 has several interesting consequences.

Corollary 2. *$\overline{u} \in L_o^\infty$ is a local minimum for ψ_∞ if and only if it is a local minimum of ψ.*

Proof. If $\overline{u} \in L_o^\infty$ is a local minimum for ψ, it clearly is a local minimum for ψ_∞. Conversely, assume $\overline{u}$ is not a local minimum for ψ. Then ψ^γ has points in L_o^β arbitrarily close to $\overline{u}$. It then follows from Lemma 1 that it contains points in L_o^∞ arbitrarily close to $\overline{u}$, which means that $\overline{u}$ is not a local minimum for ψ_∞. □

Corollary 3. *$\overline{u} \in L_o^\infty$ is an isolated critical point for ψ if and only if it is an isolated critical point for ψ_∞.*

Proof. If $u_n \in L_o^\infty$ is a sequence of critical points of ψ_∞ converging to $\overline{u}$ in L_o^∞, they also converge in L_o^β. Conversely, if $u_n \in L_o^\beta$ is a sequence of critical points of ψ converging to $\overline{u}$ in L_o^β, they are all in L_o^∞, and a simple adaptation of the proof of Lemma 1 (where the Lagrange multiplier λ disappears) shows that in fact they converge in L_o^∞. □

Corollary 4. *If $\overline{u} \in L_o^\infty$ is a mp-point for ψ, then it is a mp-point for ψ_∞.*

Proof. Assume $\overline{u}$ is not a mp-point for ψ_∞. Then, either it is a local minimum for ψ_∞, or there is a neighbourhood $\mathcal{V}$ of $\overline{u}$ in L_o^∞ such that $\mathcal{V} \cap \psi_\infty^\gamma$ is path-connected. In the first case, $\overline{u}$ will also be a local minimum for ψ, and so cannot be a mp-point for ψ.

So assume we are in the second case. Choose $\epsilon > 0$ so small that $B_\infty(\overline{u}, \epsilon) \subset \mathcal{V}$, and apply Lemma 1 to get a corresponding $\delta > 0$. Let $\mathcal{P}_1$ and $\mathcal{P}_2$ be two path components of ψ^γ such that $\mathcal{P}_i \cap B_\beta(\overline{u}, \delta) \neq \emptyset$, $i = 1, 2$. Then we can find for $i = 1, 2$:

(45) $$u_i \in \mathcal{P}_i \cap B_\infty(\overline{u}, \epsilon) \subset \mathcal{V} \cap \psi_\infty^\gamma \ .$$

Since $\mathcal{V} \cap \psi_\infty^\gamma$ is path-connected in L_o^∞, it is also path-connected in L_o^β. So u_1 and u_2 belong to the same path component of ψ^γ. It follows that $\mathcal{P}_1 = \mathcal{P}_2$, that is, there is only one path component $\mathcal{P}$ in ψ^γ intersecting $B_\beta(\overline{u}, \delta)$. Set:

(46) $$\mathcal{W} := B_\beta(\overline{u}, \delta) \cup \mathcal{P} \ .$$

Then $\mathcal{W}$ is a neighbourhood of $\overline{u}$ in L_o^β, and $\mathcal{W} \cap \psi^\gamma = \mathcal{P}$ is path-connected, so $\overline{u}$ cannot be a mp-point for $\overline{u}$. □

We are brought back to the study of the C^2 function ψ_∞. We have $\psi'_\infty(\overline{u}) = \xi \in \mathbb{R}^{2n}$. By Proposition I.4.2, the Hessian $\psi''_\infty(\overline{u}) = q_T$ induces an orthogonal splitting of L^2_o

$$L^2_o = E_- \oplus E_o \oplus E_+ \, . \tag{47}$$

Since E_- and E_o are finite-dimensional, this induces a topological splitting of L^∞_o:

$$L^\infty_o = E_- \oplus E_o \oplus E^\infty_+ \, , \tag{48}$$

where $E^\infty_+ := E_+ \cap L^\infty_o$. We denote by P_-, P_o and P_+ the corresponding projections, and by u_-, u_o and u_+ the components of u on E_-, E_o and E_+.

Lemma 5. *There is a neighbourhood $\mathcal{V}$ of $\overline{u}_- + \overline{u}_o$ in $E_- \oplus E_o$ and a C^1 map $\sigma : \mathcal{V} \to E^\infty_+$ such that*

$$\langle \psi'_\infty(v + \sigma(v)), w \rangle = 0 \quad \forall w \in E^\infty_+ \tag{49}$$

$$\sigma(\overline{u}_- + \overline{u}_o) = \overline{u}_+ \quad \text{and} \quad \sigma'(\overline{u}_- + \overline{u}_o) = 0 \, . \tag{50}$$

The composed map $\widehat{\psi}_\infty(v) := \psi_\infty(v + \sigma(v))$ is C^2. □

Proof. Define a map $S : L^\infty_o \to L^\infty_o$ by:

$$S(u) := -J\Pi u + J\nabla H^*(t, -Ju) - \frac{1}{T}\int_o^T J\nabla H^*(t, -Ju)dt \tag{51}$$

so that:

$$\langle \psi'_\infty(u), v \rangle = \int_o^T (S(u), v)\, dt \, . \tag{52}$$

Clearly S is a C^1 map. We have

$$\langle S'(\overline{u})v, w \rangle = q_T(v, w) \quad \forall v \in L^2_o \quad \forall w \in L^2_o \tag{53}$$

which implies that $S'(\overline{u})$ preserves the splitting (48). In particular, we have

$$S'(\overline{u})E^\infty_+ \subset E^\infty_+ \, . \tag{54}$$

$S'(\overline{u})$ induces an isomorphism of E_+ onto itself. Using some regularity, we can show that $S'(\overline{u})$ is a bijection of E^∞_+ onto itself, and it then follows from the open mapping theorem of Banach that $S'(\overline{u})$ is an isomorphism of E^∞_+ into itself.

Now define a map $S_+(u_- + u_o, \cdot)$ of E^∞_+ into itself, depending on the parameters $u_- \in E_-$ and $u_o \in E_o$, by:

$$S_+(u_- + u_o, u_+) := S(u_- + u_o + u_+)_+ \, . \tag{55}$$

It is C^1, and we just saw that $S'_+(\overline{u}_-, \overline{u}_o, \overline{u}_+)$ is an isomorphism. On the other hand, since $\overline{u}$ is a critical point of ψ_∞, we have $S'(\overline{u}) = 0$, which implies:

$$S_+(\overline{u}_-, \overline{u}_o, \overline{u}_+) = 0 . \tag{56}$$

It then follows from the inverse function theorem that there is a neighbourhood $\mathcal{V}$ of $\overline{u}_- + \overline{u}_o$ in $E_- \oplus E_o$ and a unique C^1 map $\sigma : \mathcal{V} \to E_+^\infty$ such that

$$S_+(v, \sigma(v)) = 0 \quad \forall v \in \mathcal{V} \tag{57}$$

$$\sigma(\overline{u}_- + \overline{u}_o) = \overline{u}_+ . \tag{58}$$

The first equation is also $S(v + \sigma(v)) \in E_+^\infty$, which is equivalent to (49) because of relation (52). Differentiating it at $\overline{u}_- + \overline{u}_o$, we get:

$$S'(\overline{u})(v + \sigma(\overline{u}_- + \overline{u}_o)v) \in E_+^\infty \quad \forall v \in E_- \oplus E_o . \tag{59}$$

We already noticed that E_-, E_o and E_+^∞ are invariant subspaces of $S'(\overline{u})$. This means that $S'(u)v$ has no component in E_+^∞, and the preceding equation amounts to:

$$S'(\overline{u})\sigma'(\overline{u}_- + \overline{u}_o)v = 0 \quad \forall v \in E_- \oplus E_o . \tag{60}$$

Since $S'(\overline{u})$ induces an isomorphism of E_+^∞ onto itself, this boils down to $\sigma'(\overline{u}_- + \overline{u}_o) = 0$, as announced.

It remains to show that $\widehat{\psi}_\infty$ is C^2. It is C^1 since σ is C^1. Compute its derivative at $v \in E_- \oplus E_o$:

$$\begin{aligned} \langle \widehat{\psi}'_\infty(v), w \rangle &= \int_o^T (S(v + \sigma(v)), w + \sigma'(w))\, dt \\ &= \int_o^T (S(v + \sigma(v)), w)\, dt \end{aligned} \tag{61}$$

since $S(v + \sigma(v)) \in E_- \oplus E_o$ and $\sigma'(w) \in E_+$. Hence

$$\widehat{\psi}'_\infty(v) = (P_- + P_+)\, S(v + \sigma(v)) \tag{62}$$

which is again C^1. □

We now need to put $\widehat{\psi}_\infty$ in standard (or normal) form near the critical point $\overline{u}$. This is the purpose of the next result, which includes the classical Morse lemma and the Gromoll-Meyer theorem as special cases:

Normal Form Theorem. *Let $\mathcal{U} \ni 0$ be an open subset in a Hilbert space V and $\varphi : \mathcal{U} \to \mathbb{R}$ a C^2 function with $\varphi'(0) = 0$. Assume that $L := \varphi''(0)$ is Fredholm, so that V splits orthogonally into positive, negative, and null subspaces relative to $\varphi''(0)$:*

(63) $$V = E_+ \oplus E_- \oplus E_o$$

with $E_o = \operatorname{Ker} L$ *and* $E_+ \oplus E_- = L(V)$. *Then there exists an open neighbourhood* $\mathcal{V}$ *of* 0 *in* $L(V)$, *and open neighbourhood* $\mathcal{W}$ *of 0 in* Ker L, *a homeomorphism* h *from* $\mathcal{V} \times \mathcal{W}$ *onto an open neighbourhood of 0 in* V, *with* $h(0,0) = 0$, *and a* C^2 *function* $f : \mathcal{W} \to \mathbb{R}$ *such that:*

(64) the restrictions of $h(\cdot, 0)$ to $\mathcal{V}$ is a C^1–diffeomorphism

(65) $$f'(0) \quad \text{and} \quad f''(0) = 0$$

(66) $$\varphi \circ h(v,w) = \frac{1}{2}(Lv, v) + f(w) \qquad \text{for } v \in \mathcal{V}, \quad w \in \mathcal{W}.$$ □

The proof will be found in the book by Mawhin and Willem. Let us just apply the normal theorem to $\varphi(u) := \widehat{\psi}_\infty(u+\overline{u})$, which is defined on the finite-dimensional space $V := E_- \oplus E_o$, and introduce the corresponding maps h and f. Note that $L(V) = E_-$, $\operatorname{Ker} L = E_o$ and $f(0) = \gamma$.

Lemma 7. *If* $\dim E_- \geq 1$, *then there exists a neighbourhood* $\mathcal{U}$ *of* $\overline{u}$ *in* L_o^∞ *with the following property: for every* $u \in \mathcal{U} \cap \psi_\infty^\gamma$ *we can find a non-zero vector* $v \in E_-$ *and some* $\rho > 0$ *such that, for* $0 < t < \rho$, *the points* $\overline{u} + tv$ *and* u *lie in the same path component of* ψ_∞^γ. □

Proof. Set $\widetilde{u} = \overline{u}_- + \overline{u}_o \in E_- \oplus E_o$. Now consider the function

(67) $$E_+^\infty \ni w \to \psi_\infty(\widetilde{u} + v + \sigma(\widetilde{u} + v) + w)$$

depending on the parameter $v \in E_- \oplus E_o$. Its second derivative at 0, for $v = 0$, is positive definite. It follows that there exist η and δ such that, when $\|v\| < \eta$, its restriction to the ball $B_\infty(\overline{u}, \delta) \cap E_+^\infty$ is strictly convex. It then follows from the definition of σ that:

(68) $$\|\sigma(v) - \overline{u}_+\| < \delta \Rightarrow \widehat{\psi}_\infty(v) = \text{Min}\,\{\psi_\infty(v + \sigma(v) + w) \,|\, \|w - u_+\| < \delta\}\ .$$

We may choose η so small that we fall within the neighbourhoods of the origin specified by the normal form theorem, and that $\|\sigma(v) - \overline{u}_+\| < \delta$ when $v = h(v')$ and $\|v'\| < \eta$. Using relations (65), we may now choose η even smaller, and get:

(69) $$|f(v_o) - \gamma| \leq \frac{1}{16}\mu\,\|v_o\|^2 \qquad \text{for} \quad \|v_o\| < \eta$$

where the constant $\mu > 0$ is defined by:

(70) $$\mu = \text{Min}\,\{-(Lv_-, v_-) | v_- \in E_-, \|v_-\| = 1\}\ .$$

We then define an open neighbourhood $\mathcal{U}$ of $\overline{u}$ in L_o^∞ by:

$$\mathcal{U} = \left\{ \overline{u} + h(v) + \sigma(\widetilde{u} + h(v)) + w \;\middle|\; \begin{matrix} v \in E_- \oplus E_o \,, & \|v\| < \eta \\ w \in E_+ \,, & \|w\| < \delta \end{matrix} \right\} . \tag{71}$$

Take any $u \in \mathcal{U} \cap \psi^\gamma$, so that

$$u = \overline{u} + h(v) + \sigma\,(\widetilde{u} + h(v)) + w \; . \tag{72}$$

Write $u \sim u'$ if u and u' belong to the same path component of ψ^γ. This defines an equivalence relation in ψ^γ. By formula (68), we have:

$$u \sim u' := \overline{u} + h(v) + \sigma\,(\widetilde{u} + h(v)) \; . \tag{73}$$

By the definition of $\widehat{\psi}_\infty$, we have:

$$\psi_\infty(u') = \widehat{\psi}_\infty\,(\overline{u} + h(v)) \; . \tag{74}$$

By the normal form theorem, this becomes:

$$\psi_\infty(u') = f(v_o) + \frac{1}{2}\,(Lv_-, v_-) \; . \tag{75}$$

Define $\widetilde{v} \in E_-$ by:

$$\begin{cases} \widetilde{v}_- = v_- & \text{if } \; \|v_-\| \geq 2 \\ \widetilde{v}_- = v_- \, \|v_-\|^{-1} \eta/2 & \text{if } \; 0 < \|v_-\| < \eta/2 \\ \widetilde{v}_- = e_- \eta/2 & \text{if } \; v_- = 0 \end{cases} \tag{76}$$

where e_- is a fixed unit vector in E_-, and set

$$u'' := \overline{u} + h\,(\widetilde{v}_- + v_o) + \sigma\,(\widetilde{u} + h(\widetilde{v}_- + v_o)) \; . \tag{77}$$

Moving along the path:

$$t \to \overline{u} + h\left(v_-^t + v_o\right) + \sigma\left(\widetilde{u} + h\left(v_-^t + v_o\right)\right) \tag{78}$$

$$v_-^t := t\widetilde{v}_- + (1-t)v_- \,, \quad 0 \leq t \leq 1 \tag{79}$$

we find that we are increasing the negative term in formula (75), and keeping the other one constant, so that the whole path lies inside $\psi_{\infty'}^\gamma$ and $u' \sim u''$.

Finally, we claim that

$$u'' \sim u''' := \overline{u} + h(\widetilde{v}_-) + \sigma\,(\widetilde{u} + h(\widetilde{v}_-)) \; . \tag{80}$$

To see this, consider the path:

$$t \to u_t := \overline{u} + h\left(\widetilde{v}_- + v_o^t\right) + \sigma\left(\widetilde{u} + h\left(\widetilde{v}_- + v_o^t\right)\right) \tag{81}$$

$$v_t^o := (1-t)v_o \,, \quad 0 \leq t \leq 1 \tag{82}$$

which brings v_o to 0, and u'' to u'''. By the normal form theorem:

$$\psi_\infty(u_t) = \widehat{\psi}_\infty\left(\overline{u} + h\left(\widetilde{v}_- + v_o^t\right)\right) = \frac{1}{2}\left(L\widetilde{v}_-, \widetilde{v}_-\right) + f\left((1-t)v_o\right) \ . \tag{83}$$

By construction, $\|\widetilde{v}_-\| \geq \eta/2$. On the other hand, $\|v_o\| < \eta$, so that estimate (69) holds, and we have:

$$\psi_\infty(u_t) \leq -\frac{1}{8}\mu\eta^2 + \gamma + \frac{1}{16}\mu\eta^2 < \gamma \tag{84}$$

which proves assertion (80).

Now consider the path $c : [0,1] \to L_o^\infty$ given by:

$$c(t) := \overline{u} + h(t\widetilde{v}_-) + \sigma\left(\widetilde{u} + h(t\widetilde{v}_-)\right) \ . \tag{85}$$

Denote by $h'(v)$ the derivative of the restriction $h(\cdot, 0)$ to $\mathcal{V} \cap E_-$, which exists by condition (64) in the normal form theorem. So c is a C^1 path in L_o^∞, and we have:

$$c'(0) = h'(0)\widetilde{v}_- \in E_- \setminus \{0\} \tag{86}$$

because $\sigma'(\widetilde{u}) = 0$, and $h'(0)$ is an isomorphism.

We have by the normal form theorem:

$$\psi_\infty\left(c(t)\right) = \frac{t^2}{2}\left(L\widetilde{v}_-, \widetilde{v}_-\right) + f(0) = \frac{t^2}{2}\left(L\widetilde{v}_-, \widetilde{v}_-\right) + \gamma < \gamma \qquad \text{for } t > 0 \ . \tag{87}$$

We have found a C^1 path c such that $c(0) = \overline{u}$, $c(t) \in \psi_\infty^\gamma$ for $t > 0$, and $c(1) = u'''$. Clearly $c(t) \sim u'''$ for every $t > 0$, and hence:

$$c(t) \sim u \quad \forall t \in (0,1] \ . \tag{88}$$

In addition, we know that

$$c'(0) \in E_- \setminus \{0\} \ . \tag{89}$$

Because of formula (47) and the definition of E_-, this implies that:

$$\langle \psi_\infty''(\overline{u})c'(0), c'(0)\rangle < 0 \ . \tag{90}$$

Now consider the path $\overline{c}$ defined by:

$$\overline{c}(t) := \overline{u} + tc'(0) \ . \tag{91}$$

We claim that there is some $\rho > 0$ such that, when $0 < t < \rho$, the line segment starting at $c(t)$ and ending at $\overline{c}(t)$ lies entirely within ψ_∞^γ.

Suppose otherwise. Then there are sequences $t_n \to 0$, and λ_n in $(0,1]$ such that for every n:

$$\psi_\infty\left[\lambda_n c(t_n) + (1-\lambda_n)\overline{c}(t_n)\right] \geq \gamma \tag{92}$$

that is

$$\psi_\infty\left[\overline{u} + t_n\left((1-\lambda_n)c'(0) + \lambda_n\frac{c(t_n)}{t_n}\right)\right] \geq \psi_\infty(\overline{u}) \ .$$

Remember that $\psi'_\infty(\overline{u}) = 0$, so that this inequality implies in the limit:

$$\langle \psi''_\infty(\overline{u})c'(0), c'(0)\rangle \geq 0 \tag{93}$$

which contradicts inequality (90). We have thus proved that there is some $\rho > 0$ such that

$$0 < t < \rho \Rightarrow c(t) \sim \overline{c}(t) \ . \tag{94}$$

Remembering relation (88), we see that $\overline{c}(t) \sim u$ for $0 < t < \rho$. This concludes the proof. □

Corollary 8. *If* $\dim E_- \geq 2$, *then there is a neighbourhood* $\mathcal{U}$ *of* $\overline{u}$ *in* L^∞_o *such that* $\psi^\gamma_\infty \cap L^\infty_o$ *is path-connected.*

Proof. Take the neighbourhood $\mathcal{U}$ defined in Lemma 7. Let u_1 and u_2 be two points in $\mathcal{U} \cap \psi^\gamma_\infty$. We have seen that $u_i \sim u'_i$, $i = 1, 2$, with

$$u'_i := \overline{u} + h(w^i_-) + \sigma\left(\widetilde{u} + h(w^i_-)\right) \tag{95}$$

$$\psi_\infty\left(u'_i\right) = \frac{1}{2}L\left(w^i_-, w^i_-\right) + \gamma \tag{96}$$

for some $w^i_- \in E_- \setminus \{0\}$.

If $\dim E_- \geq 2$, the space $E_- \setminus \{0\}$ is path-connected, so that we can find a continuous path w^t_-, $1 \leq t \leq 2$, connecting w^1_- and w^2_-, with

$$\text{Min}\left\{\|w^1_-\|, \|w^-_2\|\right\} \leq \left\{\|w^t_-\|\right\} \leq \text{Max}\left\{\|w^1_-\|, \|w^-_2\|\right\} \ . \tag{97}$$

We define accordingly:

$$u'_t := \overline{u} + h\left(w^t_-\right) + \sigma\left(\widetilde{u} + h(w^t_-)\right) \tag{98}$$

and the normal form yields immediately:

$$\psi_\infty\left(u'_t\right) = \frac{1}{2}\left(Lw^t_-, w^t_-\right) + \gamma < \gamma \tag{99}$$

whereby we have shown that

$$u'_1 \sim u'_2 \ . \tag{100}$$

Hence $u_1 \sim u_2$, and the corollary is proved. □

We now sum up the results of this section.

Theorem 9. *Let* $\overline{u} \in K(\psi, \gamma)$ *be a mp-point, and let* $\overline{x}$ *be the corresponding* T*-periodic solution of problem* (1). *Then*

$$i_T(\overline{u}) \leq 1 \ . \tag{101}$$

If $i_T(\overline{x}) = 1$, then ψ^γ has exactly two path components $\mathcal{P}_o$ and $\mathcal{P}_i$. There is an eigenvector $e \neq 0$ associated with the negative eigenvalue of $\psi''(\overline{u})$, and we can choose $t > 0$ so small that:

$$\overline{u} + te \in \mathcal{P}_o \tag{102}$$

$$\overline{u} - te \in \mathcal{P}_1 \ . \tag{103}$$

Proof. Suppose $\overline{u}$ is a mp-point for ψ. Then it is a mp-point for ψ_∞ by Corollary 4, and we must have $\dim E_- < 2$ by Corollary 8. But

$$i_T(\overline{x}) = \dim E_- \tag{104}$$

by definition. Hence formula (101).

Define ϵ and δ as in Lemma 1. Since $\overline{u}$ is a mp-point for ψ, there are two distinct path components $\mathcal{P}_o$ and $\mathcal{P}_1$ of ψ^γ intersecting $B_\beta(\overline{u}, \delta)$, and they also intersect $B_\infty(\overline{u}, \epsilon)$. Take:

$$u_i \in \mathcal{P}_i \bigcap B_\beta\,(\overline{u}, \delta) \bigcap B_\infty\,(\overline{u}, \epsilon) \ , \quad i = 0, 1 \ . \tag{105}$$

Now let $\mathcal{U}$ be the neighbourhood of $\overline{u}$ in L_o^∞ defined in Lemma 7. Since we are free to choose $\epsilon > 0$, we may assume that $B_\infty\,(\overline{u}, \epsilon) \subset \mathcal{U}$. By Lemma 7, we know that for $t > 0$ small enough,

$$u_o \sim \overline{u} + te \tag{106}$$

$$u_1 \sim \overline{u} - te \tag{107}$$

where the relation $\sim$ means "belong to the same path-component of ψ_∞^γ in L_o^∞", and hence of ψ^γ in L_o^β.

If $\overline{u} + te$ and $\overline{u} - te$ belonged to the same path-component $\mathcal{P}$ of in L_o^β, then so would u_1 and u_2 by the relation (106), (107), and we would have

$$\mathcal{P}_o = \mathcal{P} = \mathcal{P}_1 \tag{108}$$

contradicting the fact that $\mathcal{P}_o$ and $\mathcal{P}_1$ are distinct.

If there is another path-component $\mathcal{P}_2$ intersecting $B_\beta(\overline{u}, \delta)$, we get a point

$$u_2 \in \mathcal{P}_2 \bigcap B_\beta\,(\overline{u}, \delta) \bigcap B_\infty(\overline{u}, \epsilon) \tag{109}$$

which, by Lemma 7, turns out to be equivalent to u_o or u_1. So $\mathcal{P}_2 = \mathcal{P}_o$ or $\mathcal{P}_2 = \mathcal{P}_1$. □

Notes and Comments. This section is borrowed from [EkeH1] with minor improvements.

The idea of deriving information on the index of a critical point from the procedure by which it has been constructed is extremely important in modern nonlinear analysis. We will encounter it again in Sect. V.3.

If $\operatorname{Ker} L = \{0\}$, the normal form theorem reduces to the Morse lemma. The first proofs lost two orders of differentiability, so that Morse lemma held only for C^3 functions, until Cambini in [Cam] gave a C^2 proof.

If $\operatorname{Ker} L = \{0\}$, the normal form theorem was proved by Gromoll and Meyer in [GroM], assuming φ to be C^3. The improvement to C^2 was done by Hofer in [Hof2]. The statement we give here incorporates theorem 3 of [MawW] ch. 8.6 and the subsequent remark, which becomes condition (64). Mawhin and Willem provide a simple proof, following an idea of K.C. Chang [Cha].

4. Subharmonics

We go back to the situation of Sect. 2, with a few changes. We shall assume that $H(\cdot, x)$ is T-periodic in t for every fixed x, and that $H(t, \cdot)$ is C^2 in x for every fixed t, with H' and H'' continuous jointly in (t, x), and

$$H''(t,x) \quad \text{positive definite for } x \neq 0 \,. \tag{1}$$

Theorem 2.1 tells us that if:

$$H(t,x) \geq H(t,0) = 0 \tag{2}$$

$$H(t,x) \leq \frac{\omega}{2} \|x\|^2 \qquad \text{for } \|x\| < \epsilon \tag{3}$$

$$H(t,x) \geq \lambda^{\alpha} H(t,x) \qquad \text{for } \|x\| \geq r \text{ and } \lambda \geq 1 \tag{4}$$

$$\limsup_{\substack{0 \leq t \leq T \\ \|x\| \to \infty}} H(t,x) \, \|x\|^{-\alpha} < \infty \,, \tag{5}$$

where $\epsilon > 0$, $r > 0$, $\alpha > 2$ and $\omega < 2\pi/T$ are suitable constants, then the Hamiltonian system

$$\dot{x} = JH'(t,x) \tag{6}$$

has a non-constant T-periodic solution.

We note that $H(\cdot, x)$ is also kT-periodic in t for every $k \in \mathbb{N}$, so that as long as

$$k < \frac{2\pi}{\omega T} \tag{7}$$

Theorem 1 will assert the existence of a non-constant kT-periodic solution x_k for system (6). We thus get K solutions:

$$x_k(t + kT) = x_k(t) \,, \quad 1 \leq k \leq K \tag{8}$$

with $K := E\left[\frac{2\pi}{\omega T}\right]$ (recall that $E[a] = a - 1$ when a is an integer), and the question is: are they pairwise distinct?

Of course, we have to think of phase shifts first. It might happen, for instance, if $k = pq$ is not prime, that x_k is in fact a time-translate of x_p:

$$\exists r \in \mathbb{R} : \quad x_k(t) = x_p(t + rT) . \tag{9}$$

We shall say that x_{k_1} and x_{k_2} are *geometrically distinct* if:

$$x_{k_1}(t) \neq x_{k_2}(t + pT) \quad \forall t \in \mathbb{R} , \quad \forall p \in \mathbb{N} \tag{10}$$

and our question is now: are the x_k pairwise geometrically distinct?

Theorem 1 *Assumptions* (1) *to* (5). *Set* $K = E\left[\frac{2\pi}{\omega T}\right]$. *Then, for each* $k \in \{1, \ldots, K\}$, *there is a* kT*-periodic solution* x_k *of the Hamiltonian system*

$$\dot{x} = JH'(t, x) \tag{11}$$

such that the x_k, $1 \leq k \leq K$, *are pairwise geometrically distinct.* □

The proof relies on index theory. For each k with $1 \leq k \leq K$ we find a critical point u_k of the functional

$$\psi^k(u) = \int_o^{kT} \left[\frac{1}{2}(Ju, \Pi u) + H^*(t, -Ju)\right] dt \tag{12}$$

on the space

$$L_o^\beta \left(0, kT; \mathbb{R}^{2n}\right) . \tag{13}$$

The corresponding kT-periodic solution $\dot{x}_k$ of $x = JH'(t, x)$ is characterized by the relation

$$\dot{x}_k = u_k \tag{14}$$

and the index and nullity of x_k over $(0, kT)$ are given by:

$$i_{kT}(x_k) = \dim E_- \tag{15}$$

$$\nu_{kT}(x_k) = \dim E_o , \tag{16}$$

where $E_- \oplus E_o \oplus E_+$ is the splitting of L_o^2 associated with the quadratic form

$$q_{kT}(v, v) = \int_o^{kT} \left[(Jv, \Pi v) + \left(H''\left(t, x_k(t)\right)^{-1} Jv, Jv\right)\right] dt . \tag{17}$$

The results in the preceding sections give us substantial information about u_k. We know that it is either a mp-point, or that it is not a local minimum but is the limit of a sequence of local minima (Corollary 1.13).

If u_k is a mp-point, i_{kT} must be 0 or 1 (Theorem 3.9). If $i_{kT} = 0$, then $\nu_{kT} \geq 1$, otherwise u_k would be a local minimum for ψ_∞^k on L_o^∞, contradicting the fact that it is a mp-point (Corollary 3.4).

If u_k is a limit of local minima for ψ^k in L_o^β, it is also a limit of local minima for ψ_∞^k in L_o^∞ (Corollaries 3.2 and 3.3). The function ψ_∞^k being C^2, we get $i_{kT} = 0$ in the limit. On the other hand, $\nu_{kT} \geq 1$, otherwise u_k would be a local minimum for ψ_∞^k on L_o^∞, again contradicting the fact that it is a mp-point.

Finally, we have in all cases:

$$i_{kT} \leq 1 \leq i_{kT} + \nu_{kT} \ . \tag{18}$$

It will be convenient to introduce the $\mathbb{Z}$-action on $L_o^\beta\left(0, kT; \mathbb{R}^{2n}\right)$ and the isotropy group $G(u_k)$. We write:

$$(p * u)(t) := u(t + pT) \qquad \text{for } \ p \in \mathbb{Z} \ , \quad u \in L_o^\beta \tag{19}$$

$$G(u) := \{p \in \mathbb{Z} | p * u = u\} \ . \tag{20}$$

$p * x$ is defined in the same way, and $G(u_k)$ is also the isotropy group of x_k. Clearly:

$$G(u_k) \supset k\mathbb{Z} \ . \tag{21}$$

Proposition 1. *Assume $i_{kT} = 1$. If k is odd, we have $G(u_k) = k\mathbb{Z}$. If k is even, we have $G(u_k) = k\mathbb{Z}$ or $\frac{k}{2}\mathbb{Z}$.* □

Proof. Suppose $G(u_k)$ is larger than $k\mathbb{Z}$. Then $G(u_k) = p\mathbb{Z}$ for some p dividing k, say $k = pq$. Set $T_o := pT$, so $kT = qT_o$ and $i_{kT} = i_{qT_o}$. Since u_k is T_o-periodic, we may use the Bott formula (Corollary I.5.4):

$$i_{kT} = \sum_{\omega^q = 1} j_{T_o}(\omega) \ . \tag{22}$$

Since $i_{kT} = 1$ and $j_{T_o}(\omega) = j_{T_o}(\overline{\omega}) \geq 0$, we are left with two possibilities. Either q is odd and $j_{T_o}(1) = 1$, or q is even and $j_{T_o}(1) = 1$ or $j_{T_o}(-1) = 1$.

If $j_{T_o}(1) = 1$, this means that $i_{T_o} = 1$, and Theorem I.5.1 leads to

$$i_{kT} \geq q i_{T_o} \geq q \tag{23}$$

so $q = 1$ and $k = p$.

If $j_{T_o}(-1) = 1$, then, still by Bott's formula:

$$i_{2T_o} = j_{T_o}(1) + j_{T_o}(-1) \geq 1 \tag{24}$$

and Theorem I.5.1 leads to:

$$i_{kT} \geq \left(\frac{q}{2}\right) i_{2T_o} \geq \frac{q}{2} \tag{25}$$

so $q = 2$ and $p = k/2$. □

Proposition 2. *Assume $i_{kT} = 1$ and u_k is a mp-point. Then $G(u_k) = k\mathbb{Z}$.*

Proof. All we have to do is to show that, when k is even, u_k cannot be $\frac{k}{2}T$-periodic. Suppose otherwise. Denote by $e_k \neq 0$ an eigenvector associated with the negative eigenvalue.

As we noted from Bott's formula, we are in the case when $j_{T_o}(-1) = 1$, with $T_o := \frac{kT}{2}$. This means that the eigenvector of Q_{T_o} associated with the negative eigenvalue belongs to the space (see the splitting in Lemma I.5.2)

$$E_{T_o}^{-1} := \left\{y \in W^{1,2}(0, kT) \mid y(0) + y(T_o) = 0\right\} . \tag{26}$$

Differentiating this eigenfunction, we get e_k, which must therefore be a T_o-antiperiodic function:

$$e_k(t + T_o) = -e_k(t) . \tag{27}$$

By Theorem 3.9, $(\psi^k)^\gamma$ has exactly two components $\mathcal{P}_o$ and $\mathcal{P}_1$, and we can choose $h > 0$ so small that:

$$u_k + ke_k \in \mathcal{P}_o \tag{28}$$

$$u_k - he_k \in \mathcal{P}_1 . \tag{29}$$

As we just noticed, e_k is $\frac{k}{2}T$-antiperiodic

$$e_k\left(t + \frac{k}{2}T\right) + e_k(t) = 0 . \tag{30}$$

Since $(\psi^k)^\gamma = \mathcal{P}_o \cup \mathcal{P}_1$, the origin belongs to $\mathcal{P}_o$ or $\mathcal{P}_1$. Say $0 \in \mathcal{P}_o$. Let c_o be a path in $\mathcal{P}_o$ connecting 0 to $u_k + he_k$:

$$c_o(0) = 0 \quad \text{and} \quad c_o(1) = u_k + he_k \tag{31}$$

$$\psi^k\left(c_o(s)\right) < \gamma \quad \forall s \in [0, 1] . \tag{32}$$

Now act on c_o by the phase shift $s \to s + \frac{kT}{2}$. This defines a new path c_1 by:

$$c_1(s) := \frac{k}{2} * c_o(s) . \tag{33}$$

Using (30) and (31), we get

$$c_1(0) = 0 \quad \text{and} \quad c(1) = u_k - he_k . \tag{34}$$

The integral of a $\frac{k}{2}T$-periodic function over one period does not depend on where we start integrating, that is:

$$\psi^k\left(c_1(s)\right) = \psi^k\left(c_o(s)\right) \quad \forall s \in [0, 1] \tag{35}$$

so $\psi^k\left(c_1(s)\right) < \gamma$ for all s. So $u_k - he_k$ lies in the connected component of the origin in $(\psi^k)^\gamma$:

(36) $$u_k - he_k \in \mathcal{P}_o \ .$$

From (29) and (36) we get a contradiction. □

We now conclude the proof of Theorem 1. We claim that all the u_k, $1 \leq k \leq N$, are geometrically distinct. If not, there exist $j < k$ and $p \in \mathbb{Z}$ such that

(37) $$x_j = p * x_k \ .$$

Write $u_j = \dot{x}_j$ and $u_k = \dot{x}_k$, so that $u_j = p * u_k$. All the points in $\mathbb{Z} * u_k$ clearly have the same local properties, so we may assume that $p = 0$. From now on, we write $u_j = u = u_k$ and $x_j = x = x_k$.

We have

(38) $$i_{kT}(x) \leq 1 \leq i_{kT}(x) + \nu_{kT}(x)$$

(39) $$i_{jT}(x) \leq 1 \leq i_{jT}(x) + \nu_{jT}(x) \ .$$

On the other hand, using Lemma I.4.9, we have:

(40) $$i_{kT}(x) \geq i_{jT}(x) + \nu_{jT}(x) \ .$$

If $i_{kT}(x) = 0$, we get $i_{jT}(x) + \nu_{jT}(x) = 0$ and contradict (39). If $i_{jT}(x) + \nu_{jT}(x) = 2$, we get $i_{kT}(x) \geq 2$ and contradict (38). So we must have:

(41) $$i_{kT} = 1 = i_{jT}(x) + \nu_{jT}(x) \ .$$

By Proposition 2, we must have $G(u) = k\mathbb{Z}$. But this contradicts the fact that u is jT-periodic, since $j < k$. Hence the result. □

We could also ask whether $G(u_k) = k\mathbb{Z}$ for $1 \leq k \leq N$, that is, whether x_k has minimal period kT. This will not be true without further assumptions on H, which, however, turn out to be generic. It will also be true in the autonomous case, as we shall see in the next section.

Notes and Comments. Take $\omega = 0$ (that is, $H(t,x) = o(x^2)$) for the sake of simplicity. In [Rab3], Rabinowitz proved the existence of a sequence $n_k \to \infty$ of geometrically distinct subharmonics x_k, $k \in \mathbb{N}$, with x_k having period $n_k T$. The fact that we can take $n_k = k$, that is, there exists a subharmonic of every possible period, is due to Ekeland and Hofer in [EkeH2].

5. Autonomous Problems and Potential Wells

In this section, we turn to autonomous problems. We shall considerably extend the preliminary existence result of Sect. 2. Later on, we shall consider the particular case of second-order systems.

Theorem 1. *Let Ω be a convex open subset of $\mathbb{R}^{2n}$ containing the origin. Let $H \in C^2(\Omega, \mathbb{R})$ be such that:*

(1) $$\forall x \in \Omega\ , \quad H(x) \geq H(0) = 0$$

(2) $$\forall x \neq 0\ , \quad H''(x) \ \text{ is positive definite}$$

(3) $$\exists \omega > 0\ ,\ \ \exists \epsilon > 0\ :\quad H(x) \leq \frac{\omega}{2}\|x\|^2 \qquad \text{for } \ \|x\| \leq \epsilon$$

(4) $$H''(x)^{-1} \to 0 \qquad \text{when } \ \|x\| \to \infty \ \text{ or } \ x \to \partial\Omega\ .$$

Then, for every $T < \frac{2\pi}{\omega}$, the system

(5) $$\begin{cases} \dot{x} = JH'(x) \\ x(0) = x(T) \end{cases}$$

has a solution with minimal period T. □

Some comments are in order before we start the proof. By $\partial\Omega$ we denote the boundary of Ω; the theorem does not preclude the case when $\Omega = \mathbb{R}^{2n}$, so $\partial\Omega = \emptyset$. The precise meaning of assumption (4) is that for every $a > 0$, there exist $\rho > 0$ and $\eta > 0$ such that:

(6) $$\|x\| \geq \rho \ \text{ or } \ d(x, \partial\Omega) \leq \eta \Rightarrow H''(x) \geq aI\ .$$

This implies that $H(x)\|x\|^{-2} \to \infty$ when $\|x\| \to \infty$ or $x \to \partial\Omega$. More precisely, for every $a' > 0$ there exist $\rho' > 0$ and $\eta' > 0$ such that

(7) $$\|x\| \geq \rho' \ \text{ or } \ d(x, \partial\Omega) \leq \eta' \Rightarrow H(x)\|x\|^{-2} \geq a'\ .$$

Saying that x has minimal period T means that it is not T/k-periodic for any integer $k \geq 2$. As a consequence, it is not constant:

(8) $$x(t) \neq 0 \quad \forall t\ .$$

We first prove the theorem in a particular case.

Lemma 2. *Assume that $\Omega = \mathbb{R}^{2n}$, and $H \in C^2\left(\mathbb{R}^{2n}, \mathbb{R}\right)$ satisfies conditions (1) to (4). Assume in addition that there exist $\alpha > 2$ and $r > 0$ such that:*

(9) $$\limsup_{\|x\| \to \infty} H(x)\|x\|^{-\alpha} < \infty$$

(10) $$\|x\| \geq r \Rightarrow H(\lambda x) \geq \lambda^\alpha H(x) \quad \forall \lambda \geq 1\ .$$

Then, for every $T < \frac{2\pi}{\omega}$, the system (5) has a solution $\overline{x}(t)$ with minimal period T, and we must have, for some t_o:

(11) $$H''\left(x(t_o)\right) \leq \frac{2\pi}{T} I\ .$$

Proof. Once we have added assumptions (9) and (10), we are back in the situation of Theorem 2.1. We introduce the (reduced) dual action functional:

$$\psi(u) := \int_o^T \left[\frac{1}{2}(Ju, \Pi u) + H^*(-Ju)\right] dt \tag{12}$$

on the space L_o^β, and we find a critical point $\overline{u} = \frac{d\overline{x}}{dt}$ by the Ambrosetti-Rabinowitz theorem. By Hofer's Corollary 1.13, we known that we can choose $\overline{u}$ to be

either a mp − point
or a local minimum .

In the first case, $\overline{u}$, being a local minimum for ψ on L_o^β, will also be a local minimum for ψ_∞ on L_o^∞, and its index must be zero:

$$i_T(\overline{x}) = 0 \; . \tag{13}$$

In the second case, by Theorem 3.9, the index must be zero or one:

$$i_T(\overline{x}) \leq 1 \; . \tag{14}$$

We know that $\overline{u} \neq 0$; so $\overline{x}$ is non-constant and $\overline{x}(t) \neq 0$ for every t. The solution $\overline{x}$ is admissible; if it has period T/k, we have by Corollary I.6.5:

$$i_T = i_{k \cdot T/k} \geq k - 1 \; . \tag{15}$$

If $i_T(\overline{x}) = 0$, we get $k = 1$, and $\overline{x}$ has minimal period T.

If $i_T(\overline{x}) = 1$, we get $k \leq 2$, and $\overline{x}$ has minimal period T or $T/2$. We have to show that it cannot have period $T/2$. We argue as in Proposition 4.2.

Assume $i_T(\overline{x}) = 1$ and $\overline{x}$ is $T/2$-periodic. Then $\overline{u}$ is a mp-point. Denote by $e \neq 0$ an eigenvector associated with the negative eigenvalue.

By Bott's formula (Corollary I.5.4) we have:

$$i_T = i_{T/2} + j_{T/2}(-1) \tag{16}$$

while the theory of conjugate points (Corollary I.6.5) yields

$$i_T \geq i_{T/2} + 1 \; . \tag{17}$$

Since $i_T = 1$, inequalities (16) and (17) imply that $i_{T/2} = 0$ and $j_{T/2}(-1) = 1$. This means that the eigenvector of Q_T associated with the negative eigenvalue belongs to the space (see Definition I.5.3)

$$E_{T/2}^{-1} := \left\{y \in W^{1,2}\left(0, T; \mathbb{R}^{2n}\right) \mid y(0) + y(T/2) = 0\right\} \; . \tag{18}$$

Differentiating this eigenfunction, we get e, which must therefore be $T/2$-antiperiodic:

$$e(t + T/2) = -e(t) \; . \tag{19}$$

By Theorem 3.9, ψ^γ has exactly two components $\mathcal{P}_o$ and $\mathcal{P}_1$, and we choose $h > 0$ so small that $\overline{u} + he \in \mathcal{P}_o$ and $\overline{u} - he \in \mathcal{P}_1$.

Consider the S^1-action on L_o^β (phase shift):

$$(20) \qquad (\theta * u)(t) := u(t + \theta T) \qquad \text{for } \theta \in S^1 .$$

Define a continuous path $c(\theta)$, starting at $\overline{u} + he$, by:

$$(21) \qquad c(\theta) := \theta * (\overline{u} + he) , \quad 0 \leq \theta \leq 1/2$$

We have $\psi(c(\theta)) = \psi(c(0)) < \gamma$ for every θ, so this path lies inside ψ^γ. This ends at:

$$(22) \qquad c\left(\frac{1}{2}\right) = \frac{1}{2} * \overline{u} + h\left(\frac{1}{2} * e\right) = \overline{u} - he$$

since $\overline{u}$ is $T/2$-periodic and e is $T/2$-antiperiodic. So we have connected $\overline{u} + he$ to $\overline{u} - he$ by a continuous path in ψ^γ. This is a contradiction since they lie in different components, and our result is proved: $\overline{x}$ cannot have period $T/2$, so it has minimal period T in all cases.

All that remains is estimate (11). Assume it does not hold, then:

$$(23) \qquad \forall t , \quad H''(x(t)) > \frac{2\pi}{T} I$$

and equivalently:

$$(24) \qquad \forall t , \quad H''(x(t))^{-1} < \frac{T}{2\pi} I .$$

By compactness, there is some $\epsilon > 0$ such that

$$(25) \qquad \forall t , \quad H''(x(t))^{-1} \leq \left(\frac{T}{2\pi} - \epsilon\right) I .$$

Consider the quadratic form on $L_o^2(0, T)$

$$(26) \qquad q_T(v, v) := \int_o^T \left[(Jv, \Pi v) + \left(H''(\overline{u}(t))^{-1} Jv, Jv\right)\right] dt .$$

The index $i_T(\overline{u})$ is the sum of the dimensions of the negative eigenspaces of q_T (see Definition I.4.3). Inequality (23) yields

$$(27) \qquad q_T(v, v) \leq \int_o^T \left[(Jv, \Pi v) + \left(\frac{T}{2\pi} - \epsilon\right) v^2\right] dt .$$

The operator $J\Pi$ has a $2n$-dimensional eigenspace E_1 associated with the eigenvalue $\frac{T}{2\pi}$ (formula III.1.50). On this eigenspace, we have:

$$(28) \qquad v \in E_1 \Rightarrow q_T(v, v) \leq -\epsilon \|v\|^2 .$$

So $i_T(\overline{u}) \geq \dim E_1 = 2n$. On the other hand, we have seen that $i_T(\overline{u}) \leq 1$. We have a contradiction, and the lemma is proved. □

We now reduce the general case to the particular case studied in Lemma 2.

Proof of Theorem 1. Choose $h_o > 1$ such that:

$$(x \in \Omega \quad \text{and} \quad H(x) \geq h_o) \Rightarrow H''(x) > \frac{2\pi}{T} I \ . \tag{29}$$

Set:

$$\widetilde{\Omega} := \{x \in \Omega | H(x) < h_o\} \ . \tag{30}$$

$\widetilde{\Omega}$ is an open bounded convex set. Construct a function $\widetilde{H} \in C^2 \left(\mathbb{R}^{2n}, \mathbb{R}\right)$ such that

$$\forall x \in \widetilde{\Omega} \ , \quad \widetilde{H}(x) = H(x) \tag{31}$$

$$\forall x \notin \widetilde{\Omega} \ , \quad \widetilde{H}''(x) > \frac{2\pi}{T} I \tag{32}$$

$$\exists r > 0 \ , \ \exists c > 0 \ : \quad \|x\| > r \Rightarrow \widetilde{H}(x) = c \, \|x\|^4 \ . \tag{33}$$

One can, for instance, proceed as follows. Choose a number $h_1 > h_o$ such that

$$H(x) \geq h_1 \Rightarrow H''(x) \geq \left(\frac{2\pi}{T} + 1\right) I \ . \tag{34}$$

Now consider the Fenchel conjugate H^* of H. It is convex and finite everywhere. Choose some large number $A > 0$ and write

$$H_A^*(y) := H^*(y) \qquad \text{if} \quad \|y\| \leq A, +\infty \ \text{ otherwise} \ . \tag{35}$$

Take the Fenchel conjugate again, thereby defining $H_A := (H_A^*)^*$. It is convex, finite everywhere, and grows linearly at infinity. If A has been chosen large enough, we will have

$$H(x) \leq 2h_1 \Rightarrow H(x) = H_A(x) \ . \tag{36}$$

Now define

$$H_A^\delta(x) := \operatorname{Max} \left\{H_A(x), \delta \, \|x\|^2\right\} \ . \tag{37}$$

For $\delta > 0$ small enough, we have

$$H(x) \leq 2h_1 \Rightarrow H(x) = H_A(x) = H_A^\delta(x) \ . \tag{38}$$

Finally, choose an increasing C^2 convex function $\varphi : \mathbb{R}_+ \to \mathbb{R}_+$ such that

(39) $$\varphi(h) = h \quad \text{for } 0 \le h \le h_1$$

(40) $$\varphi(h) = \gamma h^2 \quad \text{for } h \ge 2h_1 \ ,$$

where $\gamma > 0$ is a large constant. Now set

(41) $$\widehat{H}(x) := \varphi\left(H_A^\delta(x)\right) \ .$$

$\widehat{H}$ is a convex function, which coincides with H on the set where $H(x) \le h_1$, and with $\gamma\delta^2 \|x\|^4$ when $\|x\|$ is large. If γ is large enough, we will have $\widehat{H}''(x) \ge \left(\frac{2\pi}{T} + 1\right) I$ at every point x where $\widehat{H}$ is C^2 and $H(x) \ge h_1$. Smoothing $\widehat{H}$ down, we get a C^2 function $\widetilde{H}$ satisfying (31) to (33).

Apply Lemma 2 to $\widetilde{H}$. We get a solution $\overline{x}$ of the problem

(42) $$\dot{x} = J\widetilde{H}'(x)$$

(43) $$x(0) = x(T)$$

such that $\overline{x}$ has minimal period T and

(44) $$\exists t_o : \widetilde{H}''\left(\overline{x}(t_o)\right) \le \frac{2\pi}{T} I \ .$$

Condition (29) then implies that $\widetilde{H}\left(\overline{x}(t_o)\right) < h_o$. Since $\widetilde{H}$ is an integral of the motion

(45) $$\widetilde{H}\left(\overline{x}(t)\right) < h_o \quad \forall t \ .$$

Condition (31) then implies that $\widetilde{H}$ coincides with H on a neighbourhood of the trajectory $\overline{x}$. So $\widetilde{H}'\left(\overline{x}(t)\right) = H'\left(\overline{x}(t)\right)$, and $\overline{x}$ in fact solves the equation

(46) $$\dot{x} = JH'(x) \ . \qquad \square$$

Theorem 1 does not apply directly to second-order system

(47) $$\begin{cases} \ddot{q} + V'(q) = 0 \\ q(0) = q(T) \\ \dot{q}(0) = \dot{q}(t) \end{cases}$$

because the corresponding Hamiltonian

(48) $$H(p,q) := \frac{1}{2}p^2 + V(q)$$

cannot satisfy assumption (4). A special statement is needed, with its own proof:

Theorem 3. *Let Ω be a convex open subset of $\mathbb{R}^n$ containing the origin. Let $H \in C^2(\Omega, \mathbb{R})$ be such that*

(49) $$\forall q \in \Omega\ , \quad V(q) \geq V(0) = 0$$

(50) $$\forall q \neq 0\ , \quad V''(q) \text{ is positive definite}$$

(51) $$\exists \omega > 0\ : \quad V(q) \leq \frac{\omega}{2} \|q\|^2 \text{ for } \|q\| \leq \epsilon$$

(52) $$V''(q)^{-1} \to 0 \qquad \text{when } \|q\| \to 0 \text{ or } q \to \partial\Omega\ .$$

Then, for every $T < \frac{2\pi}{\sqrt{\omega}}$, *problem* (47) *has a solution with minimal period* T. □

Proof. We work in the Lagrangian formulation (II.4, Example 2). The dual functional is

(53) $$\psi(q) := \int_o^T \left[-\frac{1}{2}\dot{q}^2 + V^*(\ddot{q}) \right] dt\ .$$

Setting $\ddot{q} := u \in L_o^\beta$, and defining the operator Π as usual, the dual functional becomes

(54) $$\psi(u) = \int_o^T \left[-\frac{1}{2} (\Pi u)^2 + V^*(u) \right] dt\ .$$

The connexion with the Hamiltonian formalism is as follows. Using the Hamiltonian of formula (48), we get the dual functional

(55) $$\begin{aligned} \overline{\psi}(q,p) : &= \int_o^T \left[-p\dot{q} + \frac{1}{2}\dot{q}^2 + V^*(\dot{p}) \right] dt \\ &= \frac{1}{2} \int_o^T (p - \dot{q})^2 dt + \int_o^T \left[-\frac{1}{2}p^2 + V^*(\dot{p}) \right] dt \\ &= \frac{1}{2} \|p - \dot{q}\|_2^2 + \psi(\dot{p})\ . \end{aligned}$$

So (p, q) is a critical point of $\overline{\psi}$ if and only if $u = \dot{p}$ is a critical point of ψ, and $\dot{q} = p$. Note that ψ and $\overline{\psi}$ have the same index, since they differ only by a positive definite term.

The proof now proceeds in two steps.

Step 1. Assume in addition that there exist $\alpha > 2$ and $r > 0$ such that

(56) $$\limsup_{\|q\| \to \infty} V(q) \|q\|^{-\alpha} < \infty$$

(57) $$\|q\| \geq r \Rightarrow V(\lambda q) \geq \lambda^\alpha V(q) \quad \forall \lambda \geq 1\ .$$

We then argue as in Lemma 2, by applying the Ambrosetti-Rabinowitz theorem to ψ, and showing that the critical point has index 0 or 1; details are left to the reader. So Theorem 3 obtains under the additional assumptions (56) and (57).

Step 2. In the general case, that is, when V no longer satisfies (56) and (57), we proceed as follows. With every $k \in \mathbb{N}$, we associate the set:

$$\Omega_k := \{q \in \Omega | V(q) \leq k\} \ . \tag{58}$$

Ω_k is an open bounded subset of $\mathbb{R}^n$. Proceeding as in the proof of Theorem 1, we construct a function $V_k \in C^2(\mathbb{R}^n, \mathbb{R})$ such that:

$$\forall k \in \mathbb{N} \ , \quad V_k \leq V_{k+1} \leq V \tag{59}$$

$$\forall q \in \Omega_k \ , \quad V_k(q) = V(q) \tag{60}$$

$$\forall \lambda > 0 \ , \ \exists M \ : \ \forall k \geq M \ , \ q \notin \Omega_M \Rightarrow V_k''(q) \geq \lambda I \tag{61}$$

$$\exists r > 0 \ , \ \exists c_k > 0 \ : \quad \|q\| > r \Rightarrow V_k(q) = c_k \|q\|^4 \ . \tag{62}$$

V_k satisfies the additional conditions (56) and (57). By Step 1, the problem

$$\begin{cases} \ddot{q} + V_k'(q) = 0 \\ q(t+T) = q(T) \end{cases} \tag{63}$$

has a solution q_k, with minimal period T and index ≤ 1. The latter is the index of the quadratic form $\psi_k''(u_k)$ on L_o^2, with $u_k := \frac{d^2}{dt^2} q_k$ and

$$\begin{aligned} (\psi_k''(u_k)v, v) :&= \int_o^T \left[-(\Pi v)^2 + ((V_k^*)''(u_k)v, v) \right] dt \\ &= \int_o^T \left[-(\Pi v)^2 + \left(V_k''(q_k)^{-1} v, v \right) \right] dt \ . \end{aligned} \tag{64}$$

The critical point u_k for ψ_k is found by the Ambrosetti-Rabinowitz theorem (see Sect. 2). The corresponding critical value γ_k is given by:

$$\gamma_k := \inf_{c \in \Gamma} \max_{0 \leq s \leq 1} \psi_k(c(s)) \ , \tag{65}$$

where Γ is the set of all continuous paths $c : [0,1] \to L_o^{4/3}$ such that $c(0) = 0$ and $\psi_k(c(1)) < 0$. Since $\psi_k(0) = 0$, we have $\gamma_k > 0$, and since $V_k \leq V_{k+1}$, we have $V_k^* \geq V_{k+1}^*$. It follows that the sequence γ_k is bounded:

$$0 \leq \gamma_{k+1} \leq \gamma_k \ . \tag{66}$$

By the duality Theorem II.4.2, we have:

$$\gamma_k = \int_o^T \left[\frac{1}{2} \dot{q}^2 - V_k(q_k) \right] dt \ . \tag{67}$$

We also introduce the constants:

$$h_k := \frac{1}{2}\dot{q}_k(t)^2 + V_k\left(q_k(t)\right) \ . \tag{68}$$

If the $V_k \circ q_k$, $k \in \mathbb{N}$, are uniformly bounded, say $V_k\left(q_k(t)\right) \leq b$, the problem is over. Indeed, it then follows from (60) that $q_k(t) \in \Omega_k$ as soon as $k \geq b$, so that q_k is in fact a solution of problem (47), with minimal period T.

So all we have to do is to show that the sequence $\|V_k \circ q_k\|_\infty$ is bounded. Assume otherwise. Then we may assume that

$$\|V_k \circ q_k\| \to \infty \tag{69}$$

so that $h_k \to \infty$ by formula (68).

We will now relate the h_k to the γ_k. We have, by adding (67)and (68):

$$\gamma_k + Th_k = \int_o^T \dot{q}_k^2 dt \ . \tag{70}$$

Take any $M \in \mathbb{N}$, and set

$$m_k := \text{Max}\left\{V_k\left(q_k(t)\right) | 0 \leq t \leq T\right\} \to \infty \ . \tag{71}$$

We take k so large that $m_k > M$. Denote by μ the Lebesgue measure on $[0, T]$ and set:

$$A_k := \mu\left\{t | V_k\left(q_k(t)\right) \geq M\right\} \ . \tag{72}$$

From Eqs. (68) and (70) it follows that:

$$\gamma_k + Th_k \geq \int_{V_k(q_k)<M} \dot{q}_k^2 dt \geq 2\left(h_k - M\right)\left(T - \mu(A_k)\right) \ . \tag{73}$$

Rewrite this inequality as follows

$$\begin{aligned} \gamma_k &\geq h_k\left(T - 2\mu(A_k)\right) - 2M\left(T - \mu(A_k)\right) \\ &\geq h_k\left(T - 2\mu(A_k)\right) - 2M\left(T - \mu(A_k)\right) \ . \end{aligned} \tag{74}$$

Since γ_k is bounded and $h_k \to +\infty$, we must have $T - 2\mu(A_k) \leq 0$ for all but a finite number of $k \in \mathbb{N}$. Thus we have proved that:

$$\forall M \ , \ \exists K \ : \ \forall k \geq K \ , \quad \mu(A_k) \geq \frac{T}{2} \ . \tag{75}$$

By condition (61), there is an M so large that:

$$(k \leq M \quad \text{and} \quad V_k(q) > M) \Rightarrow V_k''(q) \geq 2\left(\frac{4\pi}{T}\right)^2 \ . \tag{76}$$

Pick the corresponding K from formula (75). With K fixed in this way, pick a vector $\xi \in \mathbb{R}^n$, and set, for $t \geq 0$:

$$\mu_K(t) := 2\pi\mu\left(A_K \bigcap [0,T]\right) \Big/ \mu(A_K) \tag{77}$$

$$v_K(t) := \begin{cases} e^{J\mu_K(t)}\xi & \text{if } t \in A_K \\ 0 & \text{otherwise .} \end{cases} \tag{78}$$

The function v_K has been built to have mean value zero, so $v_K \in L_o^2(0,T)$. Since A_K is closed, $[0,T]\backslash A_K$ is a countable union of subintervals $(t_n, t_n+\delta_n)$, $n \in \mathbb{N}$, so that:

$$\frac{2\pi}{\mu(A_K)} \int_o^T v_K(s)ds = J\xi - \begin{cases} Je^{J\mu_K(t)}\xi & \text{if } t \in A_K \\ Je^{J\mu_K(t_n)}\xi & \text{if } t_n < t < t_n + \delta_n \text{ .} \end{cases} \tag{79}$$

Hence:

$$(\Pi v_K)(t) = \int_o^T v_K(s)ds - \frac{1}{T}\int_o^T \left(\int_o^t v_k(s)ds\right) dt \tag{80}$$

with:

$$\int_o^T \left(\int_o^t v_k(s)ds\right) dt = \frac{\mu(A_K)}{2\pi} J\xi T - \frac{\mu(A_K)}{2\pi} J \sum_{n=0}^{\infty} \delta_n e^{J\mu_K(t_n)}\xi \text{ .} \tag{81}$$

We have:

$$\int_o^T (\Pi v_K)^2\, dt = \int_o^T \left(\int_0^t v_K(s)ds\right)^2 dt - \frac{1}{T}\left(\int_o^T \left(\int_o^t v_k(s)ds\right) dt\right)^2 \text{ .} \tag{82}$$

Substituting (79) and (81) into the right-hand side yields:

$$\begin{aligned} \int_o^T (\Pi v_K)^2\, dt &= \frac{\mu(A_K)^2}{4\pi^2} T \|\xi\|^2 - \frac{\mu(A_K)^2}{4\pi^2}\frac{1}{T}\left(\sum_{n=0}^{\infty} \delta_n e^{J\mu_K(t_n)}\xi\right)^2 \\ &\geq \frac{\mu(A_K)^2}{4\pi^2} \|\xi\|^2 \left[T - \frac{1}{T}\left(\sum_{n=0}^{\infty}\delta_n\right)^2\right] \\ &= \frac{\mu(A_K)^2}{4\pi^2} \|\xi\|^2 \left[T - \frac{1}{T}(T-\mu(A_K))^2\right] \\ &= \frac{\mu(A_K)^2}{4\pi^2} \|\xi\|^2 \mu(A_K)\left(2 - \frac{\mu(A_K)}{T}\right) \text{ .} \end{aligned} \tag{83}$$

Finally, we use (75) to get a uniform estimate:

$$\int_o^T (\Pi v_K)^2\, dt \geq \left(\frac{T}{4\pi}\right)^2 \mu(A_K) \|\xi\|^2 \text{ .} \tag{84}$$

We now substitute this into the quadratic form $\psi_K''(u_K)$ given by formula (64). Recall also that v_K vanishes outside A_K and that estimate (76) holds on A_K:

$$(85)\qquad \begin{aligned}(\psi_K''(u_K)v_K, v_K) &\leq -\left(\frac{T}{4\pi}\right)^2 \mu(A_K)\,\|\xi\|^2 + \int_o^T \left(V_K''(q_K)^{-1}v_K, v_K\right) dt \\ &\leq \mu(A_K)\,\|\xi\|^2 \left[-\left(\frac{T}{4\pi}\right)^2 + \frac{1}{2}\left(\frac{T}{4\pi}\right)^2\right] \\ &= -\frac{1}{2}\left(\frac{T}{4\pi}\right)^2 \mu(A_K)\,\|\xi\|^2 \ .\end{aligned}$$

Since ξ varies in $\mathbb{R}^n$, we have found an n-dimensional space of functions v_K where the restriction of $\psi_K''(u_K)$ is negative definite. This means that the index of q_K, the solution of problem (63) associated with u_K, is at least n. But we know that this index is ≤ 1. If $n > 1$, we have a contradiction, so that assumption (69) cannot hold, and the theorem is proved. The case $n = 1$ is of course trivial, and can be handled directly. □

Notes and Comments. In a series of papers, [Rab2], [BenR], [Rab3], [Rab4], Rabinowitz considered Hamiltonians of the form $H(x) = Q(x) + R(x)$, where $Q(x)$ is a quadratic form and R is superquadratic in the following sense:

$$(86)\qquad R(x) \geq R(0) = 0$$

$$(87)\qquad R(x)\,\|x\|^{-2} \to 0 \quad \text{when } x \to 0$$

$$(88)\qquad \exists \alpha > 2\ ,\ \exists r\ :\quad \|x\| \geq r \Rightarrow (x, R'(x)) \geq \alpha R(x) > 0\ .$$

This type of condition was first introduced in [AmbR]. It does not imply that R is convex; if it is, then it is equivalent to $R(\lambda x) \geq \lambda^\alpha R(x)$ for $\lambda \geq 1$ and $\|x\| \geq r$, as we pointed out in Lemma 2.4.

Rabinowitz then proved that for every $T > 0$ there was a non-trivial T-periodic solution, and asked if one could find a solution with minimal period T. After substantial progress by Ambrosetti and Mancini [AmbM], and by Girardi and Matzeu [GirM1], [GirM2], the question was finally settled by Ekeland and Hofer in [EkeH1], in the case when H is convex and $\omega = 0$. They introduced index theory into the subject, and their method was extended by Girardi and Matzeu in [GirM3] to the case $H(x) = Q(x) + R(x)$, with Q positive definite.

Theorem 1 is joint work with P.L. Lions (unpublished). It contains all the previous results for the convex case, and extends them by using a method of P.L. Lions, which consists in deriving a priori estimates from the fact that the index at the critical point under consideration is known (see [Lio]). This has enabled us to replace the Ambrosetti-Rabinowitz condition (88) by the more natural condition that $H''(x)$ grows to infinity when $\|x\| \to \infty$. As a premium, our theorem holds also for Hamltionians which are defined on a subset of $\mathbb{R}^{2n}$ only, and blow up on the boundary.

Theorem 2 is joint work with Coti Zelati and P.L. Lions (unpublished). It extends Theorem 1 to the case of second-order systems. It covers the case of potential wells, that is, potentials which are defined on a bounded subset of $\mathbb{R}^n$, and got to infinity on the boundary. Such potentials were studied first by Benci [Ben2], and then by Ambrosetti and Coti-Zelati [AmbCZ], under an assumption which restricts the growth near the boundary:

$$(V'(q), q) \geq \alpha V(q) \quad \text{when } d(q, \partial\Omega) < \epsilon . \tag{89}$$

Again, in Theorem 2, this assumption is replaced by the more natural one that $V''(q)$ goes to infinity at the boundary.

It would be of interest to know whether conditions (4) or (52) can be weakened further. In Theorem 1, for instance, is it enough that the highest eigenvalue of $H''(x)$ goes to infinity when $\|x\| \to \infty$? Is it enough that $\|H'(x)\| \, \|x\|^{-1} \to \infty$ when $\|x\| \to \infty$? Is it enough that $H(x) \, \|x\|^{-2} \to \infty$?

Chapter V. Fixed-Energy Problems

Introduction

The fixed-energy problems are the most interesting (and the most difficult) in the theory, because of their geometric significance. Many are still unsolved, and we conclude this chapter by listing the most important ones.

From now on, the index theory of Chap. I will be an essential tool. In Sect. 2, we also introduce the S^1-index of Fadell and Rabinowitz in the case of a free S^1-action, with some attempt at justifying its properties. We abandon all pretense in Sect. 3, where we use the S^1-index in the non-free case, and the technical side of the proofs to the original work of Ekeland and Hofer.

1. Existence, Length, Stability

The two preceding chapters were devoted to fixed-period problems. In this last chapter, we turn to fixed-energy problems.

We are given in $\mathbb{R}^{2n}$ a convex compact set C, with non-empty interior. Denote its boundary by Σ. For $x \in \Sigma$, denote by $N_\Sigma(x)$ the normal cone to C at x:

(1) $$N_\Sigma(x) := \left\{ y \in \mathbb{R}^{2n} | (x' - x, y) \leq 0, \forall x' \in C \right\} .$$

We are interested in finding $T > 0$ and an absolutely continuous curve $x : [0, T] \to \mathbb{R}^{2n}$ such that:

(2) $$\begin{cases} x(t) \in \Sigma & \forall t \\ \dot{x}(t) \in J N_\Sigma(x) \\ x(0) = x(T) \end{cases} .$$

If Σ is smooth, that is, if it is a C^1 hypersurface, then $N_\Sigma(x)$ is a half-line depending continuously on $x \in \Sigma$. Let n_Σ be any non-vanishing continuous section of N:

(3) $$n_\Sigma \in C^o\left(\Sigma, \mathbb{R}^{2n}\right) \ : \quad 0 \neq n_\Sigma(x) \in N_\Sigma(x) \quad \forall x \in \Sigma \, .$$

Then problem (2) is equivalent to finding $T > 0$ and $x \in C^1$ such that:

(4) $$\begin{cases} x(t) \in \Sigma & \forall t \\ \dot{x} = J n_\Sigma(x) & \\ x(0) = x(T) & \end{cases} .$$

Indeed, any solution of (4) clearly is a solution of (2). Conversely, if $x(t)$ is a solution of (2), then there is a continuous map $\lambda : [0, T] \to (0, \infty)$ such that $\dot{x}(t) = \lambda(t) n_\Sigma(x(t))$. Defining a new time variable $s = \varphi(t)$ by

(5) $$\varphi(t) := \int_o^t \lambda(u) du$$

we see that $(\varphi(T), x \circ \varphi)$ is a solution of (4).

If Σ is a C^2 hypersurface, the equation $\dot{x} = Jn(x)$ defines a flow on Σ, since $Jn(x)$ is a C^1 field of tangent vectors, and the flow exists for all times since Σ is compact. Problem (4) consists of finding the closed trajectories of that flow.

Definition 1. A solution (T, x) of problem (2) is called a *closed characteristic* on Σ. Two closed characteristics (T, x) and (S, y) are *geometrically distinct* if there is no map $\varphi : \mathbb{R}/T\mathbb{Z} \to R/S\mathbb{Z}$ such that $x = y \circ \varphi$. □

In other words, two closed characteristics are geometrically distinct if you cannot deduce one from the other by rescaling time. For instance, if (T, x) is a closed characteristic, so are

$$\begin{array}{llll} (T, x_1)\,, & \text{with} & x_1(t) = x(t + t_o)\,, & \text{for} \quad t_o \in \mathbb{R} \\ (kT, x_2)\,, & \text{with} & x_2(t) = x(t)\,, & \text{for} \quad k \in \mathbb{N} \\ (T/\alpha, x_3) & \text{with} & x_3(t) = x(\alpha t)\,, & \text{for} \quad \alpha > 0 \end{array}$$

but they are not geometrically distinct from x.

The basic questions we want to answer are as follows. Are there any closed characteristics? If so, how many (geometrically distinct)? Can anything be said about their length and their (linear) stability?

To cast the problem in Hamiltonian form, we proceed as in Sect. I.7 (Example 2), assuming that $0 \in \text{Int}C$. We first introduce the gauge $j_C : \mathbb{R}^{2n} \to [0, \infty)$ of C

(6) $$j_C(x) := \text{Min}\left\{\lambda \,\middle|\, \frac{x}{\lambda} \in C\right\} \qquad \text{for} \quad x \neq 0$$

(7) $$j_C(0) := 0$$

and then the Hamiltonian $H_\alpha : \mathbb{R}^{2n} \to [0, \infty)$ defined by:

$$H_\alpha(x) := j_C(x)^\alpha \; . \tag{8}$$

H_α is convex and positively homogeneous of degree α. It is also C^1 if Σ is C^1, and C^2 if Σ is C^2 and $\alpha > 2$. We refer to Example I.7.2 for the Euler identities which hold in the C^1 case.

If Σ is C^1, then $n(x) := H'_\alpha(x)$ clearly provides a continuous non-vanishing section of $N_\Sigma(x)$. With this choice of $n(x)$, problem (4) becomes:

$$\begin{cases} H_\alpha(x) = 1 \\ \dot{x} = JH'_\alpha(x) \\ x(0) = x(S) \end{cases} . \tag{9}$$

More generally, we have:

Lemma 2. *Assume Σ bounds a convex compact set C with $0 \in \mathrm{Int}C$. Set $H_\alpha = j_C^\alpha$ with $\alpha > 1$. Then any solution (T_α, x_α) of*

$$\begin{cases} H_\alpha(x) = 1 \\ \dot{x} \in J\partial H_\alpha(x) \\ x(0) = x(S) \end{cases} \tag{10}$$

is a closed characteristic of Σ. Conversely, if (T, x) is a closed characteristic of Σ, there is a bijection $\varphi : \mathbb{R}/T_\alpha\mathbb{Z} \to \mathbb{R}/T\mathbb{Z}$ such that (T_α, x_α) with $x_\alpha = x \circ \varphi$, is a solution of (10). □

Proof. We have, for $x \in \Sigma$:

$$\partial H_\alpha(x) = \alpha \partial j_c(x) = \{x^* \in N_\Sigma(x) | \langle x, x^* \rangle = \alpha\} \tag{11}$$

so $\partial H_\alpha(x)$ generates the convex cone $N_\Sigma(x)$. Since $\partial H_\alpha \subset N_\Sigma(x)$, any solution of (10) must be a closed characteristic of Σ. Conversely, if (T, x) is a closed characteristic of Σ, there is for every s some $\lambda(s) > 0$ such that $\dot{x}(s) \in \lambda(s)\partial H_\alpha(x(s))$. The map $\dot{x}$ belongs to $L^1\left(\mathbb{R}/T\mathbb{Z}; \mathbb{R}^{2n}\right)$, and ∂H_α, which has compact convex values, bounded away from the origin, is upper semi-continuous (that is, has closed graph). Using a measurable selection theorem, we find that we can choose the map $\lambda : \mathbb{R}/T\mathbb{Z} \to (0, \infty)$ to be L^1 too. We then define a new time variable $t = \psi(s)$ by

$$\psi(s) = \int_o^s \lambda(u) du \; . \tag{12}$$

Since $\lambda(u) > 0$, the map $\psi : [0, S] \to [0, T]$, with $T := \psi(S)$, is a homeomorphism. Let φ be its inverse, and set $x_\alpha := x \circ \varphi$. We have

$$\frac{dx_\alpha}{dt} = \frac{dx}{ds}\frac{d\varphi}{dt} = \dot{x}(s)\lambda(s)^{-1} \in \partial H_\alpha\left(x(s)\right) \tag{13}$$

with $s = \varphi(t)$. So x solves problem (10). □

As we already observed in Lemma I.7.2, the solutions (T, x) of the problem

$$\begin{cases} \dot{x} \in J\partial H_\alpha(x) \\ x(0) = x(S) \end{cases} \tag{14}$$

come in continuous families, indexed by the energy level h. To be precise, choose $S = 1$, and let $x_1 \neq 0$ be a solution of:

$$\begin{cases} \dot{x} \in J\partial H_\alpha(x) \\ x(0) = x(1) \end{cases} . \tag{15}$$

Let h be its energy level:

$$H_\alpha(x_1(t)) := h . \tag{16}$$

Then (T_o, x_o) is a closed characteristic of Σ, with:

$$x_o(t) := h^{-1/\alpha} x_1\left(h^{\frac{2-\alpha}{\alpha}} t\right) \tag{17}$$

$$T_o := h^{-\frac{2-\alpha}{\alpha}} . \tag{18}$$

This trick reduces the fixed-energy problem to the fixed-period problem. It gives us our first existence result.

Theorem 3. *Let $C \subset \mathbb{R}^{2n}$ be a convex compact set with non-empty interior. Its boundary Σ always carries a closed characteristic.*

Proof. The problem is translation-invariant, so we may assume that $0 \in \operatorname{Int} C$. Choose $\alpha \in (1, 2)$ and introduce the Hamiltonian $H_\alpha := j_C^\alpha$, where j_C is the gauge of C.

Consider the boundary-value problem:

$$\begin{cases} \dot{x} \in J\partial H_\alpha(x) \\ x(0) = x(1) \end{cases} . \tag{19}$$

By Corollary III.3.6, it has a solution x_1 with minimal period 1. In particular, x_1 is non-zero, so it can be lifted to the energy level 1: the curve x_o given by formulas (17) and (18) is a closed characteristic on Σ. □

Now that the question of existence is answered, we may start looking for qualitative properties of the closed characteristics. The first one is the length – the shorter the better. We shall give two basic estimates in the C^1 case.

If $x : \mathbb{R}/T\mathbb{Z} \to \Sigma$ is a closed characteristic, a natural measure of length in this context is the *action* $A(T, x)$, defined by:

$$A(T, x) := \frac{1}{2} \int_o^T (Jx, \dot{x})\, dt . \tag{20}$$

Indeed, if $\varphi : \mathbb{R}/S\mathbb{Z} \to \mathbb{R}/T\mathbb{Z}$ is any orientation-preserving diffeomorphism, we have $A(T, x \circ \varphi) = A(T, x)$; so $A(T, x)$ is a geometric quantity, depending only on how many times one runs around the closed characteristic.

Denote by (T_α, x_α) the particular solution of problem (19) associated with the closed characteristic x. We have:

$$
\begin{aligned}
A(T,x) &= \frac{1}{2}\int_o^{T_\alpha} (Jx_\alpha, \dot{x}_\alpha)\,dt = \frac{1}{2}\int_o^{T_\alpha} (Jx_\alpha, JH'_\alpha(x_\alpha))\,dt \\
&= \frac{1}{2}\int_o^{T_\alpha} (x_\alpha, H'_\alpha(x_\alpha))\,dt = \frac{\alpha}{2}\int_o^{T_\alpha} H_\alpha(x_\alpha)dt = \frac{\alpha}{2}T_\alpha
\end{aligned} \tag{21}
$$

where we have used the Euler identity on the function H_α, which is positively homogeneous of degree α.

So $2\alpha^{-1}A(x)$ is the period of x_α. Taking into account the equation $\dot{x}_\alpha = JH'_\alpha(x_\alpha)$ on $H(x_\alpha) = 1$, it is easy to deduce from the period $\alpha^{-1}A(x)$ upper and lower bounds for the euclidian length of the trajectory of x_α, that is, of the closed characteristic x.

Theorem 4 (Croke-Weinstein). *Assume Σ is a C^1 hypersurface bounding a convex compact set C. Suppose C contains a ball of radius $r > 0$*

$$\forall x \in \Sigma\ , \quad r \le \|x - x_o\| \tag{22}$$

for some $x_o \in \mathbb{R}^{2n}$. Then, if (T,x) is a closed characteristic on Σ, we have:

$$A(T,x) \ge \pi r^2\ . \tag{23}$$

□

Proof. We shall work with $\alpha = 2$. Without loss of generality we may take $x_o = 0$, so that:

$$H_2(x) \le \|x\|^2 r^{-2}\ , \quad \forall x \in \mathbb{R}^{2n}\ . \tag{24}$$

Denoting by θ the angle between x and $H'_2(x)$, we have

$$(x, H'_2(x)) = \|x\|\,\|H'_2(x)\|\cos\theta\ , \quad \forall x \in \Sigma\ . \tag{25}$$

The tangent plane to Σ at x cannot intersect the ball $\|x\| \le r$, since the latter is contained in C. It follows that $\|x\|\cos\theta \ge r$, and hence:

$$(x, H'_2(x)) \ge r\,\|H'_2(x)\|\ , \quad \forall x \in \Sigma\ . \tag{26}$$

By the Euler identity, the left-hand side is just 2, so that $r\,\|H'_2(x)\|/2 \le 1$. Multiplying the right-hand side by this number, we get:

$$(x, H'_2(s)) \ge \frac{r^2}{2}\,\|H'_2(x)\|^2\ , \quad \forall x \in \Sigma\ . \tag{27}$$

We now look at the solution (T_2, x_2) of

$$\begin{cases} \dot{x} = JH'_2(x) \\ x(0) = x(S) \end{cases} \tag{28}$$

corresponding to the closed characteristic (T, x) by Lemma 2. We have $x_2(t) \in \Sigma$ for every t, so that inequality (27) yields:

$$\int_o^{T_2} (x_2, -J\dot{x}_2)\, dt \geq \frac{r^2}{2} \int_o^{T_2} \|\dot{x}_2\|^2\, dt \; . \tag{29}$$

Note that we can add a constant to x_2 without affecting the left-hand side. Replace x_2 by $\Pi\dot{x}_2$, the primitive with mean zero, and use Cauchy-Schwarz:

$$\|\Pi\dot{x}_2\| \geq \frac{r^2}{2} \|\dot{x}_2\| \; . \tag{30}$$

By the Wirtinger inequality (Lemma I.4.1), this yields

$$T_2(2\pi)^{-1} \geq r^2 2^{-1} \; . \tag{31}$$

Since $T_2 = A(T, x)$, this is the desired result. □

Proposition 5. *Assume Σ is a C^1 hypersurface bounding a convex compact set C. Suppose C is contained in a ball of radius $R > 0$:*

$$\forall x \in \Sigma \, , \quad \|x - x_o\| \leq R \tag{32}$$

for some $x_o \in \mathbb{R}^{2n}$. Then, there is a closed characteristic (T, x) on Σ such that

$$A(T, x) \leq \pi R^2 \; . \tag{33}$$

□

Proof. We shall work with $1 < \alpha < 2$. Without loss of generality, we may take $x_o = 0$, so that

$$H_\alpha(x) \geq \|x\|^\alpha R^{-\alpha} \, , \quad \forall x \in \mathbb{R}^{2n} \; . \tag{34}$$

Fix $T > 0$, and consider the (reduced) dual functional ψ_T associated with the fixed-period problem (10):

$$\psi_T(u) := \int_o^T \left[\frac{1}{2}(Ju, \Pi u) + H_\alpha^*(-Ju)\right] dt$$

on $L_o^\beta\left(0, T; \mathbb{R}^{2n}\right)$, with $\alpha^{-1} + \beta^{-1} = 1$. By Corollary III.3.6, problem (14) has a solution x_T such that $u_T := \dot{x}_T$ minimizes ψ_T

$$\psi_T(u_T) = \operatorname{Inf} \psi_T \; . \tag{35}$$

The left-hand side can be computed, by using the duality Theorem II.4.2 and the homogeneity of H_α:

$$\psi_T(u_T) = \int_o^T \left[\frac{1}{2}\left[(Ju_T, \Pi u_T) + H_\alpha^*(-Ju_T)\right]\right] dt$$

(36)
$$= -\int_o^T \left[\frac{1}{2}(J\dot{x}_T, x_T) + H_\alpha(x_T)\right] dt$$

$$= \int_o^T \left[\frac{1}{2}(H'_\alpha(x_T), x_T) - H_\alpha(x_T)\right] dt = -\left(1 - \frac{\alpha}{2}\right) T H_\alpha(x_T) \ .$$

We now have to adjust $T =: T_\alpha$ so that $x_T := x_\alpha$ belongs to Σ. We thereby get a closed characteristic (T_α, x_α) on Σ, and (36) yields (with $u_\alpha = \dot{x}_\alpha$)

(37)
$$\psi_{T_\alpha}(u_\alpha) = -\left(1 - \frac{\alpha}{2}\right) T_\alpha \ .$$

On the other hand, we know that $H_\alpha(x) \geq \|x\|^\alpha R^{-\alpha}$, and hence:

(38)
$$H_\alpha^*(y) \leq \frac{1}{\beta}\left(\frac{R^\alpha}{\alpha}\right)^{\beta-1} \|y\|^\beta \ .$$

This will enable us to estimate directly the right-hand side of formula (35). Indeed, setting

(39)
$$\overline{\psi}_\alpha(u) := \int_o^{T_\alpha} \left[\frac{1}{2}(Ju, \Pi u) + \frac{1}{\beta}\|u\|^\beta \left(\frac{R^\alpha}{\alpha}\right)^{\beta-1}\right] dt$$

we have by (38):

(40)
$$\mathrm{Min}\,\psi_{T_\alpha} \leq \mathrm{Min}\,\overline{\psi}_\alpha \ .$$

We compute the right-hand side explicitly. It is given by formula (36), with H_α replaced by $\|x\|^\alpha R^{-\alpha}$, namely:

(41)
$$\mathrm{Min}\,\overline{\psi}_\alpha = -\left(1 - \frac{\alpha}{2}\right) T_\alpha \|\overline{x}_\alpha\|^\alpha R^{-\alpha}$$

where $\overline{x}_\alpha$ is a solution of:

(42)
$$\begin{cases} \dot{x} = J\frac{\alpha}{R^\alpha}\frac{x}{\|x\|^{2-\alpha}} \\ x(0) = x(T_\alpha) \end{cases} .$$

We remark that, by Corollary III.3.6, the solution $\overline{x}_\alpha$ must have minimal period T_α. It is clearly of the form:

(43)
$$\overline{x}_\alpha(t) = e^{J\omega t}\overline{x}(0) \ .$$

Adjusting ω to fit the boundary-value problem (42) we get:

(44)
$$\omega = \frac{\alpha}{R^\alpha}\|\overline{x}_\alpha\|^{\alpha-2}$$

(45)
$$\omega = \frac{2\pi}{T_\alpha}$$

and hence

(46)
$$\|\overline{x}_\alpha\| = \left(\frac{T_\alpha}{2\pi}\frac{\alpha}{R^\alpha}\right)^{\frac{1}{2-\alpha}} .$$

Writing this back into Eq. (41), we get:

(47)
$$\mathrm{Min}\,\overline{\psi}_\alpha = -\left(1-\frac{\alpha}{2}\right)\frac{T_\alpha}{R_\alpha}\left(\frac{T_\alpha}{2\pi}\frac{\alpha}{R^\alpha}\right)^{\frac{\alpha}{2-\alpha}} .$$

Hence, by inequality (40):

(48)
$$\begin{aligned}\left(1-\frac{\alpha}{2}\right)T_\alpha &= -\psi_{T_\alpha}(u_\alpha) = -\mathrm{Min}\,\psi_{T_\alpha}\\ &\geq \left(1-\frac{\alpha}{2}\right)\frac{T_\alpha}{R^\alpha}\left(\frac{T_\alpha}{2\pi}\frac{\alpha}{R^\alpha}\right)^{\alpha/(2-\alpha)} .\end{aligned}$$

Simplifying, we get

(49)
$$\alpha T_\alpha \leq 2\pi R^2 .$$

Since $A(x) = \frac{\alpha}{2}T_\alpha$, by formula (21), this is the desired result. □

In the case when Σ is C^2, we have index estimates, which we recall from Sect. II.4. For ease of exposition, we introduce (compare with Definition I.8.4):

Definition 6. Let Σ be a C^2 hypersurface bounding a compact convex set C, containing 0 in its interior. We shall say that Σ is (r,R)-*pinched*, with $0 < r \leq R$, if:

(50)
$$\|y\|\,R^{-2} \leq \frac{1}{2}\left(H_2''(x)y,y\right) \leq \|y\|^2\,r^{-2}\,, \quad \forall x \in \Sigma .$$
□

More generally, if Σ bounds a compact convex set C with non-empty interior, we shall say that Σ is (r,R)-pinched if $\Sigma - x_o$ is (r,R)-pinched for some $x_o \in \mathrm{Int}C$. For instance, a sphere of radius ρ is (r,R)-pinched provided $0 < r \leq \rho \leq R$.

Definition 7. Let (T,x) be a closed characteristic and (T_2,x_2) a solution of

(51)
$$\begin{cases}\dot{x} = JH_2'(x)\\ x(0) = x(S)\\ H_2(x) = 1\end{cases}$$

following (T,x). The *index of the closed characteristic* (T,x) is the index $i_{T_2}(x_2)$ of the periodic solution x_2 over the interval $(0,T_2)$. □

By Proposition I.7.5, we also have

(52) $$i(T,x) = i_{T_\alpha}(x_\alpha) \quad \forall \alpha \in (1,2] ,$$

where (T_α, x_α) is a solution of:

(53) $$\begin{cases} \dot{x} = JH'_\alpha(x) \\ x(0) = x(S) \\ H_\alpha(x) = 1 \end{cases} .$$

We now estimate the action along a closed characteristic in terms of its index:

Proposition 8. *Let Σ be (r,R)-pinched, and (T,x) a closed characteristic with index $i(T,x)$. Then:*

(54) $$2nE\left[\frac{A(T,x)}{\pi R^2}\right] \le i(T,x) \le 2nE\left[\frac{A(T,x)}{\pi r^2}\right] .$$

Proof. We have $i(T,x) = i_{T_2}(x_2)$, where (T_2, x_2) is a minimal period solution of problem (51). From formula (21), we have $T_2 = A(T,x)$. Applying Proposition I.4.14 with $s = T_2$, and identifying the coefficients b with $R^2/2$ and a with $2/r^2$, we get the result. □

Note for instance that any closed characteristic with index less than $2n$ (that is $i(T,x) < 2n$) must have action less than $\pi \mathbb{R}^2$, that is $A(T,x) \le \pi R^2$. Conversely, $A(T,x) \le \pi r^2$ implies $i(T,x) = 0$.

Note also that closed characteristics with index zero always exist:

Proposition 9. *Let Σ be a compact C^2 hypersurface bounding a convex set with non-empty interior. Then Σ carries at least one closed characteristic with index zero.*

Proof. In Theorem 3 we found a closed characteristic on Σ by minimizing the dual action functional for problem (18). The corresponding solution x_1 must have index zero, and the closed characteristic inherits this property by homothety. □

We conclude this section by a stability result. As before, we associate with a closed characteristic x a periodic solution x_2 of $\dot{x} = JH'_2(x)$, with minimal period $T_2 = A(x_2)$. Consider the matrizant R_2 of the linearized system:

(55) $$\begin{cases} \dot{R}_2 = JH''_2\left(x_2(t)\right) R_2 \\ R_2(0) = I \end{cases} .$$

We shall say that x is *elliptic* if all the Floquet multipliers of (T_2, x_2), i.e. all the eigenvalues of $R_2(T_2)$, are on the unit circle. We shall say that it is *strictly elliptic* if all the eigenvalues $\neq 1$ are Krein-definite. Recall that, by Proposition I.6.12, $R_2(T_2)$ and $R_\alpha(T_2)$ have the same Floquet multipliers with the same multiplicity, and the same Krein sign, so that ellipticity or strict ellipticity can also be seen on $R_\alpha(T_\alpha)$.

Theorem 10. *Assume Σ is (r,R)-pinched with $\frac{R}{r} < \sqrt{2}$. Then any closed characteristic with index zero is strictly elliptic.* □

There are two main steps in the proof.

Step 1. $R_2(s)$ is strongly stable when $\pi\frac{R^2}{2} < s < \pi r^2$.

Since Σ is (r,R)-pinched, we have:

$$\|y\|^2 R^{-2} \le \frac{1}{2}\left(H_2''(x)y,y\right) \le \|y\|^2 r^{-2}\,, \quad \forall x\in\Sigma\,. \tag{56}$$

Consider the three linear Hamiltonian system:

$$\dot{y} = 2JR^{-2}y \tag{57}$$

$$\dot{y} = JH_2''\left(x(t)\right)y \tag{58}$$

$$\dot{y} = 2Jr^{-2}y \tag{59}$$

and the three corresponding quadratic forms on $L_o^2\left(0,s;\mathbb{R}^{2n}\right)$:

$$q_s^o(u,u) := \frac{1}{2}\int_o^s\left[(Ju,\Pi u) + \frac{R^2}{2}\|u\|^2\right]dt \tag{60}$$

$$q_s(u,u) := \frac{1}{2}\int_o^s\left[(Ju,\Pi u) + \left(H_2''\left(x(t)\right)^{-1}Ju,Ju\right)\right]dt \tag{61}$$

$$q_s^1(u,u) := \frac{1}{2}\int_o^s\left[(Ju,\Pi u) + \frac{r^2}{2}\|u\|^2\right]dt\,. \tag{62}$$

Denote by i_s^o, i_s and i_s^1 the indices of q_s^o, q_s and q_s^1. Since $q_s^o \ge q_s \ge q_s^1$, we have $i_s^o \le i_s \le i_s^1$ (see Proposition I.4.12). On the other hand, by Lemma I.4.13, we have

$$\begin{aligned} i_s^1 &= 2nE\left[\frac{s}{\pi r^2}\right] \\ &= 0 \quad \text{since } s < \pi r^2\,. \end{aligned} \tag{63}$$

Hence:

$$i_s^o = i_s = i_s^1 = 0\,. \tag{64}$$

Now let $h\in[0,1]$ be a parameter, and consider the quadratic form q_s^h on $L^2(0,s;\mathbb{R}^{2n})$ defined by:

$$q_s^h = \begin{cases} 2hq_s + (1-2h)q_s^o & \text{for } 0\le h\le\frac{1}{2} \\ (2-2h)q_s + (2h-1)q_s^1 & \text{for } \frac{1}{2}\le h\le 1 \end{cases}\,. \tag{65}$$

We have

$$q_s^o \geq q_s^h \geq q_s \quad \text{for } 0 \leq h \leq \frac{1}{2} \tag{66}$$

$$q_s^h = q_s \quad \text{for } h = \frac{1}{2} \tag{67}$$

$$q_s \geq q_s^h \geq q_s^1 \quad \text{for } \frac{1}{2} \leq h \leq 1 . \tag{68}$$

It follows that q_s^h has index zero for every h. The corresponding linear Hamiltonian system, on the interval $(0, s)$, is $\dot{y} = JA_h(t)y$, with:

$$A_h(t) = \begin{cases} 4hR^{-2}I + (1-2h)H_2''(x(t)) & \text{for } 0 \leq h \leq \frac{1}{2} \\ (2-2h)H_2''(x(t)) + (2h-1)r^{-2}I & \text{for } 0 \leq h \leq \frac{1}{2} \end{cases} . \tag{69}$$

Denote by $M_h(s)$ the corresponding matrizant – or rather its value at time s. Note that $M_{1/2} = R_2$. For $h = 0$, we find that $M_o(s) = \exp\left(2JR^{-2}s\right)$ has two eigenvalues, $\exp\left(\pm 2iR^{-2}s\right)$, both with multiplicity n and both Krein-definite (see Proposition I.2.11). The Krein-positive one is $\exp\left(2iR^{-2}s\right)$, which is on the lower half-circle since:

$$\pi < 2R^{-2}s < 2\frac{r^2}{R^2}\pi < 2\pi . \tag{70}$$

Arguing as in Sect. II.3 (the results of which, unfortunately, do not apply directly), we find that the Krein-positive eigenvalues move counterclockwise as h increases. Starting from the situation at $h = 0$, we conclude that all the eigenvalues of $M_h(s)$ remain on the unit circle until a Krein-positive and a Krein-negative eigenvalue meet at 1, that is, until s becomes conjugate to 0, which means that q_s^h degenerates.

On the other hand, we have see that $q_s^h \geq q_q^1$, which is nondegenerate and has index zero. In other words, q_s^1 is positive definite, and so is q_s^h. It follows that q_s^h cannot degenerate.

The Krein-positive eigenvalues of $M_h(s)$ must therefore remain on the lower half-circle, moving counterclockwise from $\exp\left(2iR^{-2}s\right)$ to $\exp\left(2ir^{-2}s\right)$. For $h = \frac{1}{2}$, we get n eigenvalues of $R_2(s)$ on this arc, all Krein-positive, and Step 1 is concluded.

Step 2. $R_2(s)$ is strongly stable when $\pi r^2 \leq s < T_2$.

Now let s increase through πr^2. By the results in Sect. II.3, the Krein-positive eigenvalues (which all lie on the lower half-circle) move counterclockwise on the unit circle until they meet at Krein-negative eigenvalue. This can only happen at 1, that is, when s becomes conjugate to 0; until then, $R_2(s)$ remains strongly stable.

Remember now that the periodic solution x_2 has index zero. This means that the first time $s > 0$ which is conjugate to 0 is $s = T_2$. This concludes the proof; in fact, $R_2(T_2)$ has $m < n$ Krein-positive eigenvalues on the lower half-circle (counting multiplication), m Krein-negative eigenvalue on the upper half-circle, and 1 as an eigenvalue of multiplicity $2(n-m)$. □

Notes and Comments. The trick of reducing the fixed-energy problem to the fixed-period problem by using homogeneous Hamiltonians was introduced in this context by Rabinowitz, who claims it is classical in celestial mechanics. He used it to show that a compact C^1 hypersurface bounding a star-shaped set always carries a closed characteristic (see [Rab2]). Simultaneously, Weinstein proved in [Wei3] that a C^2 compact hypersurface bounding a convex set, with non-vanishing curvature, always carries a closed characteristic. Theorem 3, with no regularity assumption whatsoever, is due to Clarke in [Cla1], [Cla2] (see also [Cla3]).

Theorem 4 is due to Croke and Weinstein in [CroW]. Proposition 5 is due to [Eke8], and so is the material in the rest of the section. Theorem 10 was inspired by a result of [BalTZ] in Riemannian geometry, but there is no supertheorem covering both.

It would of course be extremely interesting to know whether a convex Σ always carries an elliptic trajectory, even in the absence of the $\sqrt{2}$ pinching. Some work in this direction has been done in [dOnE].

2. Multiplicity in the Pinched Case

Now that the basic question of existence is answered, we may turn to the question of multiplicity: how many geometrically distinct closed characteristics are there on Σ?

In Sect. II.7 we studied with care the case when Σ is an ellipsoid (harmonic oscillator)

$$\Sigma = \left\{ x \,\middle|\, \sum_{i=1}^{n} \frac{\alpha_i}{2} x_i^2 = 1 \right\} \qquad \alpha_i > 0 \ \ \forall i \tag{1}$$

and we showed that, if the α_i are linearly independent over $\mathbb{Q}$ (strongly non-resonant case), then Σ carries only n closed characteristics, corresponding to the n fundamental modes of the harmonic oscillator.

It has long been conjectured that this is the worst possible case. That is, any C^1 compact hypersurface $\Sigma \subset \mathbb{R}^{2n}$, bounding a convex set, should carry at least n geometrically distinct closed characteristics. In other words, any mechanical system with convex Hamiltonian and n degrees of freedom has at least n nonlinear modes of vibration on every bounded energy level.

At the moment, we have no idea whether this conjecture is true. Ekeland and Lasry have shown the existence of n closed characteristics, provided Σ is "pinched" between two Euclidian balls, the radii of which are sufficiently close.

Theorem 1. *Let $C \subset \mathbb{R}^{2n}$ be a convex compact set with non-empty interior, and let Σ be its boundary. Assume there is a point $x_o \in \mathbb{R}^{2n}$ and numbers $0 < r \leq R$ such that:*

$$\forall x \in \Sigma\ , \quad r \leq \|x - x_o\| \leq R \tag{2}$$

$$R/r < \sqrt{2}\ . \tag{3}$$

Then Σ carries at least n geometrically distinct closed characteristics $(T_i x_1), \ldots, (T_n, x_n)$, where T_i ist the minimal period of x_i, and the actions $A(T_i, x_i)$ satisfy:

$$\pi r^2 \leq A(T_i, x_i) \leq \pi R^2\ , \quad \forall i\ . \tag{4}$$

□

Without loss of generality, we assume that $x_o = 0$. Proceeding as in the preceding section, we choose some $\alpha \in (1, 2)$ and we associate with C a convex function H_α such that

$$H_\alpha(\lambda x) = \lambda^\alpha H(x)\ , \quad \forall \lambda \geq 0 \tag{5}$$

$$x \in \Sigma \Rightarrow H_\alpha(x) = 1\ . \tag{6}$$

From condition (3) it follows that

$$R^{-\alpha} \|x\|^\alpha \leq H_\alpha(x) \leq r^{-\alpha} \|x\|^\alpha\ . \tag{7}$$

We are interested in the fixed-energy problem

$$\begin{cases} H_\alpha(x) = 1 \\ \dot{x} \in J\partial H_\alpha(x) \\ x(0) = x(T) \end{cases} . \tag{8}$$

Consider instead the fixed-period problem:

$$\begin{cases} \dot{x} \in J\partial H_\alpha(x) \\ x(0) = x(1) \end{cases} . \tag{9}$$

Every solution (x, T) of problem (8), where T denotes the *minimal* period of x, gives rise to a sequence x_α^k, $k \in \mathbb{N}$, of solutions of problem (9),

$$x_\alpha^k(t) = (kT)^{-\frac{1}{2-\alpha}} x(kTt)\ . \tag{10}$$

Conversely, any solution of problem (9) scales back to a closed characteristic on Σ.

Next, we introduce the (reduced) dual action functional for the fixed-period problem:

$$\psi : L_o^\beta \left(\mathbb{R}/\mathbb{Z}; \mathbb{R}^{2n}\right) \to \mathbb{R} \tag{11}$$

$$\psi(u) = \int_o^1 \left[\frac{1}{2}(Ju, \Pi u) + H_\alpha^*(-Ju)\right] dt\ . \tag{12}$$

We know that x solves problem (9) if and only if $u = \dot{x}$ is a critical point of ψ. Form our rescaling procedure, it follows that the critical points of ψ come in infinite sequences u_k, $k \in \mathbb{N}$, corresponding to a single closed characteristic x by formula (10):

$$u_k(t) := \dot{x}^k(t) = (kT)^{-\frac{\alpha-1}{2-\alpha}} \dot{x}(kTt) . \tag{13}$$

The corresponding critical values of ψ are given by:

Lemma 2.

$$\psi(u_k) = -\left(1 - \frac{\alpha}{2}\right)(kT)^{-\frac{\alpha}{2-\alpha}} = -\left(1 - \frac{\alpha}{2}\right)\left(\frac{2}{\alpha}A(T,x)\right)^{-\frac{\alpha}{2-\alpha}} k^{-\frac{\alpha}{2-\alpha}} .$$

Proof. Since H_α is α-homogeneous, its Fenchel conjugate H_α^* must be β-homogeneous with $\alpha^{-1} + \beta^{-1} = 1$:

$$H_\alpha^*(\lambda y) = \lambda^\beta H_\alpha^*(y) , \quad \forall \lambda \geq 0 . \tag{14}$$

Relation (13) gives

$$u_k(t) = k^{-\frac{\alpha-1}{2-\alpha}} u_1(kt) . \tag{15}$$

Substituting this into formula (12), and remembering that $\beta = \frac{\alpha}{\alpha-1}$, we get

$$\psi(u_k) = k^{-\frac{\alpha}{2-\alpha}} \psi(u_1) . \tag{16}$$

To compute $\psi(u_1)$, we remember that, by the duality Theorem II.4.2:

$$\begin{aligned} \psi(u_1) &= \Psi\left(x_\alpha^1\right) = -\Phi\left(x_\alpha^1\right) \\ &= -\int_o^1 \left[\frac{1}{2}\left(J\dot{x}_\alpha^1, x_\alpha^1\right) + H_\alpha\left(x_\alpha^1\right)\right] dt . \end{aligned} \tag{17}$$

Since x^1 is a solution of problem (9), we have $-J\dot{x}^1 \in \partial H_\alpha(x^1)$. Remember the Euler identity:

$$(x^*, x) = \alpha H_\alpha(x) , \quad \forall x^* \in H'_\alpha(x) . \tag{18}$$

Hence:

$$\left(-J\dot{x}^1, x^1\right) = \alpha H_\alpha(x^1) , \quad \forall t . \tag{19}$$

Writing this into formula (17), we get:

$$\psi(u_1) = \left(\frac{\alpha}{2} - 1\right) H_\alpha(x^1) . \tag{20}$$

Relation (10) gives $x^1 = T^{-\frac{1}{2-\alpha}} x$. Using the α-homogeneity of H together with $H_\alpha(x) = 1$, we get:

$$\psi(u_1) = \left(\frac{\alpha}{2} - 1\right) T^{-\frac{\alpha}{2-\alpha}} . \tag{21}$$

By formula (1.20), we have $T = \frac{2}{\alpha} A(x)$. Hence the result. □

Lemma 3.

$$-\left(1 - \frac{\alpha}{2}\right)\left(\frac{\alpha}{2\pi r^2}\right)^{\frac{\alpha}{2-\alpha}} \leq \operatorname{Min} \psi \leq -\left(1 - \frac{\alpha}{2}\right)\left(\frac{\alpha}{2\pi R^2}\right)^{\frac{\alpha}{2-\alpha}} .$$

Proof. Taking Fenchel conjugates in formula (7), we get:

$$\frac{1}{\beta}\left(\frac{r^\alpha}{\alpha}\right)^{\beta-1} \|y\|^\beta \leq H_\alpha^*(y) \leq \frac{1}{\beta}\left(\frac{R^\alpha}{\alpha}\right)^{\beta-1} \|y\|^\beta . \tag{22}$$

Hence:

$$\psi_r(u) \leq \psi(u) \leq \psi_R(u) \tag{23}$$

$$\psi_r(u) := \int_o^1 \left[\frac{1}{2}(Ju, \Pi u) + \frac{1}{\beta}\left(\frac{r^\alpha}{\alpha}\right)^{\beta-1} \|u\|^\beta\right] dt \tag{24}$$

$$\psi_R(u) := \int_o^1 \left[\frac{1}{2}(Ju, \Pi u) + \frac{1}{\beta}\left(\frac{R^\alpha}{\alpha}\right)^{\beta-1} \|u\|^\beta\right] dt . \tag{25}$$

These formulas are the same as formula (1.39) with $T_\alpha = 1$ (and R replaced by r in the first one). Relation (1.47) then becomes:

$$\operatorname{Min} \psi_r = -\left(1 - \frac{\alpha}{2}\right)\left(\frac{\alpha}{2\pi}\right)^{\frac{\alpha}{2-\alpha}} r^{-\frac{2\alpha}{2-\alpha}} \tag{26}$$

$$\operatorname{Min} \psi_R = -\left(1 - \frac{\alpha}{2}\right)\left(\frac{\alpha}{2\pi}\right)^{\frac{\alpha}{2-\alpha}} R^{-\frac{2\alpha}{2-\alpha}} . \tag{27}$$

Hence the result. □

The proof of Theorem 1 relies crucially on the fact that the function ψ is S^1-invariant. The situation deserves to be examined in some detail.

The space $L_o^\beta\left(\mathbb{R}/\mathbb{Z} : \mathbb{R}^{2n}\right)$ is endowed with an S^1-action, that is, a continuous map

$$(\theta, u) \to \theta * u \tag{28}$$

$$S^1 \times L_o^\beta \to L_o^\beta \tag{29}$$

which is compatible with the group law:

$$\theta_1 * (\theta_2 * u) = (\theta_1 + \theta_2) * u, \quad \forall (\theta, u) \tag{30}$$

$$0 * u = u , \quad \forall u . \tag{31}$$

In the case at hand, the action is a continuous representation of S^1 into the group of isometries of L_o^β, defined by:

$$(\theta * u)(t) := u(t+\theta) \tag{32}$$

A subset $\Omega \subset L_o^\beta$ is S^1*-invariant* if

$$u \in \Omega \Rightarrow \theta * u \in \Omega\ , \quad \forall \theta \in S^1\ . \tag{33}$$

A function $f : L_o^\beta \to \mathbb{R}$ is S^1*-invariant* if

$$f(\theta * u) = f(u)\ , \quad \forall (\theta, u) \in S^1 \times L^\beta\ . \tag{34}$$

Proposition 4. *The function ψ is S^1-invariant.* □

Proof. In formula (12), we are integrating a 1-periodic function ober one period. The result does not depend on the point where we start the integration. □

The S^1*-orbit* of a point $u \in L_o^\beta$ is the set

$$\mathcal{O}(u) := \left\{\theta * u | \theta \in S^1\right\}\ . \tag{35}$$

We shall say that the S^1-action is *free at* u if the map $\theta \to \theta * u$ is injective; that is, a homeomorphism of S^1 onto $\mathcal{O}(u)$. If $\Omega \subset L_o^\beta$ is an invariant subset, the S^1-action is *free in* Ω if it is free at every point $u \in \Omega$.

In our case, the S^1-action will be free at u if and only if u has minimal period 1. If u has minimal period $1/k$, then the map $\theta \to \theta * u$ wraps $S^1 k$ times over $\mathcal{O}(u)$. It follows that the S^1 action will not be free in L_o^β.

Since ψ is S^1-invariant, its level sets:

$$\psi^\gamma := \{u \in L_o^\infty | \psi(u) < \gamma\} \tag{36}$$

are S^1-invariant. The first step in the proof of Theorem 1 consists in finding a level $\gamma > \operatorname{Min} \psi$ such that the S^1-action is free in ψ^γ.

Lemma 5. *For*

$$\gamma \leq -\left(1 - \frac{\alpha}{2}\right)\left(\frac{\alpha}{4\pi r^2}\right)^{\frac{\alpha}{2-\alpha}} \tag{37}$$

the S^1-action is free in ψ^γ. □

Proof. Let $u \in L_o^\beta\left(\mathbb{R}/\mathbb{Z}; \mathbb{R}^{2n}\right)$ have minimal period $1/k$, for some integer $k \geq 2$. We have to prove that

$$\psi(u) \geq -\left(1 - \frac{\alpha}{2}\right)\left(\frac{\alpha}{4\pi r^2}\right)^{\frac{\alpha}{2-\alpha}}\ . \tag{38}$$

Define v by:

$$v(t) := k^{\frac{\alpha-1}{2-\alpha}} u\left(\frac{t}{k}\right) . \tag{39}$$

It is 1-periodic: $v \in L_o^\beta\left(\mathbb{R}/\mathbb{Z} : \mathbb{R}^{2n}\right)$. Computing $\psi(v)$, and using the β-homogeneity of H_α^*, we get:

$$\psi(v) = k^{\frac{\alpha}{2-\alpha}} \psi(u) . \tag{40}$$

We must have $\psi(v) \geq \operatorname{Min} \psi$. Using Lemma 3, this yields:

$$k^{\frac{\alpha}{2-\alpha}} \psi(u) \geq -\left(1-\frac{\alpha}{2}\right)\left(\frac{\alpha}{2\pi r^2}\right)^{\frac{\alpha}{2-\alpha}} \tag{41}$$

which we rewrite as:

$$\psi(u) \geq -\left(1-\frac{\alpha}{2}\right)\left(\frac{\alpha}{2k\pi r^2}\right)^{\frac{\alpha}{2-\alpha}} . \tag{42}$$

For $k \geq 2$, we get the desired result. □

In the following, we shall take:

$$\gamma > -\left(1-\frac{\alpha}{2}\right)\left(\frac{\alpha}{2\pi R^2}\right)^{\frac{\alpha}{2-\alpha}} \tag{43}$$

By Lemma 3, we have $\gamma > \operatorname{Min} \psi$, so ψ^γ is not empty. Since $2r^2 > R^2$, we may also assume that

$$\gamma < -\left(1-\frac{\alpha}{2}\right)\left(\frac{\alpha}{4\pi r^2}\right)^{\frac{\alpha}{2-\alpha}} \tag{44}$$

so the S^1-action on ψ^γ is free by Lemma 4.

We shall see that ψ^γ contains a copy of the Euclidian sphere S^{2n-1} with the standard S^1-action. It is described as follows. Write

$$S^{2n-1} = \left\{x \in \mathbb{R}^{2n} \mid x^2 = 1\right\} . \tag{45}$$

The standard S^1-action is defined by:

$$\theta * x = e^{2\pi J\theta} x . \tag{46}$$

It is free on S^{2n-1}, and the S^1-orbits are circles, which constitute the so-called *Hopf fibration* of S^{2n-1}. Setting $x \sim x'$ if $x' = e^{2\pi J\theta}x$ for some θ, we get an equivalence relation on S^{2n-1}. The equivalence classes are the S^1-orbits, and the quotient is the complex projective space $\mathbb{CP}^{n-1}$ (real dimension $2n-2$).

A map $f : \mathbb{R}^{2n} \to L_o^\beta$ will be called *equivariant* if

$$f\left(e^{2\pi J\theta}x\right) = \theta * f(x) . \tag{47}$$

Lemma 6. *There is an equivariant isometry* $f : \mathbb{R}^{2n} \to L^{\beta}_{o}$ *and a number* $\rho > 0$ *such that*

$$f\left(\rho S^{2n-1}\right) \subset \psi^{\gamma} . \tag{48}$$

□

Proof. Define $u_\xi = f(\xi)$ by

$$u_\xi(t) = e^{2\pi J t}\xi \tag{49}$$

f is an isometry. It is also equivariant:

$$f\left(e^{2\pi J\theta}\xi\right)(t) = e^{2\pi Jt}e^{2\pi J\theta}\xi = e^{2\pi J(t+\theta)}\xi = \theta * f(\xi) . \tag{50}$$

Now compute $\psi(u_\xi)$, using inequality (22)

$$\begin{aligned}\psi\left(u_\xi\right) &= \int_o^1 \left(\frac{1}{2}\left(Ju_\xi, \Pi u_\xi\right) + H^*_\alpha\left(-Ju_\xi\right)\right) dt \\ &= -\frac{1}{2}\frac{1}{2\pi}\|\xi\|^2 + \int_o^t H^*_\alpha\left(-Je^{2\pi Jt}\xi\right) dt \\ &\le -\frac{1}{2}\frac{1}{2\pi}\|\xi\|^2 + \frac{1}{\beta}\left(\frac{R^\alpha}{\alpha}\right)^{\beta-1}\|\xi\|^\beta .\end{aligned} \tag{51}$$

The right-hand side is minimum for:

$$-\frac{1}{2\pi}\|\xi\| + \left(\frac{R^\alpha}{\alpha}\right)^{\beta-1}\|\xi\|^{\beta-1} = 0 \tag{52}$$

which yields:

$$\|\xi\| = \left(\frac{1}{2\pi}\right)^{\frac{\alpha-1}{2-\alpha}}\left(\frac{\alpha}{R^\alpha}\right)^{\frac{1}{2-\alpha}} \tag{53}$$

$$\psi(u_\xi) \le -\left(\frac{\alpha}{2\pi R^2}\right)^{\frac{\alpha}{2-\alpha}}\left(1 - \frac{\alpha}{2}\right) < \gamma \tag{54}$$

by formula (43). □

To draw the consequences of this situation, we must appeal to the theory of fiber bundles. There is a contractible space E, called the *classifying space* for the group S^1, endowed with a free S^1-action. It is universal for free S^1-actions, in the sense that if Ω is any other space with a free S^1-action, there is an equivariant continuous map $f : \Omega \to E$. If f_o and f_1 are two such maps, then there is a homotopy f_t, $0 \le t \le 1$, connecting them, such that each f_t is equivariant.

In Ω, consider the equivalence relation $u \sim v$ if $\theta * u = v$ for some $\theta \in S^1$, and denote the quotient space by Ω/S^1. In other words, Ω/S^1 is the space of S^1-orbits. If $\Omega = S^{2n-1}$ with the standard action (Hopf fibration), we have $\Omega/S^1 := \mathbb{CP}^n$. There is a sequence of inclusions:

$$\begin{array}{ccccccc} S^{2n-1} & \subset & S^{2n+1} & \subset & \dots & \subset & E \\ \downarrow & & \downarrow & & & & \downarrow \\ \mathbb{CP}^{n-1} & \subset & \mathbb{CP}^n & \subset & \dots & \subset & E/S^1 \end{array} . \tag{55}$$

In fact, E can be constructed explilcitely as a limit of the S^{2n+1}; for this reason, it is often denoted by S^∞, and the quotient E/S^1 by $\mathbb{CP}^\infty$.

The cohomology ring of $\mathbb{CP}^\infty$ plays a crucial role in the study of S^1-actions. Taking rational coefficients, we find that the cohomology algebra $H^*(\mathbb{CP}^\infty;\mathbb{Q})$ is generated by an element ω_2 of degree two:

$$H^{2k}(\mathbb{CP}^\infty) = \mathbb{Q}\omega_2^k \tag{56}$$

$$H^{2k+1}(\mathbb{CP}^\infty) = 0\ . \tag{57}$$

If Ω is a space with a free S^1-action, we take a classifying map $f : \Omega \to E$ and its quotient $\widetilde{f} : \Omega/S^1 \to E/S^1$. We get a homeomorphism $\widetilde{f}^*$ of the cohomology algebras:

$$\widetilde{f}^* : H^*(\mathbb{CP}^\infty) \to H^*\left(\Omega/S^1\right)\ . \tag{58}$$

Since classifying maps are unique up to homotopy, this homomorphism does not depend on the particular choice of f. Of course, it need not, be onto; it may even be trivial. The cohomology classes which belong to the image of $\widetilde{f}^*$ are called *characteristic classes.*

To identify them, introduce:

$$\alpha_2 := \widetilde{f}^*(\omega_2) \in H^2\left(\Omega/S^1\right)\ . \tag{59}$$

Since $\widetilde{f}^*$ is an algebra homomorphism, α_2 generates all characteristic classes.

We define the S^1-index of Ω to be the largest integer k such that $\alpha_2^{k-1} \neq 0$, that is, such that $H^{2(k-1)}\left(\Omega/S^1\right)$ contains a non-trivial characteristic class. It is ∞ if there are infinitely many such k. Formally:

Definition 7. Let Ω be a space with a free S^1-action and $f : \Omega \to S^\infty$ a classifying map. The S^1*-index* $I(\Omega)$ is defined by:

$$\begin{aligned} I(\Omega) :&= \operatorname{Sup}\left\{k \mid \widetilde{f}^*(\omega_2)^{k-1} \neq 0\right\} \\ &= \operatorname{Inf}\left\{p \mid \widetilde{f}^*\left(\omega_2^k\right) = 0 \quad \forall k \geq p\right\}\ . \end{aligned} \tag{60}$$

From the definition of $I(\Omega)$, and standard arguments in cohomology, one can deduce the following properties. We denote by Ω spaces with free S^1-action.

$$\text{(monotonicity)} \quad \begin{cases} \text{if there is an equivariant continuous map} \\ f : \Omega_1 \to \Omega_2\ , \ \text{ then } \ I(\Omega_1) \leq I(\Omega_2) \end{cases}$$

$$\text{(subadditivity)} \quad I(\Omega_1 \cup \Omega_2) \leq I(\Omega_1) + I(\Omega_2)$$

$$\text{(continuity)} \quad \begin{cases} \text{if } K \subset \Omega \text{ is a closed invariant subset,} \\ \text{then it has a closed invariant neighbourhood} \\ N \subset \Omega \quad \text{with } \ I(K) = I(N) \end{cases}$$

If the cohomology theory is appropriately chosen, for instance if one uses cohomology with compact support, one has the additional property

$$\text{(compact carriers)} \quad \begin{cases} I(\Omega) = \operatorname{Sup} I(K) & \text{for all compact} \\ \text{invariant subsets } K \subset \Omega \ . \end{cases}$$

We have $I(\Omega) > 0$ whenever $\Omega \neq \emptyset$. The index of S^{2n-1} with the standard S^1-action turns out to be n:

$$I\left(S^{2n-1}\right) = n \ . \tag{61}$$

All this applies to ψ^γ, which, by Lemma 5, is endowed with a free S^1-action. By Lemma 6, ψ^γ contains a copy of S^{2n-1}, so that, by the monotonicity property and formula (61), we have:

$$I\left(\psi^\gamma\right) \geq n \ . \tag{62}$$

This enables us to define n numbers c_i, $1 \leq i \leq n$, as follows:

$$c_i := \operatorname{Inf}\left\{\delta \in \mathbb{R} | I\left(\psi^\delta\right) \geq i\right\} \ . \tag{63}$$

It is clear from the definition that $c_{i-1} \leq c_i$ for $1 \leq i \leq n$. We have:

$$c_1 = \operatorname{Min} \psi \tag{64}$$

$$c_n \leq \operatorname{Max}\left\{\psi(u) | u \in f\left(\rho S^{2n-1}\right)\right\} < \gamma \tag{65}$$

by Lemma 6, taking into account the fact that $f\left(\rho S^{2n-1}\right)$ is compact. Summing up, we have

$$\operatorname{Min} \psi = c_1 \leq c_2 \leq \ldots \leq c_n \leq \gamma < 0 \ . \tag{66}$$

The c_i are the value of δ for which the S^1-index of the level sets $\psi^\delta := \{u | \psi(u) < \delta\}$ changes. If ψ is C^1, that is, if H_α^* is differentiable, it is standard procedure to show that the c_i are critical levels of ψ. We begin by:

Lemma 8. *If H_α^* is differentiable, then ψ is C^1 and satisfies condition (PS) on L_o^β.*

Proof. If $H_\alpha^* : \mathbb{R}^{2n} \to \mathbb{R}$ is differentiable, it has to be C^1 by convexity, and the functional $\psi : L_o^\beta \to \mathbb{R}$ is C^1 by Corollary II.3.5.

Assume now $|\psi(u_p)|$ is bounded and $\psi'(u_p) \to 0$ in $\left(L_o^\beta\right)^*$. This means that, for some $c \in \mathbb{R}$, $\xi_p \in \mathbb{R}^{2n}$ and $\epsilon_p \in L^\alpha$, we have

$$-c \leq \int_o^1 \left[\frac{1}{2}\left(Ju_p, \Pi u_p\right) + H_\alpha^*\left(-Ju_p\right)\right] dt \leq c \tag{67}$$

$$-J\Pi u_p + J\left(\nabla H^*\right)\left(-Ju_p\right) - \xi_p = \epsilon_p \to 0 \ . \tag{68}$$

Since H_α^* is β-homogeneous, $H_\alpha^*(\lambda y) = \lambda^\beta H_\alpha^*(y)$, we have an estimate of the form $H_\alpha^*(y) \leq k \|y\|^\beta$ and formula (67) implies that the sequence u_n is bounded in L_o^β. We extract a weakly convergent subsequence, still denoted by u_p:

$$u_p \rightharpoonup u \ . \tag{69}$$

Writing all this into relation (68), we first see that ξ_p is bounded in $\mathbb{R}^{2n}$, and hence convergent. Since $\Pi : L_o^\beta \to L_o^\alpha$ is a compact operator, $J\Pi u_p$ converges strongly in L_p^α, and so $\nabla H^* (-Ju_p)$ must also converse strongly in L^α. Applying the Legendre transform as in the proof of Lemma IV.2.10, we conclude that u_p converges strongly in L^β. So condition (PS) is satisfied. □

Until very recently, the only way to proceed would have been to use the so-called deformation lemma, which had become the standard tool of critical point theory since the pioneering word of R. Palais, and of D. Clark in the equivariant case. But in 1987 A. Szulkin, and as this book was going to press N. Ghoussoub, found a way to eliminate the deformation lemma and rely exclusively on Ekeland's variational principle. It is the latter's approach that we have chosen here.

The setting is the one of Lemma 8, namely the S^1-invariant map $\psi : L_o^\beta \to \mathbb{R}$ associated with a differentiable H_α^*, so that ψ is C^1 and satisfies condition (PS). We consider a class $\mathcal{F}$ of compact subsets of L_o^β which is stable by equivariant homotopies, that is:

$$\begin{cases} \text{if } \ A \in \mathcal{F}\,, \ \text{ and if } \ h \in C^o\left([0,1] \times L_o^\beta; L_o^\beta\right) \\ \text{is such that } \ h(s,\cdot) \ \text{ is equivariant } \ \forall s \ \text{ and} \\ h(0,\cdot) \ \text{ is the identity, then } \ h(1,A) \in \mathcal{F} \end{cases} \tag{70}$$

and we associate with it the number

$$c(\mathcal{F}) := \inf_{A \in \mathcal{F}} \max_{u \in A} \psi(u) \ . \tag{71}$$

Theorem 9 (Ghoussoub). *Suppose $F \subset L_o^\beta$ is a closed invariant subset of L_o^β such that:*

$$A \cap F \neq \emptyset \quad \forall A \in \mathcal{F} \tag{72}$$

$$\sup_{A \cap F} \psi \geq c(\mathcal{F}) \quad \forall A \in \mathcal{F} \tag{73}$$

Suppose moreover $c(\mathcal{F}) > -\infty$. Then there exists a sequence u_n in L_o^β such that:

$$\psi(u_n) \to c(\mathcal{F}) \tag{74}$$

(75) $$(1+\|u_n\|)\,\|\psi'(u_n)\|_* \to 0$$

(76) $$\delta\,(u_n, F) \to 0\ .$$ □

Note that (73) is implied by the stronger condition:

(77) $$\inf_F \psi \geq c(\mathcal{F})\ .$$

Proof. We adapt the argument in Theorem III.1.7 to a different setting, but the general trend remains the same. Recall that δ is the geodesic distance in L^β_o, defined by formulas (21) to (23) in Sect. IV.1, and that it is equivalent to the standard distance on bounded subsets.

Choose $\epsilon \in \left(0, \frac{1}{2}\right)$, and define

(78) $$F_\epsilon := \{u | \delta(u, F) < \epsilon\}\ .$$

Let $A \in \mathcal{F}$ be such that

(79) $$c(\mathcal{F}) \leq \sup_A \psi \leq c(\mathcal{F}) + \frac{\epsilon^2}{4}$$

and consider the set Γ of all equivariant homotopies of the identity which leave $A \setminus F_\epsilon := \{u \in A | u \notin F_\epsilon\}$ fixed and are uniformly bounded from the identity. In other words, $h \in C^o\left([0,1] \times L^\beta_o; L^\beta_o\right)$ belong to Γ if and only if:

(80) $$\forall s\ ,\quad h(s,\cdot) \quad \text{is equivariant}$$

(81) $$h(s,u) = u \qquad \text{whenever} \quad s=0 \quad \text{or} \quad u \in A \setminus F_\epsilon$$

(82) $$\sup_{s,u} \delta(u, h(s,u)) < \infty\ .$$

It follows from our assumptions that $h(1, A) \in \mathcal{F}$.

Endow Γ with the uniform metric:

(83) $$d\,(h_1, h_2) := \sup_{s,u} \delta\,(h_1(s,u), h_2(s,u))\ .$$

The reader will check that (Γ, d) is a complete metric space. Now define a function $I : \Gamma \to \mathbb{R}$ by

(84) $$I(h) := \max\{(\psi + \varphi) \circ h(1,u) | u \in A\}$$

where:

(85) $$\varphi(u) := \max\left\{0, \epsilon^2 - \epsilon\delta(u, F)\right\}\ .$$

For every fixed u, the function $h \to (\psi + \varphi) \circ h(1,u)$ is continuous on Γ, so I is lower semi-continuous. It is also bounded from below, since:

$$
\begin{aligned}
I(h) &= \sup\{\psi(v)+\varphi(v)|v \in h(1,A)\} \\
&\geq \sup\{\psi(v)+\varphi(v)|v \in h(1,A)\cap F\} \\
&\geq \epsilon^2 + \sup\{\psi(v)|v \in h(1,A)\cap F\}
\end{aligned} \tag{86}
$$

Since the class Γ is stable by equivariant homotopies, $h(1,A) \in \Gamma$ and condition (73) implies

$$
\inf I \geq \epsilon^2 + c(\mathcal{F})\ . \tag{87}
$$

Conversely, consider the function $\overline{h} \in \Gamma$ defined by $\overline{h}(s,u) = u$ for all (s,u). Estimate (79) yields:

$$
\inf I \leq I(\overline{h}) = \sup\{\psi(u)+\varphi(u)|u \in A\} < c(\mathcal{F}) + \frac{\epsilon^2}{4} + \epsilon^2\ . \tag{88}
$$

So $\inf I \leq I(\overline{h}) < \inf I + \frac{\epsilon^2}{4}$. This means we are in a position to apply Theorem III.1.1, and find some $\widehat{h} \in \Gamma$ such that

$$
I\left(\widehat{h}\right) \leq I\left(\overline{h}\right) \tag{89}
$$

$$
d\left(\widehat{h},\overline{h}\right) \leq \frac{\epsilon}{2} \tag{90}
$$

$$
I(h) \geq I(\widehat{h}) - \frac{\epsilon}{2} d(h,\widehat{h})\ , \quad \forall h \in \Gamma\ . \tag{91}
$$

We now introduce the subsets M and N of L_o^β defined by:

$$
M = \left\{u \in A|\,(\varphi+\psi)\circ\widehat{h}(1,u) \geq \inf I\right\} \tag{92}
$$

$$
N = \left\{u \in L_o^\beta \,\middle|\, \begin{array}{l} \delta\left(u,\widehat{h}(1,M)\right) \leq \epsilon/2 \\ -\epsilon^2 \leq \varphi(u)+\psi(u) - \inf I \leq \dfrac{\epsilon^2}{4} \end{array}\right\}\ . \tag{93}
$$

M is compact non-empty because of the definition (84) of I, and N because of relation (88). In fact:

$$
\inf I \leq I(\widehat{h}) = \sup\left\{\varphi+\psi|\widehat{h}(1,A)\right\} \leq I(\overline{h}) < \inf I + \frac{\epsilon^2}{4} \tag{94}
$$

so that:

$$
\widehat{h}(1,M) \subset N^o\ . \tag{95}
$$

It will be useful later on to note that:

$$M \subset A \cap F_\epsilon \ . \tag{96}$$

Indeed, since $\varphi = 0$ on $A \setminus F_\epsilon$, we have by (78)

$$\inf I \geq \epsilon^2 + c(\mathcal{F}) > \frac{3\epsilon^2}{4} + \sup_A \psi \geq \frac{3\epsilon^2}{4} + \sup_{A\setminus F_\epsilon} (\psi + \varphi) \ . \tag{97}$$

Since $\widehat{h}(1,\cdot)$ is the identity on $A \setminus F_\epsilon$, we also have:

$$\sup_{\widehat{h}(1,A\setminus F_\epsilon)} (\psi + \varphi) = \sup_{A\setminus F_\epsilon} (\psi + \varphi) < \inf I - \frac{3\epsilon^2}{4} \ . \tag{98}$$

Comparing with formula (92), we find that $M \cap (A \setminus F_\epsilon) = \emptyset$. But this amounts to the inclusion (96).

We now claim that:

$$\exists u_\epsilon \in N \ : \quad (1 + \|u_\epsilon\|) \, \|\psi'(u_\epsilon)\|_* \leq 3\epsilon \ . \tag{99}$$

The theorem will follow. Indeed, since $u_\epsilon \in N$, we will have

$$-\epsilon^2 \leq \varphi(u_\epsilon) + \psi(u_\epsilon) - \inf \leq \frac{\epsilon^2}{4} \tag{100}$$

while $0 \leq \varphi \leq \epsilon^2$ and $\epsilon^2 \leq \inf I - c(\mathcal{F}) \leq 5\epsilon^2/4$. So

$$-\epsilon^2 \leq \psi(u_\epsilon) - c(\mathcal{F}) \leq \frac{3\epsilon^2}{2} \ . \tag{101}$$

Letting $\epsilon = \frac{1}{n} \to 0$, we get relation (74). Of course, (75) is nothing else than (99). As for (76), we note that by formula (93), there must be a point $u_o \in M$ with $\delta\left(u_\epsilon, \widehat{h}(1, u_o)\right) \leq \epsilon/2$. Using (90), we find that $\delta\left(\widehat{h}(1, u_o), u_o\right) \leq \epsilon/2$. Finally, we have $\delta\left(u_o, F \cap A\right) \leq \epsilon$ by formula (96). Putting everything together, we find $\delta\left(u_\epsilon, F \cap A\right) \leq 2\epsilon$, and (76) follows.

So everything amounts to proving claim (99). This we do by contradiction. Suppose it is not true, so that:

$$\forall u \in N \ , \quad (1 + \|u\|) \, \|\psi'(u)\|_* > 3\epsilon \ . \tag{102}$$

Introduce an equivariant pseudo-gradient vectorfield for ψ, that is a continuous map ξ from L_o^β into itself, which is locally Lipschitz on the open set $\{u | \psi'(u) \neq 0\}$, and satisfies

$$\|\xi(u)\| \leq 2 \, \|\psi'(u)\|_* \tag{103}$$

$$\langle \psi'(u), \xi(u) \rangle \geq \|\psi'(u)\|_*^2 \tag{104}$$

$$\xi\left(\theta * u\right) = \theta * \xi(u) \quad \forall \theta \in S^1 \ . \tag{105}$$

The existence of ξ follows from a classical argument using partitions of unity. We now use ξ to construct a homotopy in the class Γ. In fact, for every $\lambda > 0$, we set

$$h_\lambda = \widehat{h} - s\lambda(\rho \circ \widehat{h})(\eta \circ \widehat{h}) \tag{106}$$

where $\rho : L_o^\beta \to [0,1]$ is a continuous S^1-invariant function such that

$$\rho(u) = \begin{cases} 1 & \text{if } (\varphi + \psi)(u) \geq \inf I \\ 0 & \text{if } (\varphi + \psi)(u) \leq \inf I - \frac{3\epsilon^2}{4} \end{cases} \tag{107}$$

and $\eta : L_o^\beta \to L_o^\beta$ is a continuous S^1-equivariant map defined by

$$\eta(u) = \begin{cases} \left[1 + \dfrac{\delta\left(u, \widehat{h}(1,M)\right)}{\delta\left(u, L_o^\beta \setminus N\right)}\right]^{-1} \dfrac{\xi(u)}{\|\xi(u)\|} & \text{if } u \in N \\ 0 & \text{if } u \in N\,. \end{cases} \tag{108}$$

If $\xi(u) = 0$, then $\psi'(u) = 0$ and $u \notin N$ by formula (102). So formula (108) is meaningful. Since $\widehat{h}(1, M)$ and N are S^1-invariant subsets, η is equivariant. We have

$$\|\eta(u)\| \leq 1\,, \quad \forall u\,. \tag{109}$$

The homotopy h_λ belongs to the class Γ. Indeed, conditions (80) and (82) are clearly satisfied by formula (106), and so is condition (81) for $s = 0$. If $u \in A \setminus F_\epsilon$, we have $\rho(u) = 0$ by formula (98), and condition (91) is again true.

So $h_\lambda \in \Gamma$, and we may write $h = h_\lambda$ in inequality (91):

$$I(h_\lambda) \geq I(\widehat{h}) - \frac{\epsilon}{2} d\left(h_\lambda, \widehat{h}\right)\,. \tag{110}$$

Since M is compact non-empty, there is for every $\lambda > 0$ some $u_\lambda \in M$ such that, by formula (84):

$$I(h_\lambda) = \psi\left(h_\lambda(1, u_\lambda)\right) + \varphi\left(h_\lambda(1, u_\lambda)\right)\,. \tag{111}$$

Taking into account the compactness of M, we find that there is a function $O(\lambda)$, with $O(\lambda) \to 0$ when $\lambda \to 0$, such that:

$$\begin{aligned} d(h_\lambda, \widehat{h}) &= \sup_{s,u} \delta\left(h_\lambda(s,u), \widehat{h}(s,u)\right) \\ &\leq \lambda\left(1 + \left\|\widehat{h}(s,u)\right\|^{-1} + O(\lambda)\right) \end{aligned} \tag{112}$$

since $0 \leq \rho \leq 1$ and $\|\eta\| \leq 1$. Writing this into formula (110) yields:

(113)

$$\begin{aligned} \psi\left(h_\lambda(1, u_\lambda)\right) + \varphi\left(h_\lambda(1, u_\lambda)\right) &\geq I(\widehat{h}) - \frac{\epsilon}{2}\lambda\left(1 + \left\|\widehat{h}(s,u)\right\|^{-1} + O(\lambda)\right) \\ &\geq \psi\left(\widehat{h}(1, u_\lambda)\right) + \varphi\left(\widehat{h}(1, u_\lambda)\right) - \frac{\epsilon}{2}\lambda\left(1 + \left\|\widehat{h}(s,u)\right\|^{-1} + O(\lambda)\right)\,. \end{aligned}$$

By formula (85), the function φ is ϵ-Lipschitz so that:

$$
\begin{aligned}
\left|\varphi\left(h_\lambda(1,u_\lambda)\right)-\varphi\left(\widehat{h}(1,u_\lambda)\right)\right| &\le \epsilon\delta\left(h_\lambda(1,u_\lambda),\widehat{h}(1,u_\lambda)\right) \\
&\le \epsilon\lambda\left(1+\left\|\widehat{h}(1,u_\lambda)\right\|\right)^{-1}+O(\lambda)\,.
\end{aligned} \tag{114}
$$

Writing this into inequality (113) yields:

$$
\psi\left(h_\lambda(1,u_\lambda)\right)-\psi\left(\widehat{h}(1,u_\lambda)\right)\ge -\frac{3}{2}\epsilon\lambda\left(1+\left\|\widehat{h}(1,u_\lambda)\right\|\right)^{-1}+O(\lambda)\,. \tag{115}
$$

Since A is compact, there is a subsequence $\lambda\to 0$ such that u_λ converges to some $u_o\in A$. Dividing both sides of (115) by $\lambda>0$, and letting $\lambda\to 0$, we get

$$
\langle\psi'\left(\widehat{h}(1,u_o)\right),\rho\left(\widehat{h}(1,u_o)\right)\eta\left(\widehat{h}(1,u_o)\right)\rangle\le\frac{3\epsilon}{2}\left(1+\left\|\widehat{h}(1,u_\lambda)\right\|\right)^{-1}+O(\lambda)\,. \tag{116}
$$

It follows from the compactness of A and formula (111) that

$$
(\varphi+\psi)\left(\widehat{h}(1,u_o)\right)=\lim_{\lambda\to 0}I(h_\lambda)\ge\inf I \tag{117}
$$

so that $\rho\left(\widehat{h}(1,u_o)\right)=1$. By formula (108), we also have

$$
\eta\left(\widehat{h}(1,u_o)\right)=\xi\left(\widehat{h}(1,u_o)\right)\Big/\left\|\xi\left(\widehat{h}(1,u_o)\right)\right\|\,. \tag{118}
$$

Writing all this in formula (116) and using condition (104) on the pseudo-gradient yields:

$$
\left\|\psi'\left(\widehat{h}(1,u_o)\right)\right\|_*^2\left\|\xi\left(\widehat{h}(1,u_o)\right)\right\|^{-1}\le\frac{3\epsilon}{2}\left(1+\left\|\widehat{h}(1,u_o)\right\|\right)^{-1}\,. \tag{119}
$$

Using condition (103) on the pseudo-gradient, this becomes:

$$
\left\|\psi'\left(\widehat{h}(1,u_o)\right)\right\|_*\le 3\epsilon\left(1+\left\|\widehat{h}(1,u_o)\right\|\right)^{-1} \tag{120}
$$

contradicting (102). The result is proved. □

We now use Theorem 9 to prove the results we need: the c_i defined by formula (63), $1\le i\le n$, are critical values of ψ, and if two of them coincide there must be infinitely many critical S^1-orbits on that level.

Proposition 10. *Each c_i, $1\le i\le n$, is a critical value of ψ. Denote by K_i the corresponding critical set:*

$$
K_i:=\{u|\psi(u)=c_i\quad\text{and}\quad\psi'(u)=0\} \tag{121}
$$

If $c_i=c_j$ for some $j\ge i$, then $K_i=K_j$ satisfies:

$$
I(K_i)\ge j-1+1\,. \tag{122}
$$

Proof. Fix i, and set:

$$\mathcal{F}_i := \left\{ A \subset L_o^\beta | A \text{ compact, invariant, } I(A) \geq i \right\} . \tag{123}$$

Define

$$c(\mathcal{F}_i) := \inf_{A \in \mathcal{F}_i} \sup_{u \in A} \psi(u) . \tag{124}$$

If we use a suitable cohomology theory, we have $c(\mathcal{F}_i) = c_i$, as defined in formula (63). Applying Theorem 9 with $F = L_o^\beta$, we get a sequence u_n such that $\psi(u_n) \to c_i$ and $\|\psi'(u_n)\|_* \to 0$. Since ψ satisfies condition (PS), there is a subsequence converging to a point u where

$$\psi(u) = c_i \quad \text{and} \quad \psi'(u) = 0 . \tag{125}$$

The set K_i therefore is non-empty. It is clearly invariant, and by condition (PS) it is compact. So $I(K_i) \geq 1$, and formula (122) holds when $c_i < c_{i+1}$, that is, $j = i$.

If $j > 1$, we argue by contradiction. Assume (122) does not hold. Using the continuity property for the index, we find a closed invariant neighbourhood N of K_i such that

$$I(N) = I(K_i) \leq j - i . \tag{126}$$

Let $\overset{\circ}{N}$ be its interior. It is an open invariant set with $K_i \subset \overset{\circ}{N} \subset N$, and by the monotonicity property of the index, we also have

$$I\left(\overset{\circ}{N}\right) \leq j - 1 . \tag{127}$$

Denote by F the complement of $\overset{\circ}{N}$ in L_o^β. It is a closed invariant subset. For any $A \in F_j$ it follows from the subadditivity property of the index that

$$I\left(A \cap F\right) \geq I(A) - I(N) \geq j - (j-1) = i . \tag{128}$$

In particular, $A \cap F \neq \emptyset$ and

$$\sup_{A \cap F} \overline{\psi} \geq c(\mathcal{F}_i) = c_i = c_j . \tag{129}$$

We may therefore apply Theorem 9 with $\mathcal{F} = \mathcal{F}_j$. Using condition (PS) again, we find a point $u \in F$ where $\psi(u) = c_j$ and $\psi'(u) = 0$. So $u \in K_j \cap F = \emptyset$, a contradiction. □

Proposition 10 gives $I(K) \geq 2$ whenever $j > 1$. It then follows from the definition of the index that $H^1\left(K/S^1\right)$ is non-trivial, so K/S^1 must be an uncountable set. In other words, if $c_i = c_j$ for some $i \neq j$, there must be uncountably many S^1-orbits of critical points on the level $c_i = c_j$. So Lemma 9 and Proposition 10 together guarantee the existence of a least n distinct critical S^1-orbits for ψ in all cases.

We now conclude the proof of Theorem 1. Assume first that C is strictly convex. Then H_α^* is C^1, and the preceding arguments apply. We have thus found at least n critical points $u_1, \dots, u_n$ of ψ with:

$$\forall i\ ,\quad \psi(u_i) < \gamma \tag{130}$$

$$i \neq j \Rightarrow \forall \theta \in S^1\ ,\quad \theta * u_i \neq u_j \tag{131}$$

$$\theta \neq 0 \Rightarrow \forall i\ ,\quad \theta * u_i \neq u_i\ . \tag{132}$$

There are n solutions x_i^1, $1 \le i \le n$, of the fixed-period problem (9), such that $\dot{x}_i^1 = u_i$. Each of them has minimal period 1, and is therefore related to a closed characteristic (x_i, T_i) on Σ by formula (10), with $k = 1$:

$$x_i^1(t) = (T_i)^{-\frac{1}{2-\alpha}}\, x_i(T_i t) \tag{133}$$

the minimal period T_i being related to the critical value $\psi(u_i)$ by Lemma 2, with $k = 1$:

$$\psi(u_i) = -\left(1 - \frac{\alpha}{2}\right)(T_i)^{-\frac{\alpha}{2-\alpha}}\ . \tag{134}$$

We have $\operatorname{Min}\psi \le \psi(u_i) < \gamma$. Using Lemma 3 and taking the infimum of γ in formula (43), we get:

$$-\left(1 - \frac{\alpha}{2}\right)\left(\frac{\alpha}{2\pi r^2}\right)^{\frac{\alpha}{2-\alpha}} \le \psi(u_i) \le -\left(1 - \frac{\alpha}{2}\right)\left(\frac{\alpha}{2\pi R^2}\right)^{\frac{\alpha}{2-\alpha}}\ . \tag{135}$$

By the preceding equation, this implies that:

$$\frac{2\pi r^2}{\alpha} \le T_i \le \frac{2\pi R^2}{\alpha}\ . \tag{136}$$

Formula (1.20) gives us T_i in terms of the action $A(x_i)$, by $T_i = \frac{2}{\alpha} A(T_i, x_i)$. The inequality now becomes:

$$\pi r^2 \le A(T_i, x_i) \le \pi R^2 \tag{137}$$

which is the desired estimate. The characteristics (x_i, T_i) must be geometrically distinct. Indeed, the T_i are all minimal periods, so if (x_i, T_i) and (x_j, T_j) are not geometrically distinct, the only possibility is that $\theta * x_i = x_j$ for some $\theta \in S^1$. But this would imply $\theta * u_i = u_j$, contradicting (86).

If H is not strictly convex, ψ is the sum of two functions, one of which is smooth and the other convex continuous (and hence locally Lipschitz). One can then either adapt Theorem 8 to this situation as in Szulkin [Szu1], or replace ψ by a C^1 function with the same critical points, as in Ekeland and Lasry [EkeL1].

Notes and Comments. In [Wei1], Weinstein proved that a hypersurface Σ which is close enough to an ellipsoid carries at least n closed characteristics. If the ellipsoid is nonresonant, this result is a consequence of the inverse function theorem, and was already known to Liapounov. But for resonant ellipsoids, such as the sphere, it is a sophisticated bifurcation result (see [Mos4] for an analytic proof). It raised the suspicion that every compact convex Σ would carry n closed characteristics.

Theorem 1 is a step in this direction. It is due to Ekeland and Lasry in [EkeL1], and says how far away from the sphere one can go and keep n closed characteristics. It was extended by many authors, notably [BerLRM] who treated the star-shaped case, but there has been no improvement on $\sqrt{2}$ (unless more symmetries are present).

The S^1-index theory is due to Fadell and Rabinowitz in [FadR]. For a more classical approach to Proposition 9, based on the so-called deformation lemma, see the book [MawW2]. The idea of using Ekeland's variational principle in Liusternik-Schnirelman theory is due to Szulkin [Szu1], [Szu3]. The idea of localizing the critical points on a closed subset F defined a priori by separation properties is due to Ghoussoub [Gho].

3. Multiplicity in the General Case

In this section, we shall work with a compact C^2 hypersurface $\Sigma \subset \mathbb{R}^{2n}$, with non-vanishing Gaussian curvature, bounding a strictly convex set C with non-empty interior. We shall give a multiplicity criterion for closed characteristics on Σ. As a consequence, it will follow that

– Σ always carries at least two closed characteristics

– if Σ is C^∞-generic, it carries infinitely many closed characteristics.

As before, we assume that 0 belongs to the interior of C. Choosing some $\alpha \in (1,2)$, we associate with C a convex function H_α such that

$$H_\alpha(\lambda x) = \lambda^\alpha H(x)\ , \quad \forall \lambda \geq 0 \tag{1}$$

$$x \in \Sigma \Leftrightarrow H_\alpha(x) = 1\ . \tag{2}$$

We shall use the framework (and the notations) of Sect. 2, without the benefit of the pinchedness condition. Formula (2.5) to (2.13) are still valid, and so is Lemma 2.2. We can always put Σ between two concentric balls, with radii $r \leq R$, so formula (2.7) will be valid as well:

$$R^{-\alpha}\left\|x\right\|^\alpha \leq H_\alpha(x) \leq r^{-\alpha}\left\|x\right\|^\alpha \tag{3}$$

but we no longer assume that $Rr^{-1} \leq \sqrt{2}$.

The (reduced) dual action functional is C^2 and satisfies condition (PS) (Lemma 2.7). It is still S^1-invariant, but we are no longer able to find a level γ below which the S^1-action is free.

We must therefore extend the S^1-index to non-free actions. The basic idea in this direction is due to A. Borel. It consists, when dealing with a space Ω

with non-free S^1-action, of introducing the classifying space S^∞ and lifting the given S^1-action Ω to a diagonal action on the product $\Omega \times S^\infty$:

$$\theta * (u, z) = (\theta * u, \theta * z) \ . \tag{4}$$

This diagonal action is free, since the S^1-action on S^∞ is free. So the space $\Omega \times S^\infty$, with the diagonal action, has a well-defined S^1-index $I(\Omega \times S^\infty)$, which we take as a definition for the S^1-index of Ω.

Definition 1. Let Ω be a space with a (non-free) S^1-action. Its S^1-*index* $\widehat{I}(\Omega)$ is defined by:

$$\widehat{I}(\Omega) := I(\Omega \times S^\infty) \ , \tag{5}$$

where $\Omega \times S^\infty$ is endowed with the diagonal action. □

The projections $pr_1 : \Omega \times S^\infty \to \Omega$ and $pr_2 : \Omega \times S^\infty \to S^\infty$ induce quotient maps

$$\widetilde{pr}_1 : (\Omega \times S^\infty)/S^1 \to \Omega/S^1 \tag{6}$$

$$\widetilde{pr}_2 : (\Omega \times S^\infty)/S^1 \to \mathbb{CP}^\infty \ . \tag{7}$$

It can be proved that $\widetilde{pr}_2$ is always a fibration with fiber Ω, and that if S^1 acts freely on Ω then $\widetilde{pr}_1$ is a homotopy equivalence. It follows that:

$$I(\Omega) = \widehat{I}(\Omega) \tag{8}$$

provided $I(\Omega)$ is well-defined, that is, the S^1-action on Ω is free. We have thus extended the earlier definition.

From the relation $\widehat{I}(\Omega) = I(\Omega \times \Omega^\infty)$, it follows that $\widehat{I}$ satisfies the same properties as I, namely:

$$\text{(monotonicity)} \quad \begin{cases} \text{if there is an equivariant continuous map} \\ f : \Omega_1 \to \Omega_2 \ , \quad \text{then } \widehat{I}(\Omega_1) \leq \widehat{I}(\Omega_2) \end{cases}$$

$$\text{(subadditivity)} \quad \widehat{I}(\Omega_1 \cup \Omega_2) \leq \widehat{I}(\Omega_1) + \widehat{I}(\Omega_2)$$

$$\text{(continuity)} \quad \begin{cases} \text{if } K \subset \Omega \text{ is a closed invariant subset} \\ \text{then it has a closed invariant neighbourhood} \\ N \subset \Omega \ \text{ with } \ \widehat{I}(K) = \widehat{I}(N) \ . \end{cases}$$

We now investigate some situations when the S^1-action is not free. The simplest one is the case of fixed points. Recall that $u_o \in \Omega$ is a fixed point of the S^1-action if $\mathcal{O}(u_o) = \{u_o\}$, that is $\theta * u_o = u_o$ for every $\theta \in S^1$. In that case, we have

$$(\{u_o\} \times S^\infty)/S^1 = \{u_o\} \times \mathbb{CP}^\infty \tag{9}$$

which clearly has the cohomology of $\mathbb{CP}^\infty$. It follows that

$$\widehat{I}(\{u_o\}) = I(\{u_o\} \times S^\infty) = \infty . \tag{10}$$

In the sequel, we shall have to deal with certain kinds of non-free S^1-actions on spheres. They are described as follows. Consider $\mathbb{R}^{2np}$, the space of p-uples $(x_1, \dots, x_p)$ with entries $x_i \in \mathbb{C}^n$. Given a p-uple $(n_1, \dots, n_p)$ of non-zero integers, the formula

$$\theta * (x_1, \dots, x_p) := \left(e^{2\pi n_1 J\theta} x_1 \dots, e^{2\pi n_p J\theta} x_p\right) \tag{11}$$

defines an S^1-action on $\mathbb{R}^{2np}$. The Euclidian sphere

$$S^{2np-1} := \left\{ (x_1, \dots, x_p) | \sum_{i=1}^{p} x_1^2 = 1 \right\} \tag{12}$$

is invariant for this action, and contains no fixed point. On the other hand, the action is not free unless $(n_1, \dots, n_p) = (1, \dots, 1)$, in which case it is just the Hopf fibration, and its index is np. It turns out that we always have

$$\widehat{I}\left(S^{2np-1}\right) = np \tag{13}$$

We now go back to the case at hand. Consider the set $\psi^o \subset L_o^\beta$ defined by:

$$\psi^o := \left\{ u \in L_o^\beta | \psi(u) < 0 \right\} . \tag{14}$$

It is S^1-invariant, and contains no fixed point of the action. We claim that it has infinite index; in fact

Lemma 2. *For every integer $p \neq 0$, we can find in ψ^o a copy of S^{2np-1} with the S^1-action associated with the p-uple $(1, 2, \dots, p)$:*

$$\theta * (x_1, \dots, x_p) := \left(e^{2\pi J\theta} x_1, \dots, e^{2\pi pJ\theta} x_p\right) . \tag{15}$$

Proof. The negative eigenvalues of $-J\Pi$ or L_o^β are the $-\frac{1}{2k\pi}$, $k \in \mathbb{N}$, each of them with multiplicity $2n$. The $2n$-dimensional eigenspace associated with $-\frac{1}{2k\pi}$ consists of the functions $e^{2k\pi Jt}\xi$, for $\xi \in \mathbb{R}^{2n}$.

For given $p \in \mathbb{N}$, define a map $f : \left(\mathbb{R}^{2n}\right)^p \to L_o^\beta$ by:

$$f(\xi_1, \dots, \xi_p) = \sum_{k=1}^{p} e^{2k\pi Jt} \xi_k . \tag{16}$$

The S^1-action on L_o^β, defined by $\theta * u(t) = u(t + \theta)$, pulls back on the S^1-action of formula (5). We define as before by S^{2np-1} the unit sphere of $\left(\mathbb{R}^{2n}\right)^p$, and we consider the image of ρS^{2np-1} by f; the radius $\rho > 0$ is to be determined later on. For $u \in f\left(\rho S^{2np-1}\right)$, we have:

$$\begin{aligned}\psi(u) &= \int_o^1 \left[\frac{1}{2}(Ju, \Pi u) + H_\alpha^*(-Ju)\right] dt \\ &= \int_o^1 \left[-\frac{1}{2}(Ju, \Pi u) + \frac{1}{\beta}\left(\frac{R^\alpha}{\alpha}\right)^{\beta-1} \|u\|^\beta\right] dt\,.\end{aligned} \tag{17}$$

Indeed, we have $H_\alpha(x) \geq \|x\|^\alpha / R^\alpha$ by condition (3), from which the above inequality follows by convex conjugation. We now replace u by $f_p(\xi_1, \ldots, \xi_p)$, with $\sum \xi_1^2 = \rho^2$:

$$\psi(u) \leq -\frac{1}{2}\sum_{k=1}^{p} \frac{1}{2k\pi}\|\xi_k\|^2 + \frac{1}{\beta}\left(\frac{R^\alpha}{\alpha}\right)^{\beta-1} \rho^\beta\,. \tag{18}$$

Hence:

$$\psi(u) \leq -\frac{1}{2}\frac{1}{2p\pi}\rho^2 + \frac{1}{\beta}\left(\frac{R^\alpha}{\alpha}\right)^{\beta-1} \rho^\beta\,. \tag{19}$$

Taking $\rho = \left(\frac{1}{2p\pi}\right)^{\frac{\alpha-1}{2-\alpha}} \left(\frac{\alpha}{R^\alpha}\right)^{\frac{1}{2-\alpha}}$ yields

$$\psi(u) \leq -\left(1 - \frac{\alpha}{2}\right)\left(\frac{\alpha}{2p\pi R^2}\right)^{\frac{\alpha}{2-\alpha}} < 0\,. \tag{20}$$

So, for this choice of ρ, we get $f\left(\rho S^{2np-1}\right) \subset \psi^o$ which is desired result.□

We now proceed as in the preceding section. Define an infinite sequence c_i, $i \in \mathbb{N}$, by

$$c_i := \operatorname{Inf}\left\{\delta \in \mathbb{R} | \widehat{I}\left(\psi^\delta\right) \geq i\right\}\,. \tag{21}$$

We have

$$\operatorname{Min}\psi = c_1 \leq c_2 \leq \ldots \leq c_i \leq c_{i+1} \leq \ldots < 0\,. \tag{22}$$

Proposition 3. *Every c_i is a critical value of ψ. If $c_i = c_j$ for some $i \leq j$, we must have:*

$$\widehat{I}(K) \geq i - j + 1 \tag{23}$$

with $K := \{u | \psi(u) = c_i = c_j \text{ and } \psi'(u) = 0\}$. □

The proof is the same as in the preceding section (Proposition 2.9). Only the three properties of monotonicity, subadditivity, continuity are used, and since they also hold for $\widehat{I}$, the proof carries over verbatim.

Since ψ^o contains no fixed point of the action, condition (23) with $i < j$ implies $\widehat{I}(K) \geq 2$, so K contains infinitely many S^1-orbits. Indeed, the index of a single S^1-orbit is 1 (this is a particular case of the S^1-actions we investigated on S^{2np-1}, with $n = p = 1$), and so is the index of a finite union of S^1-orbits (because if $\Omega = \cup_{i=1}^{N} \Omega_i$, with $\Omega_i = S^1 * u_i$, then $(\Omega \times S^\infty)/S^1$ is the disjoint union of the $(\Omega_i \times S^\infty)/S^1$, none of which carries characteristic classes in dimension 2 or higher).

The upshot is that ψ has an infinite sequence of critical points u_i, whith $\psi(u_i) = c_i$ for $i \in \mathbb{N}$ and $u_i \neq u_j$ whenever $i \neq j$. The u_i correspond to pairwise distinct periodic solutions of the fixed-period problem (2.9), but not necessarily to geometrically distinct closed characteristics on Σ because of the rescaling. So, if we want a multiplicity result, we need extra information. Much in the spirit of Sect. IV.3, this information is provided by the Morse index.

Recall in this setting the definitions of Chap. II. If u is a critical point of ψ, it is in fact smooth, and we may define a quadratic form q on L_o^2 by

$$q(v,v) := \frac{1}{2} \int_o^1 \left[(Jv, \Pi v) + \left((H_\alpha^*)'' (-Ju_i) Jv, Jv \right) \right] dt \tag{24}$$

which defines in turn an orthogonal splitting

$$L_o^2 = E_- \oplus E_o \oplus E_+ \tag{25}$$

of L_o^2 into negative, zero, and positive subspaces. The index of u_i is defined by:

$$i(u) := \dim E_- \tag{26}$$

and its nullity by:

$$\nu(u) := \dim E_o \ . \tag{27}$$

We have $i(u) < \infty$, and

$$1 \leq \nu(u) \leq 2n \ . \tag{28}$$

The following result is the heart of the argument. In the spirit of Sect. IV.3, it takes advantage of the particular way that the critical level c_i has been constructed (namely, at $c = c_i$ the S^1-index of the subset ψ^c becomes less than i) to show that on that level there exists a critical point u_i with a certain index. The proof, however, is even more technical, and relies on tools from algebraic topology, so it will not be given here.

Theorem 4. *For every i there exists a point $u \in L_o^\beta$ such that:*

$$\psi'(u) = 0 \quad \text{and} \quad \psi(u) = c_i \tag{29}$$

$$i(u) \leq 2(i-1) \leq i(u) + \nu(u) - 1 \ . \tag{30}$$

If $c_i = c_j$ for some $j \neq i$, then, given any $N \in \mathbb{N}$ and $\epsilon > 0$, we can choose u in such a way that an ϵ-ball around u in L_o^β contains at least N critical points on different S^1-orbits at the level $c_i = c_j$. □

Corollary 5. *Assume* $c_{i-1} < c_i = \ldots = c_j < c_{j+1}$ *for some* $i < j$. *There exists a family* u_k, $i \le k \le j$, *such that*

$$\psi'(u_k) = 0 \quad \text{and} \quad \psi(u_k) = c_k \tag{31}$$

$$|i(u_k) - 2(k-1)| \le 2n + 1 \tag{32}$$

$$k \neq p \Rightarrow u_k \notin S^1 * u_p \ . \tag{33}$$

□

Proof. Define u_i by Theorem 4; conditions (31) and (32) then follow from (29) and (30), together with inequality (28). We then proceed by induction.

Take some $p \in \{i+1, \ldots, j\}$ and suppose that for $k \le p$ we have found points u_k satisfying (31) to (33). We will now construct u_{p+1}.

Choose $\epsilon > 0$ so small that when u is a critical point of ψ belonging to an ϵ-neighbourhood of u_k in L_o^β, for some $k \in \{i, \ldots, p\}$, then

$$i(u_k) \le i(u) \le i(u_k) + \nu(u_k) - 1 \ . \tag{34}$$

Choose $N > p - i + 1$ and apply Theorem 4 to $p+1$. We get a critical point v_{p+1}, with

$$\psi\,(v_{p+1}) = c_i = c_j \tag{35}$$

$$i\,(v_{p+1}) \le 2p \le i\,(v_{p+1}) + \nu\,(v_{p+1}) - 1 \ . \tag{36}$$

Using inequality (28), this yields immediately

$$|i\,(v_{p+1}) - 2p| \le 2n - 1 \ . \tag{37}$$

If $v_{p+1} \notin S^1 * u_k$, for all $k \in \{i, \ldots, p\}$, we set $u_{p+1} := v_{p+1}$ and we are done. Assume then that $v_{p+1} \in S^1 * u_k$ for some $k = k_o$. By the second part of Theorem 4, the ϵ-ball in L_o^β around v_{p+1} contains at least N critical S^1-orbits on the level $c_i = c_j$. There are at most $p - i + 1$ critical S^1-orbits of type $S^1 * u_k$, for some $k \in \{i, \ldots, p\}$. Since $N > p - i + 1$, there must be in that ball some critical orbit $S^1 * w_{p+1}$, distinct from the $S^1 * u_k$. We have

$$w_{p+1} \notin S^1 * u_k \qquad \text{for} \ \ i \le k \le p \tag{38}$$

$$i\,(u_{k_o}) \le i\,(w_{p+1}) \le i\,(u_{k_o}) + \nu\,(u_{k_o}) - 1 \ . \tag{39}$$

On the other hand, since $u_{k_o} \in S^1 * v_{p+1}$, it must satisfy relation (36)

$$i\,(u_{k_o}) \le 2p \le i\,(u_{k_o}) + \nu\,(u_{k_o}) - 1 \ . \tag{40}$$

Combining inequalities (39) and (40) we get

$$|i\,(w_{p+1}) - 2p| \le \nu\,(u_{k_o}) + 1 \le 2n + 1 \ . \tag{41}$$

Setting $u_{p+1} := w_{p+1}$, we are done. □

It is now time to recall that every solution (x, T) of the fixed-energy problem (2.8) (where T is taken to be the minimal period) gives rise, by rescaling, to an infinite sequence u_k^x, $k \in \mathbb{N}$, of critical points of ψ. They are given by:

$$u_k^x(t) := (kT)^{-\frac{\alpha-1}{2-\alpha}} \dot{x}(kTt) \; . \tag{42}$$

Their index is denoted by $i_{kT}(x)$ and their nullity by $\nu_{kT}(x)$:

$$i_{kT}(x) := i\left(u_k^x\right) \tag{43}$$

$$\nu_{kT}(x) := \nu\left(u_k^x\right) \; . \tag{44}$$

The corresponding critical value was computed in Lemma 2.2.:

$$\psi\left(u_k^x\right) = -\left(1 - \frac{\alpha}{2}\right)\left(\frac{2}{\alpha} kA(x)\right)^{-\frac{\alpha}{2-\alpha}} \tag{45}$$

where $A(x) := A(T, x)$ is the action along x.

Definition 6. The solution (x, T) is called *i-essential* if there exists some $k \in \mathbb{N}$ such that:

$$\psi\left(u_k^x\right) = c_i \tag{46}$$

$$\left|i\left(u_k^x\right) - 2i\right| \leq 2n + 1 \; . \tag{47}$$

It is *essential* if it is i-essential for some $i \geq 1$. □

From now on, every closed characteristic x on Σ will be endowed with the parametrization that turns it into a solution of the fixed-energy problem, and we will denote by T the corresponding minimal period. For instance, a closed characteristic x will be called essential if (x, T) is essential according to Definition 6. We shall denote by $\boldsymbol{C}$ the family of essential closed characteristics on Σ.

Proposition 7. *There is a map $i \to x^i$ of $\mathbb{N}^*$ into $\boldsymbol{C}$ such that*

$$\forall i \geq 1 \, , \quad x^i \text{ is } i\text{-essential} \tag{48}$$

$$[c_i = c_j, i \neq j] \Rightarrow x^i \quad \text{and} \quad x^j \text{ geometrically distinct} \; . \tag{49}$$

Proof. If $c_{i-1} < c_i < c_{i+1}$, we apply Theorem 4 to get u_i. If $c_{i-1} < c_i = \ldots = c_j < c_{j+1}$, we apply Corollary 5. The sequence u_i being constructed in this way, we associate with each u_i the corresponding closed characteristic on Σ by formula (42), which we denote by x^i. There is no reason why x^i should be geometrically distinct from x^j when $i \neq j$ – except if $c_i = c_j$.

In that case, writing $x^i = x^j = x$, we get $u_i = u^x_{k_i}$ and $u_j = u^x_{k_j}$. We have $k_i \neq k_j$ since $u_i \notin S^1 * u_j$ by condition (33), and it then follows from formula (45) that $u^x_{k_i}$ and $u^x_{k_j}$ must belong to different energy levels, thereby contradicting the assumption that $c_i = c_j$. □

It is now time for an example. Consider the case when Σ is a sphere of radius R:

$$\Sigma_R = \left\{x \in \mathbb{R}^{2n} \mid \|x\| = R\right\} . \tag{50}$$

The closed characteristics are the great circles of the sphere. Taking $H_\alpha(x) := \|x\|^\alpha R^{-\alpha}$, the fixed energy problem $\dot{x} = JH'(x)$ on $H_\alpha(x) = 1$ has the solutions

$$x(t) = \exp\left(J\frac{\alpha}{R^2}t\right) x_o \tag{51}$$

which are all periodic, with minimal period $T = \frac{2\pi R^2}{\alpha}$.

Lemma 8. *$\Sigma = \Sigma_R$ is the sphere (50), the critical values c_i are given by:*

$$c_i = -\left(1 - \frac{\alpha}{2}\right)\left(\frac{\alpha}{2\pi R^2}\right)^{\frac{\alpha}{2-\alpha}} p^{-\frac{\alpha}{2-\alpha}} , \tag{52}$$

where $p = E\left[\frac{i}{n}\right] + 1$.

Every closed characteristic on Σ_R is i-essential for every $i \in \mathbb{N}$ □

In other words, $c_1 = \ldots = c_n = -\left(1 - \frac{\alpha}{2}\right)\left(\frac{\alpha}{2\pi R^2}\right)^{\frac{\alpha}{2-\alpha}}$, and so on, $c_{np+1} = \ldots = c_{(n_1)p}$. We recall that, in this book, we define $E[a]$ to be $a - 1$ when a is an integer.

Proof. All the critical values of ψ on L^β_o are known. Indeed, the 1-periodic solutions of $\dot{x} = JH'(x)$ are the $x_k(t) = \exp\left(2\pi Jtk^{-1}\right) x(0)$. The solution x_k lies on the sphere $\|x\|^{\alpha-2} = 2\pi R^\alpha (k\alpha)^{-1}$ and its index is $i_{kT}(x)$, where (T, x) are defined in formula (51). This index has been computed in Example I.7.1 formula (21):

$$i_{kT}(x) = 2n(k-1) \tag{53}$$

and its nullity has been found to be maximal:

$$\nu_{kT}(x) = 2n . \tag{54}$$

Theorem II.4.2 gives

$$\psi\left(u^x_k\right) = -\int_o^1 \left[\frac{1}{2}\left(J\dot{x}_k, x_k\right) + \|x\|^\alpha R^{-\alpha}\right] dt \tag{55}$$

$$\psi\left(u_k^x\right) = \frac{1}{2}\frac{2\pi}{k}\left(\frac{2\pi R^\alpha}{k\alpha}\right)^{\frac{2}{\alpha-2}} - \left(\frac{2\pi R^\alpha}{k\alpha}\right)^{\frac{\alpha}{2-\alpha}} R^{-\alpha} - \left(1-\frac{\alpha}{2}\right)\left(\frac{\alpha}{2k\pi R^2}\right)^{\frac{\alpha}{2-\alpha}} . \tag{56}$$

Now take some $i \in \mathbb{N}^*$, and set $p := E[\frac{i}{n}] + 1$. We have $i = n(p-1) + r$, $1 \leq r \leq n$.

By Theorem 4, there is some closed characteristic x on Σ and some $k \in \mathbb{N}^*$ such that

$$\psi\left(u_k^x\right) = c_i \tag{57}$$

$$i\left(u_k^x\right) \leq 2(i-1) \leq i\left(u_k^x\right) + \nu\left(u_k^x\right) - 1 . \tag{58}$$

Using relations (53) and (54) yields:

$$2n(k-1) \leq 2(i-1) \leq 2n(k-1) + 2n - 1 . \tag{59}$$

It follows that

$$k \leq p + \frac{r-1}{n} \leq k + 1 - \frac{1}{n} . \tag{60}$$

But $0 \leq \frac{r-1}{n} < 1$, so these inequalities boil down to $k = p$. Hence:

$$c_i = \psi\left(u_p^x\right) = -\left(1-\frac{\alpha}{2}\right)\left(\frac{\alpha}{2p\pi R^2}\right)^{\frac{\alpha}{2-\alpha}} . \tag{61}$$

□

We now return to the general case. With the hypersurface $\Sigma \subset \mathbb{R}^{2n}$ we associate two numbers $\gamma_\alpha^+(\Sigma)$ and $\gamma_\alpha^-(\Sigma)$ defined by:

$$\gamma_\alpha^+(\Sigma) := \limsup_{i\to\infty}\left[(-c_i)^{\frac{2-\alpha}{\alpha}} i\right]^{-1} \tag{62}$$

$$\gamma_\alpha^-(\Sigma) := \limsup_{i\to\infty}\left[(-c_i)^{\frac{2-\alpha}{\alpha}} i\right]^{-1} \tag{63}$$

and we also set:

$$\gamma^+(\Sigma) := \frac{\alpha}{4}\left(1-\frac{\alpha}{2}\right)^{\frac{2-\alpha}{\alpha}} \gamma_\alpha^+(\Sigma) \tag{64}$$

$$\gamma^-(\Sigma) := \frac{\alpha}{4}\left(1-\frac{\alpha}{2}\right)^{\frac{2-\alpha}{\alpha}} \gamma_\alpha^-(\Sigma) . \tag{65}$$

If Σ_R is the sphere of radius R, these numbers are easily computed from Lemma 8, yielding:

$$\gamma^+\left(\Sigma_R\right) = \gamma^-\left(\Sigma_R\right) = \frac{\pi R^2}{2n} . \tag{66}$$

Lemma 9. *We always have*

$$0 < \gamma^-(\Sigma) \le \gamma^+(\Sigma) < \infty \,. \tag{67}$$

Proof. Denote by ψ_r and ψ_R the (reduced) dual action functionals associated with the Hamiltonians $r^{-\alpha}\|x\|^\alpha$ and $R^{-\alpha}\|x\|^\alpha$ respectively. Denote by $c_i(r)$ and $c_i(R)$ the critical values associated to ψ_r and ψ_R by Proposition 3

$$c_i(r) := \operatorname{Inf}\left\{\delta \mid \widehat{I}\left(\psi_r^\delta\right) \ge i\right\} \tag{68}$$

$$c_i(R) := \operatorname{Inf}\left\{\delta \mid \widehat{I}\left(\psi_R^\delta\right) \ge i\right\} \,. \tag{69}$$

Estimate (3) on H implies that

$$\psi_r \le \psi \le \psi_R \,. \tag{70}$$

Hence, for every $\delta \in \mathbb{R}$

$$\psi_r^\delta \supset \psi^\delta \supset \psi_R^\delta \,. \tag{71}$$

It then follows from the definition of the c_i that:

$$c_i(r) \le c_i \le c_i(R) \,, \quad \forall i \,. \tag{72}$$

Hence by the definition of γ^+ and γ^- (note the minus sign in formulas (62) and (63)):

$$\gamma^+(\Sigma_r) \le \gamma^+(\Sigma) \le \gamma^+(\Sigma_R) \tag{73}$$

$$\gamma^-(\Sigma_r) \le \gamma^-(\Sigma) \le \gamma^-(\Sigma_R) \tag{74}$$

Applying formula (66) then yields:

$$\frac{\pi r^2}{2n} \le \gamma^-(\Sigma) \le \gamma^+(\Sigma) \le \frac{\pi R^2}{2n} \,. \tag{75}$$

Denote by $\boldsymbol{S}$ the set of all compact C^2 hypersurface $\Sigma \subset \mathbb{R}^{2n}$, which bound a convex set C and whose Gaussian curvature does not vanish. We endow $\boldsymbol{S}$ with the Hausdorff metric, defined by:

$$\delta(\Sigma_1, \Sigma_2) := \operatorname{Max}\left\{\operatorname*{Max}_{x\in\Sigma_1} d(x, \Sigma_2), \operatorname*{Max}_{x\in\Sigma_2} d(x, \Sigma_1)\right\} \tag{76}$$

and we order it by inclusion

$$\Sigma_1 \gtrsim \Sigma_2 \Leftrightarrow C_1 \supset C_2 \,. \tag{77}$$

Proposition 10. *The maps* γ^- *and* $\gamma^+ : \boldsymbol{S} \to]0, \infty[$ *are non-decreasing:*

$$C_1 \supset C_2 \Rightarrow \gamma^-(\Sigma_1) \ge \gamma^-(\Sigma_2) \quad \text{and} \quad \gamma^+(\Sigma_1) \ge \gamma^+(\Sigma_2) \tag{78}$$

positively homogeneous of degree two:

$$\lambda > 0 \Rightarrow \gamma^-(\lambda\Sigma) = \lambda^2\gamma(\Sigma) \quad \text{and} \quad \gamma^+(\lambda\Sigma) = \lambda^2\gamma(\Sigma) \tag{79}$$

and continuous in the Hausdorff metric.

Proof. Without loss of generality, we may assume that C_1 and C_2 contain 0 in their interior. Denote by H_1 and H_2 the corresponding α-homogeneous Hamiltonians, by ψ_1 and ψ_2 the (reduced) dual action functionals. Since $C_1 \supset C_2$, with $C_i := \{x | H_i(x) \leq 1\}$, we have $H_1 \leq H_2$, and hence $H_1^* \geq H_2^*$ by Fenchel conjugation. So $\psi_1 \leq \psi_2$, and the critical values $(c_i)_1$ and $(c_i)_2$ are related by $(c_i)_1 \leq (c_i)_2$ for every i. Formula (78) then follows from the definition of γ^- and γ^+.

Relation (79) follows from a computation which we leave to the reader. For the last part of the proposition, we fix some $\Sigma \in \boldsymbol{S}$ and we take a sequence $\Sigma_n \to \Sigma$ in $\boldsymbol{S}$. Assuming, as usual, that 0 belongs to the interior of C and of the C_n, we have, for every $\epsilon > 0$, some $N \in \mathbb{N}$ such that:

$$n \geq N \Rightarrow (1-\epsilon)C \subset C_n \subset (1+\epsilon)C\ . \tag{80}$$

The monotonicity and homogeneity of γ^+ and γ^- then yield, for $n \geq N$:

$$(1-\epsilon)^2\gamma^+(C) \leq \gamma^+(C_n) \leq (1+\epsilon)^2\gamma^+(C) \tag{81}$$

$$(1-\epsilon)^2\gamma^-(C) \leq \gamma^-(C_n) \leq (1+\epsilon)^2\gamma^-(C) \tag{82}$$

and continuity follows immediately. □

In the first main result of this section, we will relate $\gamma^-(\Sigma)$ and $\gamma^+(\Sigma)$ to the closed characteristics on Σ. We recall that the set of all essential closed characteristics on Σ has been denoted by $\boldsymbol{C}$. With every $x \in \boldsymbol{C}$, we associate its action $A(x)$ and its mean index per period $\hat{i}(x)$:

$$A(x) := \frac{1}{2}\int_o^T (Jx, \dot{x})\, dt \tag{83}$$

$$\hat{i}(x) := \lim_{k\to\infty} i_{kT}(x) \tag{84}$$

where T is taken to be the minimal period. We recall that (Theorem I.7.8)

$$\hat{i}(x) > 2 \quad \text{if} \quad n \geq 2 \tag{85}$$

and that (Theorem 1.4)

$$A(x) \geq \pi r^2\ . \tag{86}$$

Theorem 11. *We have*

$$[\gamma^-(\Sigma), \gamma^+(\Sigma)] \subset \overline{\left\{ \frac{A(x)}{\widehat{i}(x)} \,\middle|\, x \in \boldsymbol{C} \right\}} . \tag{87}$$

□

In other words, for every γ in the interval $[\gamma^-(\Sigma), \gamma^+(\Sigma)]$ there is a sequence x_n in $\boldsymbol{C}$ (possibly even a constant sequence) such that

$$\frac{A(x_n)}{\widehat{i}(x_n)} \to \gamma , \qquad \text{when } n \to \infty . \tag{88}$$

Of course, relation (87) can also be written

$$[1/\gamma^+(\Sigma), 1/\gamma^-(\Sigma)] \subset \overline{\left\{ \frac{\widehat{i}(x)}{A(x)} \,\middle|\, x \in \boldsymbol{C} \right\}} \tag{89}$$

the advantage being that the quantity $\widehat{i}/A$ is readily interpreted as the mean index per unit of action.

The proof relies on the following lemma:

Lemma 12. *There is a constant d, depending only on Σ, such that, whenever $x \in \boldsymbol{C}$ is i-essential, we have:*

$$\left| \frac{1}{C_\alpha} \frac{\widehat{i}(x)}{A(x)} - i \,|c_i|^{\frac{2-\alpha}{\alpha}} \right| \le d \,|c_i|^{\frac{2-\alpha}{2\alpha}} \tag{90}$$

with $C_\alpha := \frac{4}{\alpha}\left(1 - \frac{\alpha}{2}\right)^{-\frac{2-\alpha}{\alpha}}$.

Proof. We have, for some $k \in \mathbb{N}^*$, using formula (45):

$$|c_i|^{\frac{2-\alpha}{\alpha}} = 2\,(C_\alpha k A(x))^{-1} \tag{91}$$

$$|i_{kT}(x) - 2i| \le 2n + 1 . \tag{92}$$

By Proposition I.5.21, we have:

$$\left| i_{kT}(x) - k\widehat{i}(x) \right| \le 2(n+1)\,(i_T(x) + 2n + 1) . \tag{93}$$

Recall also from the same section formula (158)

$$\left| i_{kT} - k\widehat{i}(x) \right| \le 2nk \tag{94}$$

and the fact that $|j(\omega) - i_T| \le 2n$, from which it follows by formula (143) that

$$\left| \widehat{i} - i_T \right| \le 2n . \tag{95}$$

Writing Eq. (91) into inequality (93) yields

$$\frac{2}{C_\alpha A} \left| \frac{i_{kT}}{k} - \widehat{i} \right| \le 2(n+1)\,(i_T + 2n + 1)\,|c_i|^{\frac{2-\alpha}{\alpha}} . \tag{96}$$

Estimates (93) and (94) imply that $\left|\frac{1}{k}i_{kT} - i_T\right| \le 4n$. Write this into the preceding inequality, and use relation (92). We get:

$$\frac{2}{C_\alpha A}\left|\frac{i_{kT}}{k} - \widehat{i}\right| \le 2(n+1)\left(\frac{2i+2n+1}{k} + 6n + 1\right)|c_i|^{\frac{2-\alpha}{\alpha}} \ . \tag{97}$$

Since $\gamma_\alpha^-(\Sigma) > 0$, it follows by formula (64) that the sequence $i\,|c_i|^{\frac{2-\alpha}{\alpha}}$ is bounded by a constant depending only on Σ. Hence the inequality:

$$\frac{2}{C_\alpha A}\left|\frac{i_{kT}}{k} - \widehat{i}\right| \le d_1\left(\frac{1}{k} + |c_i|^{\frac{2-\alpha}{\alpha}}\right) \tag{98}$$

for some suitable constant d_1. By inequality (91), there is another constant d_2 such that:

$$\frac{2}{C_\alpha A}\left|\frac{i_{kT}}{k} - \widehat{i}\right| \le d_2\frac{1}{A} \ . \tag{99}$$

We now turn to estimating the left-hand side of formula (90). We have:

$$\left|\frac{1}{C_\alpha}\frac{\widehat{i}}{A} - i\,|c_i|^{\frac{2-\alpha}{\alpha}}\right| \le \frac{1}{C_\alpha A}\left|\widehat{i} - \frac{i_{kT}}{k}\right| + \left|\frac{1}{C_\alpha A}\frac{i_{kT}}{k} - i\,|c_i|^{\frac{2-\alpha}{\alpha}}\right| \ . \tag{100}$$

We estimate the last term on the right by relations (91) and (92). We get:

$$\left|\frac{1}{C_\alpha A}\frac{i_{kT}}{k} - i\,|c_i|^{\frac{2-\alpha}{\alpha}}\right| = \frac{1}{C_\alpha k A}\left(i_{kT} - 2i\right) \tag{101}$$

and hence:

$$\left|\frac{1}{C_\alpha A}\frac{i_{kT}}{k} - i\,|c_i|^{\frac{2-\alpha}{\alpha}}\right| \le \frac{2n+1}{C_\alpha k A} = d_3\,|c_i|^{\frac{2-\alpha}{\alpha}} \ . \tag{102}$$

Writing this in inequality (100) yields:

$$\left|\frac{1}{C_\alpha}\frac{\widehat{i}}{A} - i\,|c_i|^{\frac{2-\alpha}{\alpha}}\right| \le \frac{1}{C_\alpha A}\left|\widehat{i} - \frac{i_{kT}}{k}\right| + d_3\,|c_i|^{\frac{2-\alpha}{\alpha}} \ . \tag{103}$$

Combining this with inequalities (98) and (99), we get

$$\left|\frac{1}{C_\alpha}\frac{\widehat{i}}{A} - i\,|c_i|^{\frac{2-\alpha}{\alpha}}\right| \le \frac{d_4}{2}\left(\frac{1}{k} + |c_i|^{\frac{2-\alpha}{\alpha}}\right) \tag{104}$$

$$\left|\frac{1}{C_\alpha}\frac{\widehat{i}}{A} - i\,|c_i|^{\frac{2-\alpha}{\alpha}}\right| \le \frac{d_5}{2}\left(\frac{1}{A} + |c_i|^{\frac{2-\alpha}{\alpha}}\right) \ . \tag{105}$$

Multiplying these inequalities together yields:

$$
\begin{aligned}
(106)\qquad \left|\frac{1}{C_\alpha}\frac{\widehat{i}}{2} - i\left|c_i\right|^{\frac{2-\alpha}{\alpha}}\right|^2 &\leq \frac{d_4 d_5}{4}\left(\frac{1}{kA} + \left|c_i\right|^{\frac{4-2\alpha}{\alpha}} + \left(\frac{1}{k}+\frac{1}{A}\right)\left|c_i\right|^{\frac{2-\alpha}{\alpha}}\right) \\
&= \frac{d_4 d_5}{4}\left(\left[\frac{C_\alpha}{2}+\frac{1}{k}+\frac{1}{A}\right] + \left|c_i\right|^{\frac{2-\alpha}{\alpha}}\right)\left|c_i\right|^{\frac{2-\alpha}{\alpha}} ,
\end{aligned}
$$

where we have used Eq. (91). Since $k \geq 1$ and $A \geq \pi r^2$ by formula (86), we get the desired result. □

Proof of Theorem 11. Multiplying both sides by C_α, we rewrite relation (87) as follows:

$$
(107)\qquad \left[\gamma_\alpha^-(\Sigma), \gamma_\alpha^+(\Sigma)\right] \subset C_\alpha \overline{\left\{ \frac{A(x)}{\widehat{i}(x)} \,\middle|\, x \in \mathcal{C} \right\}} .
$$

Take any $\gamma \in [\gamma_\alpha^-(\Sigma), \gamma_\alpha^+(\Sigma)]$. We shall construct an increasing sequence i_p, $p \in \mathbb{N}$, such that:

$$
(108)\qquad \lim_{p\to\infty} i_p \left|c_{i_p}\right|^{\frac{2-\alpha}{\alpha}} = \gamma^{-1} .
$$

Theorem 4 then implies that there is some $x^p \in \mathcal{C}$ which is i_p-essential and by Lemma 12:

$$
(109)\qquad \left|\frac{1}{C_\alpha}\frac{\widehat{i}\left(x^p\right)}{A\left(x^p\right)} - i_p\left|c_{i_p}\right|^{\frac{2-\alpha}{\alpha}}\right| \leq d\left|c_{i_p}\right|^{\frac{2-\alpha}{\alpha}} .
$$

Letting $i_p \to \infty$, so that $\left|c_{i_p}\right| \to 0$, we get:

$$
(110)\qquad \frac{1}{C_\alpha}\frac{\widehat{i}\left(x^p\right)}{A\left(x^p\right)} \to \gamma^{-1}
$$

and formula (87) is proved.

To construct the sequence i_p, assume first that $\gamma_\alpha^-(\Sigma) = \gamma_\alpha^+(\Sigma) = \gamma$. Then, by formula (62) and (63), we have $i\left|c_i\right|^{\frac{2-\alpha}{\alpha}} \to \gamma$ when $i \to \infty$, and we are done. So assume that

$$
(111)\qquad \gamma_\alpha^-(\Sigma) < \gamma_\alpha^+(\Sigma) .
$$

We proceed by induction. Assume that we have constructed i_p, with $i_p > i_{p-1}$ and

$$
(112)\qquad \left|\gamma^{-1} - i_p\left|c_{i_p}\right|^{\frac{2-\alpha}{\alpha}}\right| < \frac{1}{p} .
$$

We shall now construct $i_{p+1} > i_p$ with

$$
(113)\qquad \left|\gamma^{-1} - i_{p+1}\left|c_{i_{p+1}}\right|^{\frac{2-\alpha}{\alpha}}\right| < \frac{1}{p+1} .
$$

Find first $j > i_p$ such that

$$j\,|c_j|^{\frac{2-\alpha}{\alpha}} < \gamma_\alpha^+(\Sigma)^{-1} + \frac{1}{p+1} \tag{114}$$

$$|c_j|^{\frac{2-\alpha}{\alpha}} < \frac{1}{(p+1)}\,. \tag{115}$$

This is possible by the definition of $\gamma_\alpha^+(\Sigma)$. Now use the fact that the sequence $i \to |c_i|$ decreases to 0, so that, for every integer k:

$$\begin{aligned}(j+k+1)\,|c_{j+k+1}|^{\frac{2-\alpha}{\alpha}} - (j+k)\,|c_{j+k}|^{\frac{2-\alpha}{\alpha}} &\le |c_{j+k}|^{\frac{2-\alpha}{\alpha}} \\ &\le |c_j|^{\frac{2-\alpha}{\alpha}} \le \frac{1}{(p+1)}\,.\end{aligned} \tag{116}$$

By the definition of $\gamma_\alpha^-(\Sigma)$, we must have

$$(j+k_j)\,|c_{j+k_j}|^{\frac{2-\alpha}{\alpha}} > \gamma_\alpha^-(\Sigma)^{-1} - \frac{1}{p+1} \tag{117}$$

for some $k_j > 0$. Putting (114), (116) and (117) together, we see that the intervals

$$I_k := \left[(j+k)\,|c_{j+k}| - \frac{1}{p+1}, (j+k)\,|c_{j+k}| + \frac{1}{p+1}\right] \tag{118}$$

for $0 \le k \le k_j$, cover $\left[\gamma_\alpha^-(\Sigma)^{-1}, \gamma_\alpha^+(\Sigma)^{-1}\right]$. So one of them must contain γ^{-1}; say it is $I_{\bar{k}}$. We then set $i_{p+1} := j + \bar{k}$, and we are done. □

As an immediate consequence of Theorem 11, we have:

Corollary 13. *If $\gamma^-(\Sigma) < \gamma^+(\Sigma)$, then Σ carries infinitely many closed characteristic, and so does every $\Sigma' \in \boldsymbol{S}$ which is close enough to Σ in the Hausdorff metric.*

Proof. If Σ carries finitely many closed characteristics, the set of all $\frac{A(x)}{\hat{i}(x)}$ is finite and its closure cannot be an interval unless the latter reduces to a point, which means that $\gamma^-(\Sigma) = \gamma^+(\Sigma)$.

By Proposition 10, γ^- and γ^+ are continuous in the Hausdorff metric. If $\gamma^-(\Sigma) < \gamma^+(\Sigma)$, we will have $\gamma^-(\Sigma') < \gamma^+(\Sigma')$ for all Σ' in a neighbourhood of Σ. □

At first glance, Corollary 13 seems quite powerful. The Hausdorff continuity is surprising: a small C^o deformation of H translates into large deformations of the differential equation $\dot{x} = JH'(x)$. Unfortunately, we do not know whether the strict inequality actually occurs.

Question 14. *Does there exist a $\Sigma \in \boldsymbol{S}$ with $\gamma^-(\Sigma) < \gamma^+(\Sigma)$?* □

The only cases when we have been able to compute $\gamma(\Sigma)$ are spheres (formula (66)) or ellipsoids (formula (130)), and $\gamma^-(\Sigma) = \gamma^+(\Sigma)$ in both cases. Until Question 14 is answered, the interest of Corollary 13 is a moot point.

We have, however, another result for the finite case. Recall that $\boldsymbol{C}$ is the family of essential closed characteristics on Σ.

Theorem 15. *Assume $\boldsymbol{C}$ is finite. Then:*

$$\gamma^-(\Sigma) = \gamma^+(\Sigma) =: \gamma(\Sigma) \tag{119}$$

$$\forall x \in \boldsymbol{C}\,, \quad \frac{\widehat{i}(x)}{A(x)} = \frac{1}{\gamma(\Sigma)} \tag{120}$$

$$\sum_{x\in \boldsymbol{C}} \frac{1}{\widehat{i}(x)} \geq \frac{1}{2}\,. \tag{121}$$

□

Recall that, by Theorem 1.14, $\widehat{i}(x)$ is a geometric quantity, which does not depend on the Hamiltonian chosen for Σ (as long as it is convex).

Proof. The two first relations are direct consequences of Theorem 11, as we just saw. For the last one, apply Proposition 7. For every $i \geq 1$, we have some k_i such that:

$$\left| i_{k_i T_i}(x^i) - 2i \right| \leq 2n + 1\,. \tag{122}$$

Define the density of a sequence a_i in $\mathbb{N}$ to be the number (if it exists)

$$\delta := \lim_{N\to\infty} \frac{1}{N} \{i | a_i \leq N\}\,, \tag{123}$$

where A denotes the number of points in A. Relation (122) clearly implies that the sequences $i_{k_i T_i}(x^i)$ and $2i$ have the same density δ, namely $\frac{1}{2}$.

Now consider the sequence $k \to i_{kT}(x)$ for $x \in \boldsymbol{C}$. By formula (93) they have the same density as the sequences $k\widehat{i}(x)$, namely

$$\delta(x) := \frac{1}{\widehat{i}(x)}\,. \tag{124}$$

Each point in the sequence $i_{k_i T_i}(x^i)$ belongs to one of the sequences $i_{kT}(x)$ and no point in the sequences $i_{kT}(x)$ can be used twice. So its density must be no more than the sum of their densities:

$$\delta \leq \sum_{x\in \boldsymbol{C}} \delta(x) \tag{125}$$

which is precisely formula (121). □

Corollary 16. *If $n \geq 2$, any $\Sigma \in \boldsymbol{S}$ carries at least two geometrically distinct closed characteristics.*

Proof. If $\mathcal{C}$ is infinite, the corollary is true. If $\mathcal{C}$ is finite, we apply Theorem 15. Since $n \geq 2$, we have $\widehat{i}(x) > 2$ by formula (85), and the result follows. □

So Theorem 15 has enabled us to state a multiplicity result in the case of a general C^2 compact hypersurface with positive Gaussian curvature. It also gives us a criterion to find additional closed characteristics in certain situations.

Corollary 17. *Assume $n \geq 2$, and let $\mathcal{E}$ be a finite family of closed characteristics on Σ, such that*

$$[x^1 \in \mathcal{E}, x^2 \in \mathcal{E}, x^1 \neq x^2] \Rightarrow \frac{\widehat{i}(x^1)}{A(x^1)} \neq \frac{\widehat{i}(x^2)}{A(x^2)} . \tag{126}$$

Then there is a closed characteristic x on Σ which does not belong to $\mathcal{E}$.□

Proof. If there are infinitely many closed characteristics on Σ, the result is clear. Otherwise, apply formula (120). □

We conclude with two examples.

Example 1. The Harmonic Oscillator

It has already been studied in Sect. II.7. Consider the ellipsoid $\Sigma_{\alpha_1,\ldots,\alpha_n}$ defined by $H(x) = 1$, with $x = (p, q)$ and

$$H(x) := \sum_{i=1}^{n} \frac{1}{2} \alpha_i \left(p_i^2 + q_i^2\right) . \tag{127}$$

We know (Theorem I.6.13) that the mean index per (minimal) period does not depend on the Hamiltonian chosen to represent Σ. Using this particular choice of H, and assuming the system to be weakly nonresonant, we have found

$$\widehat{i}(x^i) = 2 \sum_j a_j / \alpha_i ,$$

where x^i is the closed characteristic lying in the x^i-plane.

On the other hand, the action along x^i is readily computed:

$$A(x_i) = \frac{1}{2} \oint \left(Jx^i, dx^i\right) = \frac{1}{2} \int_o^{T_i} \alpha_i \left(p_i^2 + q_i^2\right) dt = T_i = 2\frac{\pi}{\alpha_i} . \tag{128}$$

Hence the mean action per unit of action:

$$\frac{\widehat{i}(x^i)}{A(x^i)} = \frac{1}{\pi} \sum_{j=1}^{n} \alpha_j . \tag{129}$$

There are finitely many closed characteristics all of which have the same $\frac{\widehat{i}(x)}{A(x)}$. This is in accordance with Theorem 15, and shows that

$$\gamma\left(\Sigma_{\alpha_1,\ldots,\alpha_n}\right) = \pi \left(\sum_{j=1}^{n} \alpha_j\right)^{-1} . \tag{130}$$

By continuity (Proposition 10), the same formula must also hold in the general (resonant) case.

Example 2

Consider the two-dimensional system

$$\begin{cases} \ddot{q}_1 + q_1 = 3q_1^2 q_2^2 \\ \ddot{q}_2 + q_2 = 2q_1^3 q_2 \end{cases} . \tag{131}$$

It can be written as $\ddot{q} + V'(q) = 0$, with potential V given by

$$V(q_1, q_2) = \frac{1}{2}q_1^2 + \frac{1}{2}q_2^2 - q_1^3 q_2^2 \tag{132}$$

The Hamiltonian is:

$$H(q_1, q_2, p_1, p_2) = \frac{1}{2}p_1^2 + \frac{1}{2}p_2^2 + V(q_1, q_2) \tag{133}$$

For h small enough, the energy level $H(q,p) = h$ will be convex. We look for periodic solutions of system (2) such that

$$\frac{1}{2}\dot{q}_1^2 + \frac{1}{2}\dot{q}_2^2 + \frac{1}{2}q_1^2 + \frac{1}{2}q_2^2 + V(q_1, q_2) = h . \tag{134}$$

Two solutions are found immediately. They are (up to a phase shift):

$$q_1(t) = 0 , \quad q_2(t) = \sqrt{2h}\, \cos t \tag{135}$$

$$q_1(t) = \sqrt{2h}\, \cos t , \quad q_2(t) = 0 . \tag{136}$$

We claim that there is a third one. To see this, denote by x_1 the solution (p, q) of Hamilton's equations associated with $q(t) = \left(0, \sqrt{2h}\, \cos t\right)$, and by x_2 the solution associated with $q(t) = \left(\sqrt{2h}\, \cos t, 0\right)$. In other words, $x = (\dot{q}, q)$. Let us compute $\widehat{i}/A$ for x_1 and x_2.

The linearized equations around x_1 are:

$$\ddot{q}_1 + q_1 = 0 \tag{137}$$

$$\ddot{q}_2 + q_2 = 0 . \tag{138}$$

The linearized equations around x_2 are:

$$\ddot{q}_1 + q_1 = 0 \tag{139}$$

$$\ddot{q}_2 + q_2 = 2(2h)^{3/2} \cos^3 t q_2 . \tag{140}$$

We have:

$$\oint (Jx_1, dx_1) = 4\pi = \oint (Jx_2, dx_2) \ . \tag{141}$$

To have $\widehat{\imath}/A_1 \neq \widehat{\imath}_2/A_2$, it is enough that $\widehat{\imath}_1 \neq \widehat{\imath}_2$. Both $\widehat{\imath}_1$ and $\widehat{\imath}_2$ are given by formula (11), so that it is enough to show that $j_1(\omega) \leq j_2(\omega)$ for all ω on the unit circle, with strict inequality on a subset with positive measure.

By definition, $j(\omega)$ is the index of the Hermitian form

$$\int_o^T \left[(J\dot{y}, y) + \left(H^{*''}(-J\dot{x})\, J\dot{y}, J\dot{y}\right)\right] dt \tag{142}$$

on the complex Hilbert space

$$E_T^\omega = \left\{y \in L^2(0,T;\mathbb{C}) |\ \dot{y} \in L^2 \quad \text{and} \quad y(T) = \omega y(0)\right\} \ . \tag{143}$$

In the case when $H(x)$ splits, $H(q,p) = \frac{1}{2}p^2 + V(q)$, the Hermitian form (142) can be rewritten as:

$$\int_o^T \left[(\dot{p}, q) - (\dot{q}, p) + |\dot{q}|^2 + \left(V^{*''}(\dot{p}_o(t))\, \dot{p}, \dot{p}\right)\right] dt \tag{144}$$

where $x = (q_o, p_o)$ and $y = (q, p)$. This can again be rewritten as:

$$\int_o^T |(p - \dot{q})|^2\, dt + \int_o^T \left[\left(V^{*''}(\dot{p}_o)\, \dot{p}, \dot{p}\right) - |p|^2\right] dt \ . \tag{145}$$

The first term is positive definite, and does not contribute to the index. So $j(\omega)$ is the index of the Hermitian form

$$\int_o^T \left[\left(V^{*''}(\dot{p}_o)\, \dot{p}, \dot{p}\right) - |p|^2\right] dt \tag{146}$$

on the complex Hilbert space

$$F_T^\omega = \left\{p \in L^2(0,T;\mathbb{C}) |\dot{p} \in L^2 \quad \text{and} \quad p(T) = \omega p(0)\right\} \ . \tag{147}$$

When computing $j_1(\omega)$ and $j_2(\omega)$, the space F_T^ω is unchanged, with $T = 2\pi$. The Hermitian form associated with $j_1(\omega)$ is:

$$\int_o^{2\pi} \left(|\dot{p}|^2 - |p|^2\right) dt = Q_o(p) \tag{148}$$

while the quadratic form associated with $j_2(\omega)$ is:

$$\int_o^{2\pi} \left(\frac{1}{1 - 2(2h)^{3/2}\cos^3 t} |\dot{p}_2|^2 + |\dot{p}_1|^2 - |p|^2\right)^2 dt = Q_h(p) \ . \tag{149}$$

Clearly $j_2(\omega) \leq j_1(\omega)$ for $0 \leq h < 1/4\sqrt{2}$.

Note that for $h = 0$, $Q_h = Q_o$. A detailed analysis near $\omega = 1$ shows

$$j_1(1) = 0 \quad \text{and} \quad j_1(\omega) = 3 \quad \text{for} \quad \omega \neq 1 \tag{150}$$

$$j_2(1) = 0 \quad \text{and} \quad j_2(\omega) = 2 \quad \text{for} \;\; \omega \text{ close enough to } 1\ . \tag{151}$$

Hence

$$\widehat{i}_2 = \frac{1}{2\pi} \int j_2(\omega) d\omega < \frac{1}{2} \int j_1(\omega) d\omega = \widehat{i}_1 \tag{152}$$

and the result follows:

Proposition 18. *If $h < 1/4\sqrt{2}$ is such that the energy level $H(q,p) = h$ is convex, then problem* (131) *has at least three closed trajectories lying on that energy level.*

Notes and Comments. The results in this section are due to Ekeland and Hofer in [EkeH3]. The setting is slightly different from the one we have adopted here.

The crucial point is of course Theorem 4. The first proof was given by Viterbo in [Vit1], which unfortunately was limited to the case of an invariant C^2 function on Hilbert space. More work was necessary to bridge the gap to our case; it was done in [Szu2] and [EkeH3].

Using these results, the existence of two closed characteristics (Corollary 16) was proved by Ekeland and Lassoued in [EkeL4], [EkeL5] and then by Szulkin in [Szu2]. No such result is known for non-convex Σ, even in the star-shaped case.

This line of investigation was initiated by Ekeland, who proved in [Eke9] that if a convex compact Σ carries finitely many closed characteristics, then there must be a resonance condition relating them. Corollary 17 can be understood as making this resonance condition explicit: two of the closed characteristics must have the same $\widehat{i}/A$.

The approach in [Eke9] relied on Morse theory. It was carried to the star-shaped case by Viterbo, wo showed in [Vit2] that when all the closed characteristics are nondegenerate inequality (121) can be replaced by an equality:

$$\sum_{x \in \mathcal{C}_2} \frac{1}{\widehat{i}(x)} - \sum_{x \in \mathcal{C}_1} \frac{1}{\widehat{i}(x)} + \frac{1}{2} \sum_{x \in \mathcal{C}_o} \frac{(-1)^{i_1(x)}}{\widehat{i}(x)} = \frac{1}{2} \tag{153}$$

where $\mathcal{C}_2$ (resp. $\mathcal{C}_1$ or $\mathcal{C}_o$) is the set of all $x \in \mathcal{C}$ such that the sequence $i_{kT}(x)$, $k \in \mathbb{N}$, is even (resp. odd or alternating). We refer to I.5 for this classification of closed characteristics.

In [Eke9], it was also claimed that the property of carrying infinitely many closed characteristics is C^∞-generic among all C^∞ compact $\Sigma \subset \mathbb{R}^{2n}$ with positive curvature. This is correct, the idea being that resonance relations such as $\frac{\widehat{i}(x^1)}{A(x^1)} = \frac{\widehat{i}(x^2)}{A(x^2)}$ or Viterbo's formula (153) can be perturbed away by Thom's transversality theorem. Unfortunately, the transversality argument provided in [Eke9] is wrong; one must use more sophisticated ones, such as those in [Rob2] or [Tak2].

4. Open Problems

This chapter has been devoted to the study of periodic solutions for the fixed-energy problem in the convex case. We have shown that there is existence (Sect. 1) and multiplicity (Sect. 3), but an old question remains open since the time of Liapounov:

Problem 1. *Does every compact $\Sigma \subset \mathbb{R}^{2n}$ bounding an open convex set carry at least n closed characteristics ?* □

In Sect. 4, we ran into another question.

Problem 2. *Does there exist a compact C^2 hypersurface Σ, bounding a convex set and having positive curvature, such that $\gamma^+(\Sigma) > \gamma^-(\Sigma)$?* □

A major difficulty in answering these questions lies in the paucity of examples. The only case when we can actually compute $\gamma^+(\Sigma)$ and $\gamma^-(\Sigma)$ is when Σ is an ellipsoid, and it is pretty well the only known case when there are finitely many closed characteristics. Michael Herman, however, has shown me how to construct a $\Sigma \subset \mathbb{R}^4$ which carries exactly two closed characteristics and is arbitrarily close to (but distinct from) an ellipsoid. Even with this new example, there are no known cases when Σ carries p closed characteristics, with $n < p < \infty$.

Still in the convex framework, there is another important question:

Problem 3. *Does there exist a C^2 compact hypersurface $\Sigma \subset \mathbb{R}^{2n}$, bounding a convex open set, such that every closed characteristic has at least one Floquet multiplier away from the unit circle ?* □

If the answer is yes, then this particular Σ will have quite interesting properties from the dynamical point of view. It will be very far from an ellipsoid, which means that the corresponding Hamiltonian flow will be strongly nonlinear, and one may entertain the hope that it will turn out to be ergodic, and that its entropy will be computed, as in Pesin's theory (see [Pes] or [Man] for references). Birkhoff's quasi-ergodic hypothesis (see [Sma2] for references and discussions) would then appear to be a property of strongly nonlinear systems (always with the provision of convexity). On the other hand, if the answer is no, then by the KAM theorem there will be in Σ a subset of positive measure filled out with invariant tori, and even this way of salvaging Birkhoff's quasi-ergodic hypothesis will be cut off.

We conclude with a famous conjecture by Poincaré ([Poi2], vol. 1, p. 82).

Problem 4. *Denote by $\boldsymbol{S}^k$ the set of all compact C^k hypersurface in $\mathbb{R}^{2n}$, with the C^k metric, and by $\boldsymbol{S}_o^k$ the subset of all $\Sigma \in \boldsymbol{S}^k$ such that closed characteristics are dense on Σ. Does $\boldsymbol{S}_o^k$ contain a dense $\mathcal{G}_\delta$-set, that is, a countable intersection of open dense subsets ?* □

The reason for this conjecture is of course Poincaré's recurrence theorem (see [Arn2]). The question has been answered positively by Robinsion in [Rob1], for the lowest value of k which makes sense, $k = 2$. It is an outgrowth of Pugh's proof of the closing lemma in general dynamical systems. The case $k > 2$ remains open. If we denote by $\boldsymbol{C}$ the set of all C^2 hypersurfaces which bound a convex set and have positive curvature, and by $\boldsymbol{S}_1^k$ the set of all $\Sigma \in \boldsymbol{S}^k$ which carry infinitely many closed characteristics, we can use index theory to show that $\boldsymbol{S}_1^k \cap \boldsymbol{C}$ contains a dense $\mathcal{G}_\delta$-set in $\boldsymbol{S}^k \cap \boldsymbol{C}$; but this is still a far cry from the Poincaré conjecture.

This problem has led us away from the convex framework, and now a host of new questions appear, beginning with the most basic one.

Problem 5. *Does every compact C^2 hypersurface $\Sigma \subset \mathbb{R}^{2n}$ carry a closed characteristic ?* □

Although the answer in the general case is not known, it was conjectured by Weinstein [Wei4] and proved by Viterbo [Vit3] that the answer is yes in the particular case when Σ is of contact type. This means that the ω, the restriction to Σ of the 2-form $\sum_{i=1}^n dp_i \wedge dq_i$ on $\mathbb{R}^{2n}$, has a primitive Ω such that:

$$[\xi \neq 0 \quad \text{and} \quad \omega_x(\xi, \eta) \neq 0\,, \quad \forall \eta \in T_x\Sigma] \Rightarrow \Omega_x(\xi) \neq 0\,.$$

Viterbo's proof (see also [HofZ]) opened a new domain, which is in full development at the time of this writing. In particular, it has led to the construction of new classes of symplectic invariants, the Ekeland-Hofer capacities, which have the property of being monotonic with respect to inclusions (see [EkeH4] and [EkeH5]), and which seem to be a basic tool in symplectic geometry.

Bibliography

[AmaZ1] Amann, H., Zehnder, E.: Nontrivial solutions for a class of nonresonance problems and applications to nonlinear differential equations. Ann. Scuola Normale Superiore di Pisa **7** (1980) 539–603

[AmaZ2] Amann, H., Zehnder, E.: Periodic solutions of asymptotically linear Hamiltonian equations. Manuscripta Mathematica **32** (1980) 149–189

[AmbCZ] Ambrosetti, A., Coti Zelati, V.: Solutions with minimal period for Hamiltonian systems in a potential well. Ann. IHP "Analyse non linéaire" (1987) 275–242

[AmbM] Ambrosetti, A., Mancini, G.: On a theorem of Ekeland and Lasry concerning the number of periodic Hamiltonian trajectories. J. Diff. Eq. **43** (1981) 1–6

[AmbM] Ambrosetti, A., Mancini, G.: Solutions of minimal period for a class of convex Hamiltonian systems. Math. Ann. **255** (1981) 405–421

[AmbR] Ambrosetti, A., Rabinowitz, P.: Dual variational methods in critical point theory and applications. J. Funct. An. **14** (1973) 349–381

[Arn1] Arnold, V.: Equations différentielles ordinaires. Editions Mir, Moscou 1974

[Arn2] Arnold, V.: Méthodes mathématiques de la mécanique classique. Editions Mir, Moscou 1976

[Arn3] Arnold, V.: Chapitres supplémentaires de la théorie des équations différentielles ordinaires. Editions Mir, Moscou 1980

[AubE] Aubin, J.P., Ekeland, I.: Applied nonlinear analysis. Wiley 1984

[BahB] Bahri, A., Berestycki, H.: Existence of forced oscillations for some nonlinear differential equations. Comm. Pure Appl. Math. **37** (1984) 403–442

[BahB] Bahri, A., Berestycki, H.: Forced vibrations of superquadratic Hamiltonian systems. Acta Mathematica **152** (1984) 143–197

[BalTZ] Ballman, W., Thorbergsson, G., Ziller, W.: Closed geodesics on positvely curved manifolds. Ann. Math. **116** (1982) 213–247

[Ben1] Benci, V.: On critical point theory for indefinite functionals in the presence of symmetries. Trans. AMS **274** (1982) 553–572

[Ben2] Benci, V.: Normal modes of a Lagrangian system constrained in a potential well. Ann. IHP "Analyse non linéaire" **1** (1984) 362–379

[Ben3] Benci, V.: Closed geodesics for the Jacobi metric and periodic solutions of prescribed energy of natural hamiltonian systems. Ann. IHP "Analyse non linéaire" **1** (1984) 380–401

[BenR] Benci, V., Rabinowitz, P.: Critical point theory for indefinite functionals. Inv. Math **52** (1979) 241–273

[BerLRM] Berestycki, H., Lasry, J.M., Ruf, B., Mancini, G.: Existence of multiple periodic orbits on star-shaped Hamiltonian surfaces. Comm. Pure Appl. Math. **38** (1985) 252–290

[Ber1] Berger, M.: On periodic solutions of second order Hamiltonian systems. J. Math. Anal. Appl. **29** (1970) 512–522

[Ber2] Berger, M.: Periodic solutions of second order dynamical systems and isoperimetric variational problems. Am. J. Math. **93** (1971) 1–10

[Bot1] Bott, R.: On the iteration of closed geodesics and Sturm intersection theory. Comm. Pure Appl. Math. **9** (1956) 176–206

[Bot2] Bott, R.: Morsee theory, old and new. Bull. AMS **7** (1982) 331–358

[BreCN] Brézis, H., Coron, J.M., Nirenberg, L.: Free vibrations for a nonlinear wave equation and a theorem of Rabinowitz. Comm. Pure Appl. Math. **33** (1980) 667–689

[Bro1] Brousseau, V.: L'index d'un système hamiltonien linéaire. C.R.A.S. Paris **8** (1986) 351–354

[Bro2] Brousseau, V.: L'index des systèmes hamiltoniens linéaires à coefficients periodiques. Ann. IHP "Analyse non linéaire", to appear 1990

[Cam] Cambini, A.: Sul lemma di Morse. Boll. UMI **7** (1973) 87–93

[Cer] Cerami, G.: Un criterio di esistenza per i punti critici su varietà illimitati. Rendiconti dell' Academia di Sc. e Lett. dell' Istituto Lombardo **112** (1978) 332–336

[Cha] Chang, K.C.: Infinite-dimensional Morse theory and its applications. Séminaire de Mathématiques Supérieures 97, Université de Montréal, 1985

[Cla1] Clarke, F.: Solutions périodiques des équations hamiltoniennes. Note CRAS Paris **287** (1978) 951–952

[Cla2] Clarke, F.: A classical variational principle for periodic Hamiltonian trajectories. Proceedings A.M.S. **76** (1979) 186–189

[Cla3] Clarke, F.: Periodic solutions of Hamiltonian inclusions. J. Diff. Eq. **40** (1981) 1–6

[Cla4] Clarke, F.: On Hamiltonian flows and symplectic transformations. SIAM J. Control and Optimization **20** (1982) 355–359

[Cla5] Clarke, F.: Periodic solutions of Hamilton's equations and local minima of the dual action. Trans. AMS **287** (1985) 239–251

[Cla6] Clarke, F.: Optimization and non-smooth analysis. Wiley 1983

[ClaE1] Clarke, F., Ekeland, I.: Solutions périodiques, de période donnée, des équations hamiltoniennes. Note CRAS Paris **287** (1978) 1013–1015

[ClaE2] Clarke, F., Ekeland, I.: Hamiltonian trajectories having prescribed minimal period. Comm. Pure Appl. Math. **33** (1980) 103–116

[ClaE3] Clarke, F., Ekeland, I.: Nonlinear oscillations and boundary-value problems for Hamiltonian systems. Arch. Rat. Mech. Annal. **78** (1982) 315–333

[ConZ1] Conley, C., Zehnder, E.: The Birkhoff-Lewis fixed-point theorem and a conjecture of V. Arnold. Inv. Math. **33** (1983) 49

[ConZ2] Conley, C., Zehnder, E.: Morse type index theory for flows and periodic solutions for Hamiltonian equations. Comm. Pure Appl. Math. **27** (1984) 211–253

[ConZ3] Conley, C., Zehnder, E.: Subharmonic solutions and Morse theory. Physica **124A** (1984) 649–658

[CroW] Croke, C., Weinstein, A.: Closed curves on convex hypersurfaces and periods of nonlinear oscillations. Inv. Math. **64** (1981), 199–202

[CusD] Cushman, R., Duistermaat, J.J.: The behaviour of the index of a periodic linear Hamiltonian system under iteration. Adv. in Math. **23** (1977) 1–21

[dOnE] d'Onofrio, B., Ekeland, I.: Hamiltonian systems with elliptic periodic orbits. To appear, J. Nonlinear Analysis TMA

[Dui] Duistermaat, J.J.: On the Morse index of variational calculus. Adv. in Math. **21** (1976) 173–195

[Eke1] Ekeland, I.: On the variational principle. J. Math. Anal. **47** (1974) 324–353

[Eke2] Ekeland, I.: Nonconvex minimization problems. Bull. AMS **1**, New Series (1979) 443–474

[Eke3] Ekeland, I.: Periodic solutions of Hamiltonian systems and a theorem of P. Rabinowitz. J. Diff. Eq. **34** (1970) 523–534

[Eke4] Ekeland, I.: Forced oscillations for nonlinear Hamiltonian systems II. Advances in Mathematics, volume in honour of L. Schwartz, Nachbin ed., Academic Press, 1981

[Eke5] Ekeland, I.: Oscillations forcées de systémes hamiltoniens non linéaires. Bull. SMF **109** (1981) 297–330

[Eke6] Ekeland, I.: A perturbation theory near convex Hamiltonian systems. J. Diff. Eq. **50** (1983) 407–440

[Eke7] Ekeland, I.: Dual variational methods in non-convex optimization and differential equations. Mathematical Theories of Optimization, Cecconi and Zolezzi ed., Springer Lecture Notes in Mathematics **979** (1983) 108–120

[Eke8] Ekeland, I.: An index theory for periodic solutions of convex Hamiltonian systems. Proceedings of Symposia in Pure Mathemathics **45** (1986) 395–423

[Eke9] Ekeland, I.: Une théorie de Morse pour les systèmes hamiltoniens convexes. Ann. IHP "Analyse non linéaire" **1** (1984) 19–78

[EkeH1] Ekeland, I., Hofer, H.: Periodic solutions with prescribed period for convex autonomous Hamiltonian systems. Inventiones Math. **81** (1985) 155–188

[EkeH2] Ekeland, I., Hofer, H.: Subharmonics of convex nonautonomous Hamiltonian systems. Comm. Pure and Applied Math. **40** (1987) 1–37

[EkeH3] Ekeland, I., Hofer, H.: Convex Hamiltonian energy surfaces and their closed trajectories. Comm. Math. Physics **113** (1987) 419–467

[EkeH4] Ekeland, I., Hofer, H.: Symplectic topology and Hamiltonian dynamics. Mathematische Zeitschrift **200** (1989) 355–378

[EkeL1] Ekeland, I., Lasry, J.M.: On the number of periodic trajectories for a Hamiltonian flow on a convex energy surface. Annals of Math. **112** (1980) 283–319

[EkeL2] Ekeland, J., Lasry, J.M.: Principes variationnels en dualité. Note CRAS Paris **291** (1980) 493–497

[EkeL3] Ekeland, I., Lasry, J.M.: Duality in non-convex variational problems. Advances in Hamiltonian Systems, Aubin, Bensoussan and Ekeland ed., (1983), Birkhauser-Boston

[EkeL4] Ekeland, I., Lassoued, L.: Un flot Hamiltonien a au moins deux trajectoires fermées sur toute surface d'énergie convexe et bornée. Note CRAS Paris **301** (1985) 162–164

[EkeL5] Ekeland, I., Lassoued, L.: Multiplicité des trajectoires fermées d'un système Hamiltonian sur une hypersurface d'énergie convexe. Annales IHP "Analyse non linéaire" **4** (1987) 1–29

[EkeT] Ekeland, I., Temam, R.: Analyse convexe et problèmes variationnels. Dunod-Gauthier-Villars, 1974; English translation, "Convex analysis and variational problems", North-Holland-Elsevier, 1976

[FadR] Fadell, E., Rabinowitz, P.: Generalized cohomological theories for Lie group actions with an application to bifurcation questions for Hamiltonian systems. Inv. Math. **45** (1978) 139–174

[GelL] Gelfand, I., Lidsky, V.: On the structure of the regions of stability of linear canonical systems of differential equations with periodic coefficients. Uspekhi Math. Naouk USSR **10** (1955) 3–40 (AMS translation 8 (1958) 143–181)

[GhoP] Ghoussoub, N., Preiss, D.: A general mountain pass principle for locating and classifying critical points. Annales IHP "Analyse non linéaire" to appear 1989

[Gho] Ghoussoub, N.: A general min-max method in critical point theory. Preprint, 1988

[Gir] Girardi, M.: Multiple orbits for Hamiltonian systems on starshaped energy surfaces with symmetries. Ann. IHP "Analyse non linéaire" **1** (1984), 285–294

[GirM1] Girardi, M., Matzeu, M.: Some results on solutions of minimal period to superquadratic Hamiltonian equations. Nonlinear Analysis TMA **7** (1983) 475–482

[GirM2] Girardi, M., Matzeu, M.: Solutions of minimal period for a class of nonconvex Hamiltonian systems and applications to the fixed energy problem. Nonlinear Analysis TMA **10** (1986) 371–382

[GirM3] Girardi, M., Matzeu, M.: Periodic solutions of convex Hamiltonian systems with a quadratic growth at the origin and superquadratic at infinity. Ann. Math. Pura ed App. **147** (1987) 21–72

[Gor1] Gordon, W.B.: A theorem on the existence of periodic solutions to Hamiltonian systems with convex potentials. J. Diff. Eq. **19** (1971) 324–335

[Gor2] Gordon, W.B.: Physical variational principles which satisfy the Palais-Smale condition. Bull. A.M.S. **78** (1972) 712–716

[Gor3] Gordon, W.B.: A minimizing property of Keplerian orbits. Am. J. Math. **99** (1977) 961–971

[GroM] Gromoll, D., Meyer, W.: On differentiable functions with isolated critical points. Topology **8** (1969) 361–369

[Hal] Hale, J.: Ordinary differential equations. Wiley-Interscience, 1969

[Hin] Hingston, N.: Equivariant Morse theory and closed geodesics. J. Diff. Geometry **19** (1984) 85–116

[Hof1] Hofer, H.: A new proof of a result of Ekeland and Lasry concerning the number of periodic Hamiltonian trajectories on a prescribed energy surface. Boll. UMI **6** (1982) 931–942

[Hof2] Hofer, H.: A geometric description of the neighbourhood of a critical point given by the mountain pass theorem. J. London Math. Soc. **31** 1985 566–570

[Hof3] Hofer, H.: The topological degree at a critical point of mountain pass type. Nonlinear Functional Analysis and its Applications, Proc. Symp. Pure Math. no. 45 (1986), 501–509

[HofZ] Hofer, H., Zehnder, E.: Periodic solutions on hypersurfaces and a result by C. Viterbo. Inv. Math. **90** (1987) 1–9

[Hor] Horn, J.: Beiträge zur Theorie der kleinen Schwingungen. Zeit. Math. Phys. **48** (1903) 400–434

[Jor] Jorna, S. (ed.): Topics in nonlinear dynamics: a tribute to sir Edward Bullard. American Institute of Physics Conference proceedings no. 46, 1978, 1978, AIP, New York

[JorS] Jordan, D., Smith, P.: Nonlinear ordinary differential equations. Clarendon Press, Oxford, 1977

[Kin] Kingman, J.F.: Subadditive ergodic theory. Ann. Prob. **1** (1973) 652–680

[KirS] Kirchgraber, U., Stiefel, U.: Methoden der analytischen Störungsrechnung und ihre Anwendungen. 1978, Teubner, Stuttgart

[Kli] Klingenberg, W.: Lectures on closed geodesics. Springer 1981

[Koz1] Kozlov, V.V.: Integrability and non-integrability in classical mechanics. Uspekhi Math. Nauk **38** (1983) 1–68; Russian Math. Surveys **38** (1983) 1–76

[Koz2] Kozlov, V.V.: Calculus of variations in the large and classical mechanics. Uspekhi Math. Nauk **40** (1985) 33–60; Russian Math. Surveys **40** (1985) a37–71

[Kra] Krasnoselskii, M.A.: Topological methods in the theory of nonlinear integral equations. English translation, Pergamon press, 1963

[Kre1] Krein, M.: Generalization of certain investigations of A.M. Liapunov on linear differential equations with periodic coefficients. Doklady Akad. Nauk USSR **73** (1950) 445–448

[Kre2] Krein, M.: On the application of an algebraic proposition in the theory of monodromy matrices. Uspekhi Math. Nauk **6** (1951) 171–177

[Kre3] Krein, M.: On the theory of entire matrix-functions of exponential type. Ukrainian Math. Journal **3** (1951) 164–173

[Kre4] Krein, M.: On some maximum and minimum problems for characteristic numbers and Liapunov stability zones. Prikl. Math. Mekh. **15** (1951) 323–348

[Kre5] Krein, M.: On criteria for stability and boundedness of solutions of periodic canonical systems. Prikl. Math. Mekh. **19** (1955) 641–680

[KreL] Krein, M., Lyubarski, G.: On analytical properties of multipliers of periodic canonical differential systems of positive type. Izv. Ak. Nauk SSSR **26** (1962) 542–572

[Lev] Levi, M.: Stability of linear Hamiltonian systems with periodic coefficients. IBM Research paper, Thomas J. Watson Research Center, 1977

[Lia] Liapunov, A.M.: Problème général de la stabilité du mouvement. Ann. Fac. Sci. Toulouse **9** (1907) 203–474. Russian original, Kharkov Math. Soc. 1892. Reedited, Princeton U. Press, 1949. Reedited, Gabay, Paris 1989

[Lio] Lions, P.L.: Solutions of Hartree-Fock equations for Coulomb systems. Comm. Math. Physics **109** (1987) 33–97

[Man] Mañe: Ergodic theory. Springer, 1987

[Mas] Maslov, V.P.: Théorie des perturbations et méthodes asymptotiques. Dunod, 1972

[MasF] Maslov, V.P., Fedoryuk. Semi-classical approximation in quantum mechanics. Reidel 1981

[Maw] Mawhin, J.: Problèmes de Dirichlet variationnels non linéaires. Séminaire de Mathématiques Supérieures **104**, Université de Montréal, 1987

[MawW1] Mawhin, J., Willem, M.: Critical points of convex perturbations of some indefinite quadratic forms and semi-linear boundary value problems at resonance. Ann. IHP "Analyse non linéaire" **6** (1986) 431–454

[MawW2] Mawhin, J., Willem, M.: Critical point theory and Hamiltonian systems. Springer 1989

[MicT] Michalek, R., Tarantello, G.: Subharmonic solutions with prescribed minimal period for nonautonomous Hamiltonian systems. J. Diff. Eq. **72** (1988) 28–55

[Mil] Milnor, J.: Morse theory. Princeton U. Press, 1963

[Mos1] Moser, J.: New aspects in the theory of stability of Hamiltonian systems. Comm. Pure Appl. Math. **11** (1958) 81–114

[Mos2] Moser, J.: Regularization of Kepler's problem and the averaging method on a manifold. Comm. Pure Appl. Math. **23** (1970) 609–636

[Mos3] Moser, J.: Stable and random motion in dynamical systems. Ann. Math. Studies **77** (1973) Princeton University Press.

[Mos4] Moser, J.: Periodic orbits near an equilibrium and a theorem by Alan Weinstein. Comm. Pure Appl. Math. **29** (1976) 727–747

[Nay] Nayfeh, A.; Perturbation methods. 1973, Wiley, New York

[Pal1] Palais, R.: Morse theory on Hilbert manifolds. Topology **2** (1963), 299–400

[Pal2] Palais, R.: Ljusternik-Schnirelman theory on Banach manifolds. Topology **5** (1966) 115–132

[PalS] Palais, R., Smale, S.: A generalized Morse theory. Bull. A.M.S. **70** (1964) 165–171

[Pes] Pesin, Y.B.: Characteristic Lyapunov exponents and smooth ergodic theory. Russian Math. Surveys **32** (1977) 55–114

[Poi1] Poincaré, H.: Mémoire sur les courbes défines par une équation différentielle; J. Math. Pures et Appl. **7** (1881) 375–422; **8** (1882) 251–296; **1** (1885) 167–244; **2** (1886) 151–217

[Poi2] Poincaré, H.: Les méthodes nouvelles de la mécanique céleste (3 vol). Gauthier-Villars, Paris, 1892–1899

[PucS1] Pucci, P., Serrin, J.: Extensions of the mountain pass theorem. J. Funct. Anal. **59** (1984) 185–210

[PucS2] Pucci, P., Serrin, J.: A mountain pass theorem. J. Diff. Eq. **60** (1985) 142–149

[PucS3] Pucci, P., Serrin, J.: The structure of the critical set in the mountain pass theorem. Trans. AMS **91** (1987) 115–132

[Rab1] Rabinowitz, P.: Free vibrations for a nonlinear wave equation. Comm. Pure and App. Math. **31** (1978) 195–184

[Rab2] Rabinowitz, P.: Periodic solutions of Hamiltonian systems. Comm. Pure Appl. Math. **31** (1978) 157–184

[Rab2] Rabinowitz, P.: Periodic solutions of a Hamiltonian system on a prescribed energy surface. J. Diff. Eq. **33** (1979) 336–352

[Rab3] Rabinowitz, P.: On subharmonic solutions of Hamiltonian systems. Comm. Pure Appl. Math. **33** (1980) 609–633

[Rab4] Rabinowitz, P.: Periodic solutions of large norm a Hamiltonian systems. J. Diff. Eq. **50** (1983) 33–48

[Rab4] Rabinowitz, P.: Minimax methods in critical point theory with applications to differential equations. C.B.M.S. **65** (1986), AMS

[RabAEZ] Rabinowitz, P., Ambrosetti, A., Ekeland, I., Zehnder, E. (eds): Periodic solutions of Hamiltonian systems and related topics. NATO ASI Series C, vol. 209, Reidel 1987

[Rob1] Robinson, R.C.: The C^1 closing lemma. Preprint

[Rob2] Robinson, R.C.: A global approximation theorem for Hamiltonian systems. Proceedings Symposia Pure Mathematics, 233-243, AMS

[Roc] Rockafellar, R.T.: Convex analysis. Princeton University Press, 1970

[Ros1] Roseau, M.: Vibrations non linéaires et théorie de la stabilité. Springer, 1966

[Ros2] Roseau, M.: Equations différentielles. Masson, Paris, 1976

[Ros3] Roseau, M.: Vibrations des systèmes mécaniques. Masson, Paris, 1984

[RouM] Rouché, N., Mahwin, J.: Equations différentielles ordinaires (2 vol). Masson, Paris, 1973

[Sei] Seifer, H.: Periodische Bewegungen mekanischer Systemen. Math. Zeitschrift **51** (1948) 197–216

[SieM] Siegel, C.L., Moser, J.: Lectures on celestial mechanics. Springer, 1971

[Skr1] Skripnik, I.V.: On the application of Morse's method to nonlinear elliptic equations. Soviet Math. Doklady **13** (1972), 202–205

[Skr2] Skripnik, I.V.: The differentiability of integral functionals. Dopovidi Ak. Nauk Ukrain. RSR Ser. A (1972) 527–529

[Sma1] Smale, S.: Morsee theory and a nonlinear generalization of the Dirichlet problem. Ann. Math. **17** (1964) 307–315

[Sma2] Smale, S.: Birkhoff's quasi-ergodic hypothesis. New York Acad. of Sci. Conference Proceedings

[Szu1] Szulkin, A.: Minimax principles for lower semi-continuous functions and applications to nonlinear boundary-value problems. Ann. IHP "Analyse non linéaire" **3** (1986) 77–110

[Szu2] Szulkin, A.: Morsee theory and existence of periodic solutions of convex Hamiltonian systems. Bull. S.M.F., to appear

[Szu3] Szulkin, A.: Ljusternik-Schnirelman theory on C^1-manifolds. Ann. IHP "Analyse non linéaire" **5** (1988) 119–140

[Tak1] Takens, F.: A note on sufficiency of jets. Inv. Math. (1971) 225–231

[Tak2] Takens, F.: Hamiltonian systems: generic properties of closed orbits and local perturbations. Math. Ann. **188** (1970) 304–312

[Tar] Tarantello, G.: Subharmonic solutions for Hamiltonian systems via a $\mathbb{Z}_p$ pseudoindex theory. Annali Scuola Normale di Pisa, to appear

[vGr] van Groesen, B.: Existence of multiple normal mode trajectories on convex energy surfaces of even, classical Hamiltonian systems. J. Diff. Eq. (1985) 70–89

[Vit1] Viterbo, C.: Indice de Morse des points critiques obtenus par minimax. Ann. IHP "Analyse non linéaire" **5** (1988) 221–226

[Vit2] Viterbo, C.: Equivariant Morse theory for starshaped Hamiltonian systems. Transactions A.M.S. **311** (1989) 621–655

[Vit3] Viterbo, C.: A proof of Weinstein's conjecture in $\mathbb{R}^{2n}$. Ann. IHP "Analyse non linéaire" **4** (1987) 337–356

[Wei1] Weinstein, A.: Normal modes for nonlinear Hamiltonian systems. Inv. Math. **20** (1973) 47–57

[Wei2] Weinstein, A.: Lagrangian submanifolds and Hamiltonian systems. Ann. Math. **98** (1973) 377–410

[Wei3] Weinstein, A.: Periodic orbits for convex Hamiltonian systems. Ann. Math. **108** (1978) 507–518

[Wei4] Weinstein, A.: On the hypothesis of Rabinowitz' periodic orbit theorem. J. Diff. Eq. 33 (1979) 353–358

[Wil] Willem, M.: Subharmonic oscillatios of convex Hamiltonian systems. J. Nonlinear Analysis TMA **9** (1985) 1303–1311

[YaS] Yakubovich, V., Starzhinskii, V.: Linear differential equations with periodic coefficients. Halsted Press, Wiley

[Zeh] Zehnder, E.: Periodic solutions of Hamiltonian equations. Lecture Notes in Mathematics **1031**, Springer 1981

[ZehM] Zehnder, E., Moser, J.: Classical mechanics. In preparation

Index

Notations and symbols
(Greek letters come last)